NUCLEAR POWER

Development in India over the last seven Decades
True Story and History

RAMASWAMI RANGANATHAN

INDIA · SINGAPORE · MALAYSIA

Copyright © Ramaswami Ranganathan 2024
All Rights Reserved.

ISBN
Paperback 979-8-89066-875-2
Hardcase 979-8-89363-967-4

This book has been published with all efforts taken to make the material error-free after the consent of the author. However, the author and the publisher do not assume and hereby disclaim any liability to any party for any loss, damage, or disruption caused by errors or omissions, whether such errors or omissions result from negligence, accident, or any other cause.

While every effort has been made to avoid any mistake or omission, this publication is being sold on the condition and understanding that neither the author nor the publishers or printers would be liable in any manner to any person by reason of any mistake or omission in this publication or for any action taken or omitted to be taken or advice rendered or accepted on the basis of this work. For any defect in printing or binding the publishers will be liable only to replace the defective copy by another copy of this work then available.

Dedication

This book is dedicated to the memory of my parents who instilled high moral values in me which have sustained me throughout my life. I am also grateful to them for building my desire to acquire more knowledge, and for encouraging me to pursue my dreams.

Contents

Acknowledgment

I sincerely thank Sarvi Sri. Y.S.R. Prasad and Sri. V.K. Chaturvedi—retired Chairmen and Managing directors (CMD's) of the Nuclear Power Corporation of India Ltd. (NPCIL)—for going through the list of contents of this book and for providing oral permission for the publication of this book.

I thank Sri. Amritesh Srivastava, Editor of Nu-Power—the international publication of the NPCIL—for his de-facto intent of permission to use the information and photos, that have been appearing frequently in Nu-Power, for this book.

I am grateful to my wife and two daughters for giving me the space and time to complete this book which spans over several decades. In particular, my eldest daughter has been functioning as an interlocutor between 'Notion Press', Smt. Megha Mehta, the DTP vendor, and me. I sincerely thank Smt. Megha Mehta for her work as a DTP vendor for this book. Both she and her team have completed this challenging work within just two months. I would also like to thank my son-in-law, Sri. Shireesh Dharap, for digitizing all the 109 photos that will go into this book. I thank my grandson, Sri. Aditya, for effectively and suitably captioning all of these photos that will form a crucial part of the entire book.

I convey my gratitude in advance to the Notion Press publishers for coming up with an excellent book both in the print and digital formats.

I thank google.com and other international and national media for my gleanings of data from their media outputs.

List of Abbreviations, Expansions and Explanations

Sl. No.	Abbreviation	Expansion	Explanation
00	&	And	
0	Hz	Frequency	A characteristic of electrical systems and of protective systems like relays
1	AC	Alternating Current	A measure of flow of electricity which changes its magnitude and polarity at regular intervals of time
2	AH	Ampere-hour	Capacity of a DC battery to deliver power in amperes/time
3	Ag	Silver	A noble metal
4	Al	Aluminium	An electricity-conducting metal
5	Bi	Bismuth	A metal having low electrical and thermal conductivity
6	Btu/B.th.U	British Thermal unit	A measure of heat energy output
7	CB	Circuit breaker	A device to break a circuit as protection against electrical fault
8	Chr	Chromium/ chromite	A mineral consisting of ferrous chromic oxide
9	Cl	Chlorine	A pungent yellow-coloured gas used for bleaching and other chemical processes
10	cm	Centimetre	A measure of length 100 mm; other units are dm, mm and metre (1000 mm)
11	CO2	Carbon dioxide	A toxic gas that is emitted from automobile exhaust, thermal power stations and human activities. It replaces oxygen in the atmosphere and causes air pollution. One of the greenhouse-gases that humans seek to curtail and save humanity

Sl. No.	Abbreviation	Expansion	Explanation
12	Cu. ft	Cubic feet	A cubic area in space made out of length, breadth and height
13	Cu. ft/sec	Cubic feet per second	Flow of fluid measure
14	DC	Direct current	A measure of flow of electricity from the operation of a generator or battery; the other major measure is Alternating current
15	DG	Diesel Generator	Generates electricity using fossil fuel
16	DM	Demineralization process	Removes minerals from water and renders it fit for use in nuclear reactors
17	Eqpt.	Equipment	Used in the erection of nuclear plants
18	GI	Gas insulated	A gas that improves insulation for electrical objects, and is used in circuit breakers and electrical conductor equipment
19	GT	Generator Transformer	Used for stepping up generated voltage to suit manifold electricity system usage particularly for High-Voltage Electricity Transmission
19(a)	GW	Giga-watt	1000 megawatt—a measure of electrical power
20	HD	High density	Refers to any material that has high weight per unit
21	H2	Frequency	An element characteristic of electricity generation which has a function of electricity relays. Value is specified as per second. It is 50 cycles per second in electricity supply systems
22	KA	Kilo-amperes	1000 amperes—a measure of flow of electrical current
23	Km	Kilo-metre	1000 metres—a measure of length
24	KW	Kilowatt	1000 watts—a measure of electricity power
25	Met	Metre	A measure of length (1000 mm)
26	mm	Millimetre	A measure of length (1/1000 of a metre)
27	Mccs	Motor control centres	They make power (electrical) available to equipment
28	Mn	Manganese	A metal used for the manufacture of electrical equipment

Sl. No.	Abbreviation	Expansion	Explanation
29	NH	National highway	A series of Trunk roads operated by the Ministry of road transport and highways. They are similar to State roads and Village roads maintained by the State and Panchayats, respectively
30	Ni	Nickel	A metal
31	OC	Over-current	A measure of electrical current that makes a relay break a circuit and thus protects equipment
32	ODC	Over-dimensioned consignment	Large-sized heavy equipment that require high-capacity cranes/jigs for movement
33	pH	A chemical characteristic	The characteristic defines the chemical nature of a liquid, particularly water
34	QA/QC	Quality assurance/ Quality control	A procedure that assures the meeting of prescribed specifications and conformance to them of equipment
35	QS	Quality Survey	Survey of measures available in a manufacturing plant to assure adherence to prescribed quality standards
36	RB	Reactor building	One of the several structures required in Nuclear Power Plants
37	SB	Service building	One of the several structures required in Nuclear Power Plants
38	SG	Steam Generator	Equipment that generates steam from hot feed water—the steam is used to run turbine generators and develop power
39	SS	Stainless Steel	The product that restricts corrosion due to oxygen and is thus used for the manufacture of critical equipment like reactors in nuclear power stations
40	sq.	Square	A product that measures and multiplies the length and breadth (diameter in case of circular cross-section) to arrive at a particular area/size (sq. mm, sq. metre, etc.)
41	TB	Turbine building	One of several critical structures required in any power station
42	TG	Turbo-generator	The equipment that generates electrical power by driving turbines with the help of steam

Sl. No.	Abbreviation	Expansion	Explanation
43	V/KV	Volts/Kilo-volts	A measure of electricity generation by windings used in generators. The other measure is current
44	VA/PF	Volt-amperes/ Power factor	A measure that is arrived at by multiplying voltage and current. This is a measure of inactive power developed by electricity generators. Real power comprises of watts, kilowatts, megawatts and gigawatts which in turn drives equipment such as motors among others. Power factor (watts divided by volt amperes) defines the ratio of effective use of power. It is at a desirable rate of 0.8
45	WH/KWH/ MWH	watt hour/kilowatt hour/megawatt hour	watthour/1000 watt hour/10 to the power of 6 watt hour : It is a measure of electricity output/energy and indicates the extent of output of a generator
46	WM	Waste management	Describes the ways in which a nuclear power station disposes off its radioactive waste produced at the station in both liquid and solid forms. It also adopts temporary and permanent measures for the storage of such waste
47	Exgrs.	Exchangers	These transfer hot/cold products from one body to another in an enclosure

About the Author

The author of this book is an Electrical Engineer who graduated from Andhra Govt. College at Kakinada, Andhra University, in 1953. He has acquired 46 years of experience—33 years in the electrical/nuclear field and 13 years in the machine lubrication field. He possesses a certificate titled 'Training in the operation and maintenance of Indian electricity systems' issued by the Power Ministry of the Govt. of India. He has a wife and two daughters with whom he lives in Pune. He has worked as a volunteer for over a decade and a half, contributing much to the upgrading of Civic life in Pune city.

Prologue

Nuclear power development in India over the last seven decades-true story and history

The objective of this book is to present before the people of this great country, a broader picture of nuclear power development in India. The book elaborates on how the programme for the harnessing of nuclear power as a front-line technology fuelled the economic and industrial development of India. The programme was initiated by the first Prime Minister of India, late Sri. Jawaharlal Nehru, and the Chairman of India's first Atomic Energy Commission of India (AEC), late Dr. H.J. Bhabha. Over the last 70 years, it has taken shape and progressed greatly.

This book was written to inform the general public about the methods adopted by the Department of Atomic Energy (DAE) of the Government of India (GOI). It also explains the strategies employed by the said authorities to achieve indigenous capability in the entire nuclear cycle, starting from the mining of nuclear fuel to the processing of spent nuclear fuel.

There are two new technical frontier fields wherein the country has scored significant successes. One of them is the Atomic energy field, particularly the power-generation field, and the other is the outer space application field.

The launching of satellites with huge volumes of orange-coloured engine exhaust gases, and the streaking rocket soaring into the sky flashed across TV screens attract instant and spectacular public attention. The running of nuclear power stations is, however, a silent and distant event that gets public attention only when the first-criticality is achieved in a nuclear unit. It may be pointed out that quite a few nuclear power-generating units have operated continuously for years (two-and-half-years and longer) without any interruption or operator intervention; some of them have been taken out only due to maintenance purposes. The above information has been pointed out only to highlight the significant successes achieved in the nuclear power-generation field and not for comparison. However, positive competition is much needed, and it does not harm anyone.

There are quite a few questions that may be asked as to why this book is needed. Some reasons have been mentioned below.

Nuclear power is well set to become an essential part of India's on-going power-generation programme.

In the 'introduction' to this book, we will be seeing how nuclear power-generation fits into this programme.

Another reason for a book of this nature is to gain more knowledge about nuclear power. At an age where knowledge is bound to be an asset, this book will truly serve its purpose. It is always easy to assume that you have to teach the reader a lot in the case of a book on science, but the truth is that the reader knows much more than you give him or her credit for. This book is based on this truth.

The book starts with the concept of the atom and the immense power path for the construction and destruction it possesses. The saga of the development of atomic power and its history over the last seven decades follows in the 39 chapters of this book. There is one more success story that we have to consider before we present our true story. The immense multiplication of telephone connections from 2 million about five decades back to about 1200 million now deserves all the praise. However, we import crucial items like switches, software, etc. whereas nuclear power development is entirely indigenous except for very high-capacity units like IGWe and 1.69 GWe. There is a need for more development and we will explore the same in this book.

Introduction

Man has always been a seeker of energy sources to fulfil his needs. Palaeolithic and Neolithic men and women might not have looked beyond the rubbing of stones to obtain sufficient heat to set up a fire and thus meet their sparse cooking needs. As men and women advanced socially and economically, his/her energy needs also expanded accordingly. As civilization advanced, humans looked for more resources to meet their increasing demands.

Coming up to the most recent centuries, we can note that steam met a large part of human energy needs in the 18th and 19th centuries. The use of steam in the energy field received a fillip with the building of the first steam-driven turbine alternator by Parsons in New Castle, England, in 1889. The workshop built by Parsons still exists today. The use of steam was then followed by hydraulic (water-driven) energy which made use of the potential energy of stored water to turn water turbines and electric alternators, thus generating Alternating current. The advent of steam and hydro-electricity turbines to produce electricity marked a significant stage in the evolution of modern means of producing electrical power on a commercial scale. The ushering in of the Industrial Revolution in the Western world about 150 years ago led to a demand for the generation of large quantities of electrical energy to fuel the rapid growth of industry and industrial applications. Thus, the concept of commercial production of electrical energy came into being.

The development and manufacture of large steam-driven turbo-electric generators played a significant role in meeting the energy needs of the rapidly developing industrial economy of the Western world over the last century and a half.

Fossil fuels played a crucial part as a primary source of input fuel to generating stations. However, the fossil fuel resources of the world are limited. The combined global energy resources of crude oil, natural gasoline, shale oil, natural gas and coal, as of 1974, were of the order of 80×10 to the power of 18 Btu (80 quads). These resources must have increased considerably since then, due to the discovery of more fossil resources and the increased exploitation of resources of lesser economic value. Notwithstanding these recent developments, these augmented resources may not be adequate to meet the ever-rising global energy needs, particularly those of the Third world.

The needs of some of these countries like India, China, Brazil, Argentina, Chile and the Republic of South Africa are burgeoning. The vast dissimilarity between the energy consumption of the Third world and the Western world may have risen from the fact that some of the more developed nations of the developing world have per capita annual electrical energy consumption of 400-600 units as against 10,000 units of the Western world. The per capita energy consumption in the Western world may stabilize at existing levels or perhaps at lower levels due to more efficient use of energy, whereas that of the developing world is bound to go up exponentially over the foreseeable future.

Global fuel resources

The fossil resources of the world have been estimated to last until the middle of the 21st century. There is, hence, an urgent need to look for alternative sources of energy. This alternative source must have, among others, the important capacity to drastically cut down greenhouse-gas emissions. As of 2005, greenhouse-gas emissions globally are of the order of 25 billion metric tonnes of carbon dioxide (CO_2) per year (800 tonnes per second). Over the last 4,00,000 years, CO_2 emission levels have been fluctuating between 200 and 300 parts particulate per million. Atmospheric temperature, on the other hand, has been fluctuating by about 10 degrees C—which is in perfect correlation with the CO_2 emissions and the melting of icecaps and glaciers from Alaska in the USA to the Antarctic.

S. No	Continent/ Country Group	2002			2010			2020			2030		
		Total	Nuclear	Nuclear Total%	Total	Nuclear	Nuclear Total%	Total	Nuclear	Nuclear Total%	Total	Nuclear	Nuclear Total%
1.	North America	4779	851	17.8	5239	884	16.5	6247	905	14.5	7299	894	12.5
2.	Latin America	1078	29	2.7	1303	34	2.6	1960	49	2.6	2993	61	1.9
3.	Western	3084	880	28.5	3481	876	255	4161	892	22	5002	827	15.5
4.	Europe	1758	299	17.0	1979	359	18.0	2521	488	19	3298	495	15
5.	Eastern	459	12	2.6	575	13.5	2.4	836	19	2.2	1203	37	2.8
6.	Europe	1176	19.6	1.7	1484	44	3.0	2200	77	3.5	3137	132	4.0
7.	Africa	600	484	–	761	–	–	1027	5.5	0.5	1372	18	1.2
8.	Middle East & South Asia South East Asia & The Pacific Far East	3157		15.3	3848	699	18	5402	990	185	7452	1161	16.5
	world Total	16091	2574.63	12.2	18670	2909.5	12.3	24354	34255	9.3	31756	3625	8.2

The energy table attached above shows the projected electricity consumption for the period 2010-2030 continent/country-wise. It can be seen from the table that the developed world with just 20% of the total population consumed 65% of the world's total electricity consumption, in terms of billions of units per year, as of 2010. This will go up to 70% as of 2020 and may go down to 60% as of 2030. It only goes to show that developing nations have to take enormous strides to catch up with the per capita energy consumption of developed countries.

Efforts made by the world to meet energy needs

We go back now to record the global efforts that were undertaken to meet the growing energy needs from the early 1900s, particularly in the nuclear field. We will go back as far as the early 1900s. The UK's Rutherford was responsible for splitting the atom in 1919. This was then followed by the crucial identification of the Neutron in 1933 at the Cavendish Laboratory in Cambridge (UK).

Otto Hahn's and Lise Meitner's predictions of the possibility of nuclear fission, and Enrico Fermi's demonstration of the fission reaction on 12.2.1942—as part of the Manhattan (USA) project—opened the floodgates of the generation of a significant amount of fairly cheap energy from uranium and other fissionable nuclear fuels.

It is relevant, though unfortunate, to record here that the first declaration of an atomic explosion was made on 16.7.1945, at Alamogordo, New Mexico State (USA).

(1) Nuclear energy

The world's first chain reaction between the nuclei of rare and heavy elements was produced in 'Fermis' at La Alamos, New Mexico, in 1942. It was only on 20.12.1945, that a nuclear reaction produced a usable amount of electricity for the first time at the US Dept. of Energy's National Engineering and Environmental Laboratory at Idaho State, USA. This nuclear reaction produced enough power to light up a string of four 100 W electric bulbs and proved that the production of electricity from nuclear reactions was a reality.

The above experiment was repeated the following day. The sixteen Engineers and Scientists associated with these experiments recorded their historic achievements by affixing their signatures on the concrete containment wall of the reactor on which the experiments were conducted. It was on the same day that an experimental breeder reactor which was housed in a small building, that stands even today in a wind-swept plain area in Eastern Idaho State (USA), saw its output touch 100 KW of power.

(2) The world's first nuclear power reactor

Hahn's and Meitner's visionary predictions became a reality when the world's first power reactor was built in Soviet Russia in 1954. This was a BMW test reactor. Calder Hall in the UK and Marcoule in France followed the Soviet's achievement in 1956. The USA's first reactor was built at a Shipping Port in 1957. It is pertinent to point out, at this point of our narrative, that the total global Nuclear Energy resources as of 1974 were 1800x10 to the power of 18 Btu against the total fossil fuel resources which were 80x10 to the power of 18 Btu.

The beginning of India's electricity generation

We will now look at how electricity generation has progressed in India since the early 1900s.

(1) Indian Electrical power-generation status

A Dynamo made by Crompton and Co. (owned by E.R.B Crompton) generated India's first unit of electricity in 1899. The first hydro-electricity generating unit of 130-KW capacity was set up in the year 1897 at Darjeeling, which is now situated in West Bengal. The first major power station, a hydro-electric one of 7.5-MWe capacity, was set up in 1902 at Shivasamudram by utilizing the waterfalls of the river Cauvery to cater to the requirements of the Kolar Gold mines. The capacity of the station was further augmented and the augmented station was active until the end of 2005.

(2) Installed Capacity

The production of appreciable quantities of electrical energy commenced only after the advent of Indian Independence in 1947. The total installed capacity as of 1947 was 2 GWe (2000 MWe). The installed capacity as of 2006-2007 was 150 GWe, of which hydro-electricity generating capacity was 34 GWe.

The estimated hydro-electric reserves of the country are 84 GWe, of which a little over 38 have been utilized thus far. These resources, especially the larger-capacity ones, are located mostly in the Central-Northern and North-Eastern parts of the country, and are unevenly distributed geographically in the rest of the country. The Northern and North-Eastern regions have a potential of 59 GWe (70 of the total potential). The single State of Himachal Pradesh is expected to possess 40of the total hydro-electric resources, while Jammu and Kashmir possess 18 of the total resources.

(3) Environmental Considerations

Enhanced environmental and bio-diversity considerations are bound to restrict the rapid growth of further installed capacity of hydro-electric plants (HEP) in the country. An idea of the environmental concern that is bound to be caused may be gauged from the fact that a 1000-MWe-capacity HEP is likely to submerge 5000 to 7500 acres of land. On the other hand, a 1320-MWe (6x220 MWe units) Nuclear Power Plant will require only 120 hectares (300 acres) of land, including housing and other related facilities. Thus, the land requirement for HEP stations is gargantuan by any standards. HEP stations are peak-hour generating ones, and the setting up of one of these stations has to conform to a desirable 40 share of the total installed capacity.

(4) Thermal Generating Capacity

Thermal energy-generating plants in India now account for 70of the total installed capacity. Limited availability of good-quality coal (low-ash content) and the concentration of coal mines in the States of Bihar, Chhattisgarh, Jharkhand and West Bengal militate against a rational geographical distribution of coal-fired thermal stations in the country.

India now imports 70% of its crude oil requirements (as of 2005-2006); this share remains the same as of 2021. The quantity of crude oil had been 85 million tonnes out of a total requirement of 120 million tonnes (as of 2005-2006); crude oil prices reached $70 per barrel during the above-mentioned period—it is expected to increase further. As a result, the use of petroleum oil as a fuel will become more unviable. The cost of generation of one unit of electricity (as of 2005-2006) varied from Rs. 0.70 paise to Rs. 1.10 for oil use; and Rs. 1.05 to Rs. 2.00 for gas use; while the nuclear power cost at that time was varying from Rs. 0.55 to Rs. 0.65 per unit.

Indian Nuclear energy activities

Nuclear energy technology is a frontier area; one decision that was taken by the Govt. of India in the early years of Indian independence was to develop this technology. The decision was made so that the country could incorporate this technology into its growing technology vista. This is what India's First Prime Minister said while committing the Government to the development of Nuclear Energy: 'I do not see any way out of our cycle of poverty except utilizing new sources of power which Science has placed at our disposal'.

F(ii) Comparative table for land and other requirements is placed below:

Some of the other advantages of Nuclear Power Plants are also listed below :

(i) Coal-fed thermal plants Capacity - Land Requirement
(a)Indigenous Fuel :   4800 MWe ------ 958 Hectares

               5800 MWe --------   1048 Hectares

(b)Imported fuel : 5 of 800 MWe---- 428.52 Hectares. -----------

(ii) Nuclear Plants : 4700 MWe -   550   Hectares-------- 6 of 1000 MWe - 650 Hectares

Fuel requirements:
(a) 1 GWe (1000 MWe) coal-fired plant: 7800 tonnes per day
(b) 1 GWe (1000 MWe) nuclear plant: 125 tonnes per annum
Giant scales of the economy are thus achievable by going in for Nuclear Power Plants.

(1) Nuclear Facilities

The stage was thus set for the development of a host of nuclear actions in the country. The Atomic Energy Act was enacted in 1948. The first Atomic Energy Commission was set up in the same year with Dr. H.J. Bhabha as its first Chairman; J.R.D. Tata was one of its members. The Department of Atomic Energy was set up in 1954. A series of nuclear facilities—wherein original research in nuclear and applied Sciences could be carried out—were also set up. The details of these facilities as well as the setting up of nuclear power stations in India are furnished in the following chapters of this book.

(2) Nuclear Power Project Cost

The cost of setting up nuclear power stations and other types of power stations was a matter of debate even as early as the 1950s due to the cost grounds. The answer of Dr. H.J. Bhabha, the father of nuclear power development in India, was the oft-quoted statement in this regard: 'No **power** is costlier than **no power**'. The truth of this statement can be seen even now in the early 21st century when daily power cuts and the enforcement of load-shedding measures have become the order of the day in India. This in turn leads to consequent losses in the manufacturing industry amounting up to thousands of crores of rupees; there is an eventual loss in export earnings and a threat of loss in industrial employment as well.

(3) An Unpleasant example

Coming back to the development of commercial nuclear power production, it is unfortunate to see that the military application of Nuclear Energy preceded its peaceful application—one such occurrence is the horror of Hiroshima and Nagasaki which has caused a severe reaction in the psyche of mankind.

(4) Balances to nuclear advantages

It would be difficult to predict today: the feelings, hopes or fears of an Oppenheimer or a Van Braeun (if either of them were alive today), who in their ways contributed to two different facets of the same parent technology. This technology, if not controlled or harnessed in the interests of the future of mankind, would result in the release of awesome, explosive power especially in the hands of maniacs, terrorists or power-crazy dictators. If, however, we look beyond these possibilities of our times and the first display of release of power as described by these Scientists/Technologists, we may be able to have a better appreciation of the significance of these discoveries. The secrets of nature or Science, or both, are constantly being discovered in myriad ways. Some of these may not always be true blessings. We have to live with their unnecessary offshoots as well. It is, however, not beyond the genius of mankind to largely diminish, if not altogether eliminate, the evil offshoots of these discoveries and to increase, to the maximum extent possible, the constructive contribution of these discoveries.

(5) Safeguards

The effects of radiation engender fear in most of mankind. Increasing environmental awareness and the necessity of preserving the globe's bio-diversity has caused many activists to take up cudgels against nuclear activities. It is essential to consider their views while making nuclear power development and related nuclear activities safer and more environmentally friendly. Safeguards against radiation risk in nuclear activities are constantly being incorporated. But despite this, the mere mention of the term 'Nuclear activity' continues to evoke a response, bordering on 'passionate', both from the proponents and opponents of nuclear power development. Shorn of rhetoric, which is often invoked by both sides of the argument, there appears to be a very large common ground in the perceived, if not stated, objectives of these two 'devoted' groups.

Although these two groups seem to be in polarized opposition, the concerns of the groups may be seen to converge on the accountability of those occupied in the 'nuclear activity regime' regarding the following:

(i) The safety, integrity, reliability, operability and maintainability of nuclear equipment and installation

(ii) The cause of no damage to the environment or bio-diversity

(iii) Other related obligations to the public on account of the result of every one of these activities

(6) Safety consideration

The anti-nuclear lobby voices its criticism of nuclear activities on two main grounds. The first is the possible hazard due to radiation caused by fission, and the second is the disposal of radioactive wastes with a long half-life. The entire philosophy of the nuclear power station design is to protect against the first-cited hazard. These two hazards have been elaborately dealt with in Chapter 4 of this book.

(7) Global nuclear Power Scenario

There are 444 operating nuclear power stations globally as of 2020 with a total installed capacity of 394 GW, generating about one-sixth of the world's total electricity generation. There are 52 more such plants, with a total capacity of about 52 GW that are currently under construction. The number of full reactor-years of operation of these 444 units was over 18,900 full-power reactor-years. Light-water reactors (LWRs) accounted for 75% of this total. These LWRs are comprised of 55% Pressurised-water reactors (PWRs) and 45% of Boiling-water reactors (BWRs). PWRs produce 60% of the world's nuclear power.

We go back to the year 2021 again to look at the units under construction globally. Fifty-two units with an installed capacity of approximately 52 GW are under construction. About thirteen of these are in India; there are four each being constructed in China and Ukraine. Japan and Russia had three units each, while Pakistan and Bangladesh had two units each. France, Germany and the USA, each of which already have a large installed capacity, had none in construction. Germany, on the other hand, had a proposal of entirely doing away with its nuclear power and substituting it with other green-energy sources.

At the end of 2020, a total of 33 countries had built up commercial nuclear reactors.

(8) Greenhouse-gas emissions: saving due to global nuclear power-generation

As of 2002, world-wide nuclear power-generation had saved 2.3 billion tonnes of CO_2. More than 50% of this quantity was saved by France and the USA. Global savings of CO_2, during the 1990s, using hydro power was 7% and by nuclear power was 9%.

(9) Further Nuclear Power Development in India

We had seen earlier that a conscious effort was made by the Central Govt. when it decided to invest in the frontier technology of nuclear power at the advent of independence. This effort gathered strength in the succeeding decades. India today (2010-2020) is counted among half a dozen countries in the world which have the facilities for a complete nuclear cycle—commencing from exploration/mining of nuclear fuel and rare ores; refining of ores; fabrication of nuclear fuel involving machining of rare metals; design, construction and commissioning of Nuclear Power Plants; processing of spent-fuel and short-term/long-term storage of both liquid and solid nuclear waste.

As of today, India has 23 power plants with a total installed capacity of 7.22 GW (7,220 MW). It has notched up a total of over 550 full-power reactor-years with no nuclear-caused fatal accidents as of 2020. Thirteen nuclear units with a total capacity of 12.3 GW are under construction.

(10) Global nuclear accidents

There have been nuclear accidents around the world. The International Atomic Energy Agency (IAEA) has classified such accidents in ascending levels of severity from 1 to 7. The Chernobyl accident which occurred in Russia was placed at level 6. The three-mile accident in the USA was also placed at level 6. These accidents have been dealt with in great detail in Chapter 19 of this book.

(11) Nuclear/non-nuclear accidents in India

There has been only one non-nuclear fatal accident in the history of nuclear power operations in India. The incident took place during a routine maintenance check and was a non-nuclear one. The details of the accident have been furnished in Chapter 2 of this book. Other accidents that have happened are listed below:

(1) A fire accident during the dome construction in RAPP 1

(2) A fire in the 220-KV Switchyards ofKAPP 1 and KAPP 2

(3) A fire in the TG set of RAPP 1

(4) A failure of the Inner Containment Dome during the construction of Kaiga 1

The details of these incidents have been explained in Chapters 2 and 3, respectively. The most severe of these incidents was the fire in the TG set of RAPP 1 which occurred on March 31st, 1993. It was classified as a level 3 accident according to the IAEA Code.

(12) The impact of Nuclear Power development on the quality of Indian life: CSR work

We have, in the respective chapters of our Nuclear Power Projects, detailed the improvements in infrastructural work carried out by the various operating stations in the vicinity of these stations. The operation and maintenance of these stations are quite expensive. Big strides in industrial development, which were triggered by the availability of quality power from these stations, have also been touched upon.

It is of great interest to record here that the NPCIL has set up a 'Tier-II' Committee to oversee Corporate Social Responsibility (CSR) in various projects. The unit was set up during 2011-2012. The committee had ear-marked initially, during 2014-2015, a fund of Rs. 52.54 crores for 157 new projects. These comprised 18 in health, 84 in education, 32 in infrastructure, 6 in skill development, 12 in sustainable development and 5 in other general projects. At the time the fund was planned, schemes amounting to Rs. 601.95 lakhs were in progress while Rs. 3961.63 lakhs were being allocated for plans. The details of the funds spent during 2014-2015 in lakhs of rupees were as follows:

1) TAPS : 509.00
2) RAPS : 643.42
3) MAPS : 341.12
4) NAPS : 242.85
5) KAPS : 374.48
6) Kaiga PS : 276.50
7) Kudankulam : 141.63

The above-listed funds work out to about Rs. 2,630.15 lakhs in total.

(13) The Impact of nuclear power development on Indian life

The rest of the planned projects were in progress. Of particular interest to environmentalists was the establishment of five artificial bays in the Bay of Bengal shores near Kalpakkam, Tamil Nadu—the site of the MAPP nuclear units. These bays are intended to stimulate marine life growth and fortify the shores.

We conclude the narration of the CSR benefits to the local population of several project sites across India, with the announcement of the NPCIL board to allot an additional CSR fund of Rs. 12,500 lakhs during 2014-2015.

Rationale for this book

In the prologue of this book, we are given the need to come up with a book of this nature. We will now come up with a much mightier reason. It has often been said by eminent persons, both in India and abroad, that India is an important country in the international scenario which has sadly been lacking in the historical account of important developments in many spheres of life. This is quite true both in our honourary past and particularly in recent centuries; the narration of events of important developments will keep the present and future generations aware of their historical roots. This book, in particular, achieves this with regard to technical developments.

It is with this important criterion in mind that this book has been compiled. Dealing with the development of nuclear power in our country over the last seven decades, this book can be titled a 'true story and history' in short. The narration has been rendered in a language that can be easily comprehended by all kinds of readers from both the technical and non-technical (general) fields and is thus recommended for regular reading.

The Genesis of Atomic Energy Development in India

1.1 The Atomic Energy Act

The Atomic Energy Act was passed in 1948. This was one of the first Acts passed by the Indian Parliament after its independence from British Rule on August 15th, 1947. The Act was piloted by Pandit Jawaharlal Nehru, who was not only the first Prime Minister of India (PM) but also its first Minister for Atomic Energy.

The Act conferred onto the Government of India: the sole and absolute rights in regards to the research, development and exploitation of India's Atomic Energy Resources. It also envisaged the creation of an Atomic Energy Commission (AEC); the commission was charged with the duty of laying down policies and exercising control through agencies under its authority over every activity related to the development of Atomic energy resources of the country.

1.2 The Constitution of the AEC

Pandit Nehru was instrumental in appointing Dr. Homi Jehangir Bhabha (H.J. Bhabha) as the first Chairman of the AEC. He was appointed to the post in 1948. Later on, he was also appointed as the Secretary to the Department of Atomic Energy (DAE), Ministry of Scientific Research in 1954. The AEC was to consist of a minimum of three and a maximum of seven members, with one member being in charge of finance. The Atomic Energy Act stipulated that the member for finance shall also be an ex-officio Secretary to the Government of India in the Dept. of Atomic Energy in financial matters except in so far as such powers have been, or may in the future, be conferred on or be delegated to the Department.

1.3   The AEC

1.3.1   General

The Chairman was empowered to carry out the day-to-day activities of the commission with the provision that the member for finance could bring issues to the notice of the Finance Minister or the Prime Minister, in the event of any disagreement with the Chairman of the AEC on matters within his charge (i.e. finance) for orders. The Chairman was empowered to overrule other members of the commission which in turn was to meet at least once in six months. The headquarters of the commission was to be in Mumbai, Maharashtra.

1.3.2   Functions of the AEC

The functions of the AEC are as set out in the Government of India (GOI) notification dated 10.08.1948 and as amended via resolutions dated 1.03.1958, 6.3.1962, 24.1.1981 and 20.3.1984, respectively. These functions are stated below.

The commission shall be responsible for:

(A) Formulating the policies of the Department of Atomic Energy for the consideration and approval of the Minister of Scientific Research/Prime Minister as the case may be

(B) Preparing a budget for the Dept. of Atomic Energy for each financial year and getting the approval of the Govt. for the same

(C) Implementing the Government's policies in all matters concerning Atomic Energy

The commission shall have powers, within the limits of the budget provision approved by the Parliament—both administrative and financial—for carrying out the work of the DAE.

1.3.3   The Administrative Officer

The Chairman in his capacity as Secretary to the GOI in the DAE shall be responsible under the concerned Cabinet Minister/PM for arriving at decisions on technical questions and advising the Govt. on matters of Atomic Energy. All recommendations of the commission on policy and allied matters shall be put up to the concerned cabinet Minister/PM through the Chairman of the AEC.

The Chairman may authorise any member of the Commission to exercise on his behalf, subject to such general or special orders as he may issue from time to time and of his powers and responsibilities as he may decide.

The administrative officers of the AEC and the Department of Atomic Energy were earlier housed in the old Yacht Club building near the Gateway of India. This building is nearly two hundred years old and was built almost entirely with teak wood. This building has been declared to be one of the heritage buildings of Mumbai City.

A modern three-storied building has since been built in the same premises. All the offices of the AEC and the DAE are now housed in this new building.

1.3.4   Initial projects of the DAE/BARC

The very first projects taken up by the DAE at the Bhabha Atomic Research Centre (BARC), Trombay, during the early fifties are listed below:

1. Rare earth factory in the South site
2. Electronics division in the South site
3. Chemical Engineering Division in the South site
4. Graphite fabrication facility in the South site
5. Uranium plant at the South site
6. Engineering halls 1 to 6 at the North site
7. Central workshop at the North site

All the above projects were set up with the basic objective of building plants where research and development in the various disciplines required for a fledging Atomic energy development could be carried out. Concurrently taken up were the modular laboratories and radiological laboratories at the North site. The Van de Graaff project and the Plutonium processing plant were the other two important projects at the North site that were taken up for construction in the early 1950s.

1.3.5 The Commissioning of Research and Development Projects

1.3.5.1 *'Apsara': The Swimming pool reactor*

The 'Apsara' reactor was the very first research reactor that was constructed and commissioned by the DAE in August 1956, in the South site at Trombay. It was also Asia's first reactor. It is a swimming pod-type reactor that uses enriched uranium as fuel. The project would later go on to provide the operating staff with valuable experience regarding Physics, Health Physics, Power levels and related operating parameters. The fuel for the reactor was supplied by the UK under an agreement signed on October 6th, 1955. As of 2005, the reactor is forty-nine years old. 'Apsara' is still operational and seems good to last for many more years of satisfactory operation.

1.3.5.2 *The 'CIRUS' Reactor*

Next to come up was the 'CIRUS' reactor (Canada–India Reactor Utility Services) that was set up in the North site, Trombay. It was a 40-MWt research reactor fuelled by natural uranium, moderated by graphite and cooled by light water. It was the first large-scale research reactor which would provide valuable experience for operators and maintenance crews. These individuals were expected to be called upon to operate and maintain commercial Nuclear Power Projects that were to be set up shortly. The reactor was commissioned in 1960.

1.3.5.3 *The Purnima Reactor*

Purnima-3 (R-5 reactor, Trombay) went critical on November 13th, 1990. It used 590 gms of Uranium (U233), Aluminium-alloy plates as fuel, with light water as coolant and Beryllium Oxide as a reflector (moderator). The reactor has only 10 litres of volume.

1.3.5.4 *Engineering Halls 1 to 6*

Developmental engineering work, required for the construction of commercial-size nuclear power reactors, was undertaken in these halls. 'Zerlina' (Zero Energy Lattice Investigation Nuclear Assembly) was housed in one of these halls and commissioned in 1962. Another hall housed the facility for trying out loops which simulate the nuclear heat-exchanger components for transferring heat from nuclear fuel to the coolant, leading to the production of steam for running the turbines.

1.3.5.5 *Modular laboratories*

The laboratory buildings, founded on Y-shaped concrete columns, are one of the longest in India—measuring about 300 metres. The laboratories are provided with all the facilities required for research and development in various disciplines of Nuclear Science. About 1500-2000 personnel work in these laboratories.

1.3.5.6 *The Central Workshop*

A Central workshop was set up in the North site, to which a foundry has been attached with all the ancillaries—essential machinery and tools required for manufacturing the nuclear plant. It fabricates fuelling machine heads, air-locks, end-fittings and other critical nuclear components required for the nuclear plants.

1.3.5.7 *Research and related facilities*

As we discussed so far, several facilities were established by the DAE for research and development in several disciplines related to the utilization of Atomic energy.

1.3.5.8 *Utilization of Nuclear Energy*

It was demonstrated in several Western countries, during the early 1950s, that the energy of the atom could be utilized for the production of commercial electric power. Soviet Russia was the first nation to set up a nuclear power station while the UK and the USA were the first Western nations to exploit this source of energy.

1.3.5.9 *India's first Nuclear Power Station: a recourse to foreign technology and 'know-how'*

It would be some years before India would be in a position to exploit nuclear power on its own. Dr. H.J. Bhabha felt that valuable time would be lost if India were to wait the time required to develop such a 'know-how' on its own. He felt that it would be a good beginning to bring in the foreign 'know-how' on a turn-key basis for constructing and commissioning commercial power reactors, and later deploy Indian Engineers and Scientists on operation and maintenance. These Scientists even went on to initiate the project process. Details of this strategy will be explored in Chapter 2.

Tarapur Atomic Power Project (TAPP) Units 1 and 2

2.1 The Genesis of the First Nuclear Power Project in Tarapur, India

2.1.1 The Intention of Dr. H.J. Bhabha

It was the intention of Dr. H.J. Bhabha, the first Chairman of the AEC, to prove the relevance of nuclear power-generation to a developing country like India. He also wanted to demonstrate the competence of Indian Engineers and Scientists to operate successful nuclear power stations.

2.1.2 Preparation of specifications and the Award of contract

The work related to the preparation of specifications and tender documents for the first nuclear power station in India was taken up in 1959. The offers received for the work were reviewed, and a decision was taken by the AEC to go for the construction and commissioning of a Boiling-Water Reactor (BWR) with 2x190 MWe units on a turn-key basis. The station was to be located at Tarapur village, near the Boisor Railway Station (Western Railway-WR) in Thane District, Maharashtra. The location decided was as per the recommendations made by a site selection committee set up by the AEC sometime earlier.

2.2 The Criteria for the Site Selection

It would be useful and proper at this stage of our narration to recount some of the essential requirements for the setting up of nuclear power stations in India. The criteria are classified below:

(1) Economic and National Security-related aspects

(2) Aspects related to natural and engineered safety features.

These aspects are further discussed in the upcoming sections.

2.2.1 Economic Criteria

2.2.1.1 *The Distance of the coal-bearing areas*

If the distance between the coal-bearing mines and the power stations is maintained at 700 km or less, it would make the transportation costs cheaper.

2.2.1.2 *The Density of Population in the vicinity of nuclear power stations*

The density of the population around the stations should be as low as possible. Usually, an exclusion zone of 1.5 km radius from the proposed site is enforced and no human habitation is allowed in this area.

2.2.1.3 *National Security*

The site should not be an easy target for attack by hostile countries or terrorist forces. Hence measures have to be taken to prevent the same.

2.2.1.4 *Cooling water for the stations*

There should be easy and economical availability of secure cooling water which is highly essential for the smooth functioning of the power stations.

2.2.1.5 *Accessibility of the site*

There should be proper and reliable accessibility by road, rail or sea/water bodies for the transportation of bulk over-dimensioned consignments for the station.

2.2.2 Safety features

2.2.2.1 *Natural Safety features*

There are safety features that need to be considered to counter problems that arise due to the following factors:

(1) Demography

(2) Meteorology

(3) Geo Hydrology

(4) Seismology

The effects of these natural phenomena on the proposed power stations are studied in advance and proper safeguards are later built into the design and construction of the said stations.

2.2.2.2 *Engineered safety features*

The engineered safety features that are to be built into the design of nuclear power stations are:

(1) Accident prevention
(2) Accident limitation
(3) Accident consequence limitation

These methods are adopted to plan the reactor design. The physical causes of reactor system failure are further analysed for pre-determination of accidents. Mathematical calculations of the probability of accidents are made to estimate the likelihood of failure. A Probabilistic Risk Assessment (PRA) method is adopted to arrive at a numeric safety value.

2.2.2.3 *Core Damage Frequency (CDF) and Large Early Release (of radioactive gases) Frequency (LERF)*

Typical releases are of two kinds:

(a) It is of the order of more than 1 multiplied by 10 to the power of 14 Becquerel of Caesium 137 into the atmosphere due to core meltdown
(b) It is expressed as a fraction of the fuel core inventory.

The USA's regulatory commission introduced PRA in 1973. It specified CDF for Nuclear Power units of the order of 10 to the power of minus 4 per reactor-year. It also adopted CDF of 1 multiplied by 10 to the power of minus 5 per reactor-year for future light-water reactors. LERF for many countries is 10 to the power of minus 5 or less. We may point out that the Fukushima (Japan) reactor crisis resulted in the release of 10% of the core inventory. The Chernobyl release was of the order of 20-40% of Caesium 137 and 50-60% of Iridium 137 of the total core inventory.

The Atomic Energy Regulatory Board (AERB) of India has adopted CDF and LERF for future designs as 1 multiplied by 10 to the power of minus 5 per reactor-year and 1 multiplied by 10 to the power of minus 6 per reactor-year, respectively.

2.2.2.5   NRC and RFNRSA study

The USA's Nuclear Regulatory Commission (NRC) and Russia's Federal Nuclear and Radiation Safety Authority (RFNRSA) studied V-338 LWR (an earlier version of VVER-104) at the Kanine nuclear plant during 1995-2004. They later conducted an internal initiation event and arrived at a CDF value of 2.39 multiplied by 10 to the power of minus 5 per reactor-year. The IAEA has categorised VVER-14 as a generation-III category of reactors.

2.2.2.6   Provision of Features in HWPTRs

The provision of features in the design of Heavy-water moderated Pressurized Tube Reactors (HWPTRs) adhering to the above-mentioned requirements will be discussed in Chapter 4. These provisions will also apply to BWRs and PWRs.

2.3   General and special features of the TAPP Site

Tarapur was a small village at the time of construction of TAPP units 1 and 2. The village, apart from fishing activities, boasted several 'Chikku' (also known as Sapodilla in English) orchards—many of which were formed by the members of the Parsi community which constitutes a fair sprinkling of the local population. Bullock carts constituted a major means of transportation at the very beginning of the construction of TAPP 1 and 2

So far as the twin nuclear units were concerned, one of the big advantages for the transportation of heavy equipment was the adjoining Arabian Sea. One of the largest pieces of heavy equipment, namely the reactor vessel, was transported by utilising barges and high-capacity cranes.

2.4   Details of TAPP Units 1 and 2

The TAPP units 1 and 2 are reportedly the first twin BWRs in the world. The power station is also the first station in the world to house two units in a single reactor building.

2.4.1 Reactor and auxiliaries

2.4.1.1 *Reactor vessel, moderator and coolant*

The reactor is a 660/707 MWt 200/210 MWe dual cycle boiling-water reactor with hemispherical top and bottom ends. It utilises light water both as a moderator and coolant.

2.4.1.2 *Reactor fuel assembly*

Each reactor contains 284 fuel bundles of 2% to 3% low-enriched uranium (LEU) in the form of uranium oxide pellets housed in Zircaloy Pins of 6×6 (36 pins) or 7×7 (49 pins) bundle material matrix serving as the fuel core. One of these pins is a light-water pin that contains a reactor-grade pin and is used for better moderation of the fission process.

2.4.1.3 *Reactor control and regulatory system*

Reactor control is affected by the bottom entry rods in a cruciform shape. This shape provides for an optimum flux distribution in the fuel core, while this arrangement leaves the top of the core free to facilitate re-fuelling. These rods consist of sheathed stainless steel (SS) tubes in a cruciform array filled with Boron Carbide powder. There are 69 control rods which control the reactor's power. The control rod is attached to a control drive system which acts as a fail-safe measure and shuts down the reactor whenever any operational parameter exceeds the design limits.

A poison injection system has been introduced as a back-up to the control system. This back-up system renders the reactor sub-critical in the event of failure of the control rods to act. The back-up system consists of sodium penta borate solution (13.5%) by weight, which is injected into the reactor by a high-pressure pump.

Low-level and high-level trips are also provided. The low-level trip is set at 152 mm while the high-level trip is set at 940 mm. Zero power corresponds to 254 mm height and full power is attained at 508 mm height. The reactor set-back is at 0.5% of reactor power per second.

The reactor control assembly consists of a central cruciform tie-bar, two end-transition castings and four U-shaped sheaths.

These components are welded together to form a structurally rigid housing for the neutron-absorber Boron Carbide-filled poison rod.

2.4.1.4 *The Steam cycle: how the reactor follows the load systems*

The initial steam cycle is a dual-cycle system. The Nuclear Steam Service Generators (NSSGs) are tubed with 304s stainless steel tube material. The primary steam is produced in the reactor vessel at 70.3 kg/cm. sq. and at 285 degrees C temperature which goes directly to the high-pressure (HP) Turbine. The by-pass valves are used instead of steam discharge valves for regulating the main steam-line pressure; the steam from the by-pass valves is condensed and routed to the fuel water cycle system. The differential temperature across the reactor is 46 degrees C. The hot coolant is also circulated through the second-system steam generators (SSGs) which are assigned 2 numbers per reactor at a pressure of 35 kg/cm. sq. This secondary system steam goes directly to the lower stages of the HP turbine. The interesting feature of the SSGs is the load-following characteristic of the reactor; an increased load demand, and hence of steam, is met by the SSG within its steam range (25%) of the total steam requirement.

Before we move to explore more on the above-mentioned characteristics, we note the details of the NSSGs. The NSSGs are vertical U-tube-type heat-exchangers with 1589 U-tubes in each SSG.

Now back to the above characteristic. Increased draw of steam from the SSGs would cause sub-cooling of the main coolant circuit of the reactor. The decrease in coolant temperature results in the acceleration of the fission process due to negative temperature. The coefficient of reactivity thus makes the reactor follow the load.

2.5 Power consumption

The power consumption of the station's auxiliaries is 2.5% that of the generated power (less than 5 MWe).

2.5.1 The Turbo-generator set

TAPP units 1 and 2 were functioning on a GE-make (General Electric, USA) with 220-MWe/248MVA, 12 KV hydrogen-cooled, 50 cycles per second frequency, 1500 rpm steam-driven turbo-generator (TG) set.

2.5.1.1 *The TG auxiliaries: the excitation system*

Shaft-driven amplidyne-controlled excitation has been provided. Apart from this, a provisory panel, automatic voltage regulatory panel and generator control panel have also been provided.

2.6 Contract for the execution of site work

The contract for the execution of the power station works on a turn-key basis and was signed in Washington on 09.05.1964. The main contractors for the project work were International General Electric Co., (USA) and the International Electric Co., (India) Pvt. Ltd. The contract was concluded as a result of an agreement for co-operation in the project by the Governments of India and the USA.

2.7 Foreign Suppliers/Contractors

Equipment for the nuclear portion of the station, including nuclear fuel, was supplied by the International General Electric Company's Atomic power equipment in San Jose, California. The reactor vessels were manufactured by Combustion Engineering (USA). The two TG units and other major electric equipment were manufactured by International General Electric Co., (USA), while Bechtel Corporation (USA) served as the Engineering contractors for the station's construction review and inspection.

2.8 Plant construction and commissioning

The International General Electric Company (USA) was charged with the construction and commissioning of TAPP units 1 and 2 because the scope for indigenous contribution to the project was very limited. It included the construction of township facilities, supply of materials for civil work and the erection of structures for the 220-KV switchyard and switchyard equipment. The switchyard work was farmed out to the Maharashtra State Electricity Board. The scope of this project has already been discussed in Chapter 1.

2.9 Materials for Civil Work

Over 100,000 tonnes of materials like sand and metal and tens of thousands of re-inforced steel were procured by Engineering contractors from the Indian industry. As discussed earlier, the tender specifications for the plant were drawn up indigenously and the site work inspection was also done similarly.

2.10 TAPP township

The township was built indigenously to house Indian and foreign personnel therein. This work was taken up during the middle of 1964. A guest house was also constructed to house single officials from India and abroad. The township and the guest house were constructed 7 km from the site.

2.11 First criticality of TAPP units 1 and 2: taking over of units by the DAE

The construction of both units was completed successfully by the end of 1968. The commissioning of various systems of both units was also completed. The first-criticality of Unit 1 was achieved on 01.02.1969, while that of the second unit took place on 27.02.1969. Unit 1 was synchronised with the Western Regional Electricity Grid (WREB) on 01.04.1969 while unit 2 achieved the same on 05.05.1969. It was a red-letter day in the history of nuclear power in India when power was fed from its first station at 20:52 hours on 01.04.1969. The capacity of 190 MWe from unit 1 was also the highest in India at that time. The operation of the station was taken over by the DAE from the International General Electric Co., (GE, USA), in November 1969. The station was declared commercial in the same month. The station was dedicated to the nation by the late PM Indira Gandhi on 19.01.1970.

TAPS 1&2 Main Control Panel

TAPS 1&2 - Main Control Panel

2.12   Re-fuelling

This is an off-load operation and is carried out during a planned shut-down. About 30% of the total fuel inventory is replaced every 18-20 months. One reactor is taken up at a time so that the continuity of power-generation at the station is maintained. Fuel management was being carried out by GE until 1976. The DAE has since taken over this responsibility as well.

2.13   Fuel supply agreement with the USA: subsequent problems and their solutions

2.13.1   Agreement with the USA

The Tarapur reactors use 2% to 3% enriched uranium oxide fuel. It can be noted that natural uranium contains only 0.7% of U235 which is required for the fission process. Hence nuclear fuel was being imported from the USA in the form of enriched uranium hexafluoride to meet the demands of the project. The enriched uranium hexafluoride was being converted to enriched uranium dioxide at the nuclear fuel complex at Hyderabad (AP) for use at Tarapur. The Government of the USA was committed—as per its agreement with the Govt. of India—to supply the fuel for 30 years from the date of the agreement, i.e. 24.19.1993. This commitment followed on the heels of the tri-partite agreement signed between the USA, the IAEA and India.

2.13.2 Change in stance of the USA

The US Government, however, was not happy when India carried out a 'Peaceful Nuclear Explosion' (PNE) at Pokhran in Rajasthan State, in May 1974. Increasing difficulty began to be experienced by India in obtaining nuclear fuel from the Government of the US. This stance of the USA got tougher as India subsequently refused to sign the 'Non-Proliferation Treaty' (NPT). The USA banned the supply of fuel in 1980 as per its Domestic Non-Proliferation Act (1976), which was passed by the US Congress.

2.13.3 Alternative arrangement for fuel supply

The Govt. of India subsequently negotiated successfully with the French Govt. for the import of enriched uranium fuel. This source also dried up in October 1993. The Government of India concluded, since then, on a commercial contract with the 'Nuclear Energy Corporation of China' under which China agreed to supply 2.4% enriched uranium under the 'material specific safeguards' as per Article III point 2 of the NPT, and not under full-scope safeguards. The first supply of fuel under this contract reached the nuclear fuel complex (NFC), Hyderabad, on 31.01.1995. This supply signified a landmark in the field of partnerships between the two countries on the peaceful use of Nuclear Energy. China supplied the fuel until 2002. Russia stepped in after 2002 and has since been continuing its supply of the fuel. The latest agreement for supplying 60 tonnes of fuel was signed on 17.03.2006.

2.13.4 Indigenous efforts towards the manufacture of enriched uranium

Two bundles of indigenously designed and manufactured mixed oxide UO2-PUO2 fuel were loaded in TAPP 1 during the 13th re-fuelling outage of 1994. Subsequently, ten more such indigenous bundles have been introduced into the core. Fuel assemblies of seven fuel assemblies of 7x7 arrays, instead of the earlier 6x6 arrays, were developed to improve performance. The plutonium of the mixed oxide consists of four isotopes, namely P239, P240, P241 and P242. The fissile plutonium is contributed by P239 and P241 and comprises 77% of the total plutonium dioxide content. It is quoted on the cards that future enriched uranium requirements of TAPPs will be met indigenously.

2.14   De-rating of TAPPs 190-MWe reactors

The 2 TAPP reactors have been de-rated to 160 MWe since 1986, as the secondary steam generators have been taken out of service following the deteriorating condition of their tubes.

2.15   Post-construction facilities provided at TAPPs

2.15.1   Spout-fuel storage

A facility designed to enable the storage of spent-fuel from the reactor for thirty years was initially built. A third facility has been augmented by 300% in recent years. The agreement with the US envisages the return of spent-fuel to that country if requested. But so far, the USA has not requested the return of spent-fuel because it did not have a commercial-size re-processing plant at that time. It may not ask for the return of spent-fuel in the future as well because India has the right to reprocess the unreturned fuel.

2.15.2   Fuel re-processing plant

A fuel re-processing plant was commissioned during the year 2011. Its main role was to reprocess the spent-fuel.

2.15.3   Waste immobilisation

A waste immobilisation plant was commissioned during the year 1985. It bituminises and vitrifies high-level wastes. The vitrified glass blocks are cast into stainless steel containers, two of which are placed in metal over-packs and kept in an interim air-cooled solid storage and surveillance facility. This facility can store up to 2,000 over-packs with air-cooling for 30 years.

2.15.4   Radioactive waste augmentation facility: phase-II

This would serve as an integrated system for the treatment of segregated streams of effluents from TAPPs, and the immobilisation of spent ion-exchanger resins and filter sludges. They would also carry out de-contamination of various equipment. This additional facility became essential as the then-existing rad-waste or radioactive waste facility was inadequate for coping with the planned rise in power demand.

2.15.5   Solid Storage Surveillance Facility

The facility was provided for purposes of surveillance of the storage of high-level radioactive waste that was solidifying in the waste immobilisation plant. This facility ensures that no radioactive leakage from the storage facility takes place.

2.16   Tarapur power station: operation and maintenance snags faced and solved life extension

2.16.1   Steam generation tubing: TAPP 1

A leak in one tube plugged in early 1975 recurred due to the previous plug being loose. This tube was then re-plugged to prevent further leakage. Another leak was detected in steam generator 1B in July 1975 which was re-plugged in early 1976.

2.16.2   SG tubing: TAPP 2

Twenty-one SG tubes were plugged into the steam generator SG 2B late in early 1974. Further leakage was detected in 1975 using radio-chemical analysis. So another eighty tubes were plugged in during the 1976 outage. The defect mechanism causing the leakage in the tubing has not yet been well understood. However, the defect has been attributed to intra-granular stress corrosion, resulting in pin-hole perforations in the tubes.

2.16.3   The Secondary Steam Generators (SSGs)

Four SSGs, from both units, presented numerous problems due to leaky tubes. All of them were isolated on the secondary side. Both the units have since been operating on a single cycle with a reduced unit capacity of 160 MWe.

2.16.4   Feed-water heaters

One HP heater and one LP heater were replaced by indigenous heaters. These heaters are now working satisfactorily.

2.16.5   Fuel channels and control blades

Zircaloy-4 fuel channels were developed jointly by the TAPPs, the NFC, the BARC and the NPCIL. The channels were manufactured at the NFC—Hyderabad, Andhra Pradesh—and are now in use at the station.

Stainless steel tubes and hollow bars are required for fabricating the control blades. These were developed by the NFC. The control blades were later fabricated at the Atomic Fuels Division (AFD) at BARC, Trombay. These blades are now in use at the station.

2.16.6   Station Performance-Review

In the year 2000, the Safety Review Committee for Operating Plants (SARCOP) mandated TAPPs to carry out a comprehensive safety review of TAPP units 1 and 2. The review was later carried out and covered the following:
(1) Operational performance
(2) Design basis
(3) Safety analysis
(4) Seismic analysis
(5) Assessment of civil structures

The above-mentioned review established the mode of safety upgradation work. The work that was proposed to be carried out can be seen below:
(1) Operation work to be done on both the units
(2) Work to be done under shut-down conditions of both the units

Major work was carried out after both units were shut down on 01.10.2005. Design support was provided by the NPCIL headquarters and the continuous review of the work was carried out by the AERB. The following work was carried out:
(1) Improvement of the shut-down cooling system and control-rod drive system

(2) Replacement of electrical cables and re-routing of cables in the cable-spreading room

(3) Relocation of major electrical equipment

(4) Installation of new programmable logic circuit-driven EMTR (Emergency Transfer)

Apart from the above, 500 design modifications were carried out at the station. The cost of upgradation was a total of Rs. 96 crores. All of the above-mentioned work will serve to extend the life of the station until 2025-2030.

2.17 Project costs: financial performance during the first-quarter century operation of the station

2.17.1 Estimated cost

The estimated cost of TAPP units 1 and 2 was Rs. 69.50 crores, exclusive of the cost of fuel which was Rs. 24 crores. The sanctioned cost had provided for an additional capital equipment cost of Rs. 103.69 crores.

2.17.2 Estimated profit from the sale of Power

TAPPs had completed 25 years of operation as of 01.04.1995. The station has cumulatively generated 44.789 billion units of electricity as of the end of the year 1994. The total generation of electricity at the end of 1998 was 52.6 billion units, while that generated as of June 2008 was 75.44 billion units. The revenue earned by way of the sale of power to the Maharashtra State Electricity Board (MSEB)—which is a part of the WERB—at a very low tariff of 63 paise per unit, as of June 2008, was Rs. 4,743 crores. The station has accumulated a sizeable net surplus of Rs. 207 crores in its reserves as of the above-mentioned period. The electricity supplied by the station has resulted in the generation of gross national product of value Rs. 110 billion as of June 2008.

2.18 Important improvements carried out to enhance station safety and reliability

2.18.1 Second start-up transformer (SUT)

TAPPs had only one start-up transformer for the start-up of both the units of the station. It is normal station practice to have one start-up power source for each unit. In line with this practice, a second SUT was procured, installed and commissioned in the year 1982.

2.18.2 Higher capacity station batteries

Two sets of improved, 250-volt DC batteries were put in place of the then-existing 'Gould batteries'. These batteries had a higher capacity of 1,100 ampere hours compared to their predecessors which had a capacity of only 800 ampere hours. The high-capacity batteries were procured, installed and commissioned satisfactorily during the years 1982-1983.

2.18.3 The Augmentation of station black-outs (SBO)

Another milestone in the improvement of the safety aspects of TAPPs was the augmentation in the form of a full-capacity 800-KW station black-out diesel generator. The installation of the generator was completed in July 1999 and was formally inducted into the main plant system on 23.08.1999.

The DG set was required as a stand-by at full capacity to supply power to the emergency load in the unlikely event of the front-line emergency DG sets failing to come on, due to common-cause or common-mode failure. These emergency loads work out to 700 KW, and hence the choice of the set is 800 KW.

The contract for supply and commissioning was awarded in February 1998. Dynamic tests such as the direct online switching of the large-capacity 380-volt 113-KW core spray pump motor, measurements of the voltage and frequency transients at various base loads and 24-hour endurance tests were carried out on the set. Other salient features of the system are as stated below:

(1) Power and control cables were laid in the dedicated trench and cable trays. The shortest possible route was arrived at to optimise voltage drops across the cable lengths

(2) Three of the 1.1-KV rated 400 mm sq. annealed flexible plain copper conductor, LT XPLE-insulated and HRPVC inner- and outer-sheathed cables were used per phase from the alternator terminals to the emergency bus

(3) The DG set has been placed in a separate room away from the main plant building

(4) Dedicated controls to enable the starting of the set and consequent loading under station black-out conditions have been provided

(5) The DG day tank of 900-litre capacity has been installed in the DG set room; the tank has been inter-connected with the main diesel storage tank so that the day tank may receive the feed by gravity from the latter, even in the case of the diesel transfer pump becoming inoperable

(6) The DG set has been supported on anti-vibration pads to reduce the impact of the vibrations on the foundation

(7) Cooling of the DG set is affected by a dedicated cooling tower consisting of a blower and a pair of cooling tower pumps which in turn is powered by the DG set

(8) The DG set has been inter-connected with the adjacent 1 MVA sub-station so that the DG set may be periodically load-tested

(9) The DG set is designed for automatic starting, on loss of Class-IV power, or emergency bus under voltage. The first preference is given to the front-line DG sets G2A, G2B and G2C to start up and synchronise with the emergency bus. In the event of these sets not synchronising successfully, these sets are isolated electrically and the SBO DG set is synchronised with the emergency buses BO4, BO5 and BO9. The SBO has significantly contributed to the reliability of the station emergency Class-III bus.

2.18.4   Other work

The work to replace the feed-water charger of unit 1 was carried out without a single injury or loss of man-days.

2.18.5   The Augmentation of station black-out measures

The independent verification and validation (IV and V) of the emergency power-transfer system of TAPP units 1 and 2 was carried out.

2.19   The Operational history of TAPP units 1 and 2 over the years

2.19.1   Modifications

TAPP units 1 and 2 are recorded as the second oldest BWRs in the world, having outlived their sister units. Some 500-odd modifications have been carried out so far on different systems and equipment in the station to ensure state-of-the-art functioning of the same. These modifications have proved to be effective in maintaining the health and safety of the plant. The station has demonstrated its ability to run at stable power levels for continuously long periods and under accident-free conditions.

2.19.2   Capacity availability factors: accident-free operation

High capacity and availability factors have been established in the station. Some of these are detailed below.

(a) Long continuous runs

S. no	Unit no.	Year	Period of continuous runs
1	1	1988	226 days
2	2	1987	215 days
3	1	1992	205 days

(b) Capacity and availability factors

S.no	Unit no.	Year	Capacity factor	Availability factor
1	1	1987	90.56	93.85
2	1	1992	81.71	86.36
3	1 and 2	1992/1993	81.71	--
4	1	1985-1994 average	65.37	72.49

S.no	Unit no.	Year	Capacity factor	Availability factor
5	2	1985-1994	64.42	70.30
6	1 and 2	2004	93.44	--
7	1 and 2	2004-2005	92.28	--

It can be observed from the table above that the performances at serial numbers 6 and 7 are the best so far since 1969.

(c) Shortest re-fuelling duration

    Unit 1 : 20 days in 2003

    Unit 2 : 26 days in 2008

(d) No. of accident-free days as of 1995 : 628 days

(e) No. of man-hours worked during these 628 days : 75,32,736

2.20   Fatal accidents in TAPP Units 1 and 2

TAPP units 1 and 2 have been functioning for the last 51 years, as of 2020. There has been one fatal accident during this long run of more than half a century. It happened when there was a maintenance/check-up of a large steel tank. Normally, large tanks are checked up for the non-presence of toxic gases and the presence of oxygen for life sustenance before entry into these tanks. It appears that this was not done in this particular instance. Two TAPP employees who went into the tank for maintenance check-ups did not emerge out of the tank for a long time. It was later found out that both of them were dead. This was the only fatal accident in the long history of TAPPs and it was a non-nuclear accident.

2.21   Social and Community Life in TAPP township

2.21.1   Facilities provided for station personnel and their families

The Tarapur township today is a well-planned colony wherein are housed all the essential operation and maintenance personnel of the station. Some of the employees live outside the township in Tarapur, in the Chinchane and Varor villages nearby. More than 800 families live in the township. A department-run primary and higher secondary school affiliated with the Central Board of Secondary Education (CBSE) is located in the colony. A Marathi-medium school,—managed by the duly registered DAE Employees' Educational Society—named 'Dr. Homi Bhabha Vidyalaya' was inaugurated in 1993.

A recreation club housed in the department-built facility provides space for playing games such as table tennis and badminton. Tennis courts have also been laid out in the vicinity along with a swimming pool.

The Tarapur CBSE school administration routinely conducts intra- and inter-school games and athletic meets. The students of this school have distinguished themselves in these events.

2.21.2   Catering to the religious requirements

A temple complex—built through private funding, and run under the aegis of a registered public trust formed by the station employees—caters to the religious requirements of the colony population.

2.21.3   Help provided to families of the diseased

Another innovative feature of the colony life is the working of the 'Antya sanskar mandal' formed by the colony residents to dispose of the mortal remains of the deceased, irrespective of their religion. The Mandal undertakes the initial expenses incurred without bothering the members of the family of the deceased. The trauma caused to the family members is thus lessened considerably. The mandal has also constructed a crematorium on the banks of the nearby Banganga River to be used by the colony residents.

2.21.4   Hospital facility

A department-run hospital caters to the health of the station personnel and their families.

2.21.5   Co-operative Store

A co-operative store caters to the day-to-day grocery requirements of the colony population.

2.21.6 Transport facility

Regular department-run buses provide transportation facilities to the residents of the colony as they commute to the station and back. Limited transport facilities between the colony and the Boisor Railway station are also available for station employees and their family members.

2.21.7 Social welfare measures of nearby villages

2.21.7.1 *Corporate social responsibility (CSR)*

The Tarapur station authorities have been fulfilling their responsibilities towards the people residing in the nearby villages. They have adopted fifty children and funded their studies up to the XII standard.

Mobile health clinic facilities for 24 nearby villages were also provided. Medicines are also being supplied at free cost to the people of these villages.

2.21.7.2 *Educational CSR Work*

The TAPP CSR unit has constructed the 'Nuclear Friends English School' with two upper floors. It was inaugurated on 03.07.2014 at Tarapur village by the Tarapur Site Director of the NPCIL. The president of the 70-year-old 'Tarapur Education Society', which will run the school, was present for the occasion. The function was also attended by 700 persons, comprising students, parents, teachers and the Tarapur villagers.

2.21.7.3 *Further educational CSR work*

The Tarapur CSR unit has also constructed a two-storey school building at Dandi village. The school named 'Anuvilas English School' was inaugurated on 17.7.14. The function was attended by 300 people including students, parents, teachers and residents of the entire village. The unit has also carried out holistic checks on 1,320 students in the nearby villages in the years 2013-2014. Feedback from the checks was received on 27.06.2014. The unit has also constructed several internal roads, developed a school playground and constructed a crematorium in the Pathroli village.

The first-phase distribution of school uniforms and notebooks in the schools of 33 neighbouring villages was carried out on 7.9.2017.

2.21.8 Tarapur town economy: then and now

Tarapur was a small village with a population of around 5000 people in the early 1960s; the main vocations then were fishing and fruit farming. When nuclear power from the Tarapur units 1 and 2 became available, the advent of this cheap power—spearheaded by the availability of local and nearby labour—triggered economic advancement in the area. The area has developed into an industrial town with a permanent estimated population of 35,000 and a large floating population. Big companies like Camlin Ltd. (Stationery specialists in Indian dye industries), Jindal Iron and Steel Company Limited, Ganesh Ben Plast, Lupin Chemicals, Vadilal trans-freight Containers Ltd. and Imperial Process and Printing (P) Ltd. are now located here.

Several medium and small-scale industries have also come into existence in the industrial estate of the Maharashtra Industrial Development Corporation (MIDC) at Tarapur. There are about 600 such units in the chemical, textile and engineering fields, employing about 20,000 workers.

A 35-bed hospital has been set up by the Tarapur Medical and research charitable trust. The hospital will serve the needs of these workers and other people residing in Boisor, Tarapur, and other nearby villages.

Where once only bullock carts served as a mode of transport, auto rickshaws are now plentiful. Private transport is also available for hire, with increasing housing facilities being provided for top executives at Tarapur and other nearby areas. Thus, the service sector is also getting a big boost.

The Tarapur town that has now come into existence may look forward to continuing industrial, social and other related developments in the years to come.

2.22 Concluding remarks on the establishment of India's first nuclear power station

2.22.1 Operational experience

Very valuable operational experience on nuclear power stations has been acquired by Indian nuclear Operations and Maintenance

(O & M) Engineers and Scientists. The design that was imported has been fully assimilated. Although additional BWRs have not yet been explored (see Chapter 3), the operational experience in nuclear power stations has provided valuable inputs for furthering the concept of indigenous Engineering installation and the commissioning of future nuclear power stations.

2.22.2   Labour Management Relationship

The relationship in TAPPs has been excellent to date. The labour leadership has offered consistent co-operation in the running of the station and has seen the full commitment of its employees. A noteworthy feature of this relationship has been that a single day has not been lost due to labour unrest throughout the 36 years of the running of the station as of the end of 2005.

2.22.3   Award for Station/Station Director

TAPP units 1 and 2 received the 'Fire Safety' award from the AERB for the year 2005. Both the units received the 'Shrestha Suraksha Puraskar' from the National Safety Council of India for the same year. The Station Director, Sri. Shashi Bhattacharya was conferred with the 'Nuclear Excellence Award' by the World Association of Nuclear Operators (WANO) for the year 2009.

2.22.4   Independent verification and validation for the EMTR system

An exercise for the independent verification and validation (IV and V) for the Emergency Power Transfer (EMTR) system of TAPP units 1 and 2 was initiated and completed.

2.23   Further improvements to the TAPP operational performance

Thus far, we have listed the improvements over the years 1982 to 1999. With the increasing emphasis on further safety measures for nuclear power stations and based on the performance of similar nuclear power stations across the world, further safety measures were taken up in early 2011. These safety measures were in line with the safety report given by the International Nuclear Safety Advisory Group (INSAG) of the IAEA, especially INSAG 8 titled 'Common basis for judging the safety of Nuclear Power Plants'. The safety measures also comply with NU-REG0800, issued by the USA's Nuclear Regulatory Commission (NRC), as the standard review plan along with the NPCIL's analytical documents. The NPCIL had also reviewed the seismic valuation of structures, systems and components as per the IAEA safety review series 28 on 'Seismic re-evaluation of existing nuclear plants'. The upgradation details were as stated below:

(1) Three new Class-III DG sets of higher capacity were installed
(2) Unit-wise segregation of power supplies was carried out to avoid common mode failures of these supplies
(3) Independent control-rod drive pumps were added to strengthen emergency feed-water cooling of the reactor system
(4) The emergency control room was installed
(5) The fire protection system was upgraded

2.23.1 Measures common to other nuclear power stations

The NPCIL had been reviewing reactor safety after the Fukushima 'tsunami' reactor accident. It also has been augmenting emergency operating procedures to ensure site self-sufficiency for seven days, after any extraordinary accident occurrences without any reliance on external help.

It may be noted that the primary containment-volume-to-power-generation-volume at TAPP units 1 and 2 was 10 times that of the Fukushima reactor. The position may be similar in other Indian nuclear power-generating stations as well.

To prevent accidents, measures were taken to:
(1) Raise 'Tsunami'-resistant walls around Class-III DG sets
(2) Provide mobile DG sets
(3) Relocate block-out Class-III DG sets at a higher elevation
(4) Provide for power-supply from CNG or gas-fed generators located outside the reactor site

Indian and Global Evolution of Nuclear Power Reactor Technology

3.1 Choice of reactors for power-generation in India: the current scene and future scenario

3.1.1 Boiling-water reactors (BWRs)

BWR was the first reactor type that was chosen by India to prove that frontier technology could be successfully accessed to trigger the economy of a developing country.

The choice of a BWR as the first reactor type was a conscious choice that was made to conform to the long-term goals of the Indian nuclear power programme. India has not had facilities—until recently—for the enrichment of nuclear fuel, and in the absence of such facilities, the viability of reactor technology based on enriched uranium fuel could not be assured. On the other hand, present reserves of uranium in India are adequate for establishing Nuclear Power Plants based on natural uranium fuel in the form of uranium dioxide to generate 10 to 15 GW of power. The use of natural uranium would require a moderator in the form of heavy water. Heavy water has the property of slowing down neutrons in the atoms of the nuclear fuel thus rendering a much higher cross-section of the nucleus for further collision and hence for continued nuclear fission. Light water which is used both as a coolant and moderator in BWRs, as in Tarapur, does not possess this property. Due to these reasons, we need enriched uranium as fuel for BWRs.

3.1.2 Heavy-water Moderated Pressurized Tube Reactors (HWPTRs)

3.1.2.1 First-generation reactors
The heavy-water moderated and cooled uranium oxide-fuelled reactors could be described as the first-generation reactors. These reactors would produce plutonium during the process of fission. This plutonium would be recovered during the re-processing of the spent-fuel.

3.1.2.2 Second-generation reactors: the first stage
The first stage of the second-generation reactors would be utilizing the combination of uranium dioxide and plutonium dioxide—that was re-processed from the spent-fuel of the first-generation HWPTRs—as fuel and depleted uranium as blanket fuel in fast-breeder reactors to breed more plutonium.

3.1.2.3 Second-generation reactors: the second stage
The second stage of the second-generation reactors would use the main fuel, as discussed earlier, but would utilize thorium as the blanket fuel instead of depleted uranium.

3.1.3 Third-generation reactors: fast-breeder reactors (FBRs)
The third-generation reactors would use U233 and sustain the future breeders by converting the blanket fuel U232 to Pu233. These reactors would produce more U233 than they burn up nuclear fuel, thus sustaining a succession of U233-fuelled reactors. India has reserves of nearly 3,00,000 tonnes of Thorium, which is the largest in the world. The reserves would serve to sustain a succession of future FBRs.

The burn-up in these FBRs would be 50,000 mega-watt days (MWDs) per tonne of fuel, as compared with a rate of 7000 to 8000 MWDs of burn-up achievable in HWPTRs. This economy of scale could thus be attained and a self-sustaining programme of nuclear power production can be ensured over the foreseeable future.

It may be mentioned that the 40-MWt fast-breeder test reactor (FBTR) at Kalpakkam, Tamil Nadu (TN), uses a mix of plutonium (re-processed from the spent-fuel of existing HWPTRs) and natural uranium U238, both in the carbide form.

The DAE insisted on the use of oxide fuel for the proposed 500 MWe prototype fast-breeder test reactor (FBTR) at the Indira Gandhi Centre for Atomic Research (IGCAR), Kalpakkam (TN), given the problem faced with carbide fuel in 40 MWt FBTR.

3.2   The Global scenario: choice of reactor systems in the USA

Nuclear power based on pressurized light-water reactors was introduced in the USA in the 1960s and 1970s. More than 90 Nuclear Power units were commissioned thereafter. The increasing cost of setting up new stations and the public concern for safety against radiation effectively stopped new constructions after the 1970s. The share of 20% of total power by nuclear stations remained fairly constant over the succeeding years. As of March 2003, ninety-four plants had permits for increasing unit output and the operating licenses of many plants had been renewed for another twenty years. The applications for the continued operation of seven plants were under review. The continuing rise in crude oil prices, which had touched $60 per barrel in 2005, had resulted in the above permits being granted.

The Dept. of Energy (DOE) of the USA had, after a long period, issued a financial solicitation on the 20th of November, 2003, as part of a future programme for 2010 and beyond. As a result, two consortia bid for this assistance solicitation. The Dominicon project envisaged an advanced Canada Deuterium Uranium (CANDU) reactor (ACR 700) at the North Anna site in Louisiana Country, sixty miles North-East of Richmond, Virginia. The Dominicon-led consortia included Atomic Energy of Canada Limited (AECL) and its subsidiary AECL Technologies; Bechtel Power Corporation in Frederick, Maryland and Hitachi America Inc. located in Tarrytown, New York. 'Dominicon' proposed to spend $61 million over the subsequent six years (up to 2009) while the other partners were willing to invest more than $61 million towards the project, with the DOE's (USA) approval. If an order had resulted, then 'Dominicon' could have had a new plant in operation as early as 2014.

The NuStart Energy Development, LLC consortium, was the other party to bid against the DOE solicitation. They said that they would evaluate the Westinghouse advanced pressurised water reactor (AP10000 PWR) and the General Electric Economic Simplified BWR (ESBWR) before coming up with a proposal. The consortium comprised Philadelphia (Pa); Entergy nuclear, Jackson (Mississippi); Southern Company, Atlanta (Georgia); Constellation Generation Group, Baltimore (Maryland); Duke Energy, Charlotte(North Carolina); Tennessee Valley Authority, Knoxville (Tennessee); ....Florida Power & Light Co., Juno Beach (Florida); Progress energy, Raleigh (North Carolina); and EDF International North America, Houston (Texas), Washington D.C Turn reactor vendors, General Electric and Westinghouse Electric Co.—Pittsburgh (Pennsylvania)—also formed part of the consortium.

The NuStart Energy consortium planned to select a final reactor technology and a site by 2007. If an order had resulted, then NuStart Energy could also have had new nuclear power stations in operation by 2014.

Co-operative agreements for each of these projects were expected to be ready by late 2004 or early 2005. A final decision on the implementation was to be taken during the project-planning stage.

The above-mentioned projects formed part of the demonstration process of USA's NRC for licensing the construction and operation involved in the generation of three or more Nuclear Power Plants.

The following table elaborates on the choice of various reactor types by some of the International nuclear power players. The burn-up rates of these types are also listed.

S. No.	Name of the Country	Type of Reactors in operation/under construction	Remarks
1	USA	Pressurised Light-water Reactors (PWRs)	
2	Japan	Boiling-Water Reactors (BWRs), Pressurised Light-water Reactors (PWRs). A fast-breeder reactor (FBR) was commissioned in August 1995.	
3	UK	Gas-cooled graphite-moderated reactors; the first-generation carbon dioxide-cooled reactors were named 'Magnox' reactors after the magnesium alloy used to ease the uranium fuel. These reactors were built between 1965 and 1971. The successors of the 'Magnox' reactors were the advanced gas-cooled reactors which are designed to operate at higher temperature and pressure in order to achieve higher thermal efficiency and power density of the core. A prototype fast-breeder reactor has also been in operation at Dounreay since 1977. A PWR had been commissioned at a cost of $3 billion in January 1995. A second PWR was proposed to be built at Suffolk but funds were tied up for the project. In 1978, a decision was made to order one or more advanced gas-cooled reactors. A subsequent decision was made to not go in for more steam-generating heavy-water reactors (SGHWRs); a SGHWR which was commissioned in 1968 is still in operation. The UK foresees FBRs getting into further use only during the middle of the 21st century. The use of Thermal Reactors and AGERs with a view of possible use of re-processed spent-fuel of these reactors in FBRs will be exercised by the UK at an appropriate time. The UK's first PWR at Sizewell was officially inaugurated on 25.3.1996. The country went in for one 1.65 MWe advanced pressurised light-water reactor (AREVA) in the mid-2010s.	
4	China	PWRs/BWRs	
5	Russia	BWRs, PWRs (Graphite-Moderated RBMK type). One FBR has also been commissioned. FBRs are intended for part power production and part de-salination water-supply purposes.	
6	Germany	PWRs, FBRs	

S. No.	Name of the Country	Type of Reactors in operation/under construction	Remarks
7	France	Mostly PWRs; four gas-cooled reactors were under construction in 1987. An FBR 1200-MWe 'Super-Phoenix' was also under construction.	
8	South Korea	PWRs (most likely)	
9	Canada	HWPTRs	
10	Taiwan	BWRs, PWRs (most likely)	

3.2.1   Indian and global evolution of nuclear power technology

The burn-up rates of nuclear power reactor types are tabulated below:

S. No.	Type of Reactors	Burn-up rates (Megawatt days per tonne) of fuel
1	Heavy-water pressurized tube-type reactors (HWPTRs)	7000 to 8000
2	Pressurized water reactor (PWRs)	6500
3	Light-water reactors	30000 to 35000
4	Magnox-type reactors	3000 to 4000
5	Fast-breeder reactor (FBRs)	50000
6	Exxon (Demonstration fuel assembly for Ginna NP 470 MWe PWR)	55,000 (expected)

Types of fuel for all of the above types of reactors play an important part in the evolution of the above-mentioned technologies. We will discuss the requirement of uranium fuel globally in the upcoming sections.

3.2.3   Current global reactor construction programmes (other than India)

As of 2002, seven Nuclear Power units with a capacity of 6.5 GWe were in construction globally. Of these, three were in Japan, two in South Korea and two in the Slovak Republic. In addition to these, eight units totalling 10 GWe were commissioned in Japan. An additional eleven units were planned: Finland - 1 unit; South Korea - 2 units and Turkey - 2 units.

3.2.4 Total Global Nuclear Energy generation including India

As of 2002, the total global Nuclear Energy generation was 2,170 tetra-watt hours. Nuclear Energy generation in 2002 was 23.8% of the total power-generation in the 'Organisation of Economic Co-operation and Development' (OECD) countries. Seventeen OECD countries have Nuclear Power Plants in operation, providing between 4% (Netherlands) and 78% (France) of the respective countries' total power-generation.

Four hundred and forty-four Nuclear Power units stood connected to international grids as of the end of 2020. These units had a total installed capacity of 394 GWe of electricity. As of 2010, the Global nuclear power-generation was expected to be 22% in total.

3.3 Uranium fuel

We will now take a look at the current production and future requirement of uranium fuel globally with particular reference to the USA's production. The USA had hitherto been a major nuclear power producer, and its production of uranium fuel had a major influence on the global production of nuclear power.

3.3.1 The USA's role in uranium fuel production

There was an annual requirement of 60,000 tonnes of uranium fuel, as of 2006, against a production of 45,000 tonnes. The use of breeder technology could reduce the world commitment of 3,00,000 tonnes by 50%, as of 2020. The deficit of 15,000 tonnes as seen above was made good through recourse to inventory. This is a stop-gap arrangement and urgent steps to increase uranium production to meet the rising demand are essential. In any case, it is impossible to reduce nuclear fuel commitment in the future without recourse to advanced fuel cycles.

As of 2010-2015, the estimated global uranium fuel resources were about 14 million tonnes with the cost of production varying from \$80 per kg to \$260 per kg. The USA's annual uranium fuel production was 11,000 tonnes to 13,000 tonnes during the early 1970s. This was expected to increase to 25,000 tonnes per annum in the early 1990s and to 60,000 tonnes per annum by 2000. It is, however, of interest to record here that no fresh orders for nuclear power stations were issued in the USA over the two decades from 1978 to 1998. This would have increased the uranium fuel production as envisaged by the USA in the early 1970s.

The General Description of the HWPTR Project: The Difference Between Fossil-Powered and Nuclear-Powered Stations

4.1 Comparison with fossil fuel-fired stations

In the upcoming chapters, we will be looking into the development and construction of mostly HWPTRs, among others. These chapters would be dealing with the design and construction details, organizational methods adopted to expedite construction, the commissioning evolution of standardized HWPTRs etc. It would be desirable to have an idea of the essential features of such HWPTR units so that construction methods etc., of the plants are better appreciated by the masses. This chapter is entirely devoted to this effort.

4.2 The difference between the working of fossil fuel-fired and nuclear power-generating stations

4.2.1 Fossil fuel-fired power-generating stations

A general awareness of the principles of operation of fossil fuel-fired power stations is prevalent among the masses. These stations use fossil fuels such as coal, furnace oil, gas and other such fossils to fire the steam boilers and generate steam for running the steam turbines. The fossil fuels we normally come across are in the form of coal, pulverized coal, furnace oil, etc. Natural gas or liquified natural gas, compressed natural gas, naphtha or other petroleum derivatives may also be used as input fuel to run gas turbines which in turn drive turbo-generators and derive electrical power.

4.2.2   HWPTRs: details of the generation of electrical power

4.2.2.1   *Nuclear fuel fission process*

U238 in the form of uranium oxide is used as fuel in HWPTRs. U238 is the most significant nuclear fuel and it is the only naturally occurring nuclear fuel in the world. U235 constitutes about 0.7% weight of naturally occurring U238 which causes nuclear fission and hence nuclear heat. One gram of UO2 contains 0.0071 gm of U235. The nucleus of the atoms in UO2 fuel participates in the fission reaction under certain conditions which are listed below:

(1) The presence of a critical mass of nuclear fuel in the form of the core in a reactor

(2) The presence of a moderator to slow down neutron speed in the negative charge so that the equilibrium of the atom is maintained

The mass number before an atom is indicated by the addition of protons and neutrons, while the atomic number is indicated by the number of electrons. Thus, for U238 the mass number is 238 and the atomic number is 92. U238 is indicated as 92U238 so its designation is complete.

A sketch of an atom and its constituents is given below:

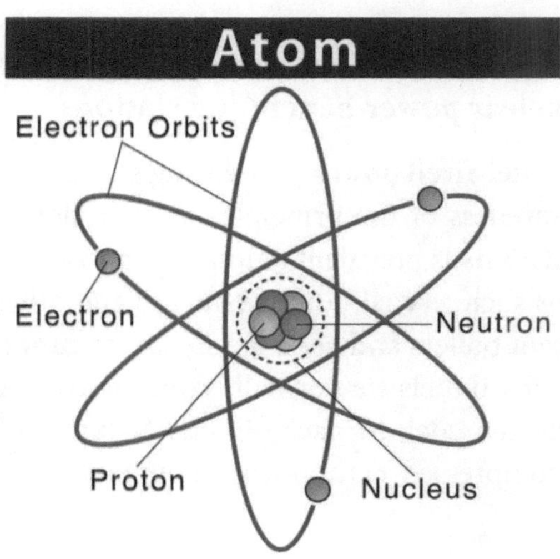

4.2.2.2   *The release of thermal energy as a result of nuclear fission*

The neutron in the nucleus of the U238 atom bombards the proton as a result of which further neutrons are released. At least one neutron of the released ones bombards another proton, thus releasing a continuous chain reaction of fissions. Immense quantities of thermal energy are released as a result of these fission chain reactions. Each fission reaction releases 200 MEV (Million Electron Volts) 6.25×10 raised to the power of 5 MEV which constitutes 1 watthour of energy.

4.2.2.3   *Nuclear fission energy*

The 200 MEV of energy released by the fissile material may be compared with the energy released by the fast neutron bombardment of non-fissile material in the nucleus of the U238 atom of 2 MEV (1% of that of the fissile material). It can be mentioned here that 3.1×10 to the power of 6 fissions are required per second to produce 1 mega-watt (1000 KW) of power. It can also be noted that each nuclear fission reaction produces 3.2×10 to the power of minus 11 watt seconds of electrical energy.

The above fission production takes place in the Calandria (pressure vessel) of a nuclear reactor. This vessel may be described as the heart of a Nuclear Power Plant.

The heat produced as a result of nuclear fissions is removed by the pressurized heavy water surrounding the nuclear fuel. The heated heavy water transfers this fission heat in turn to the light fuel water contained in the Nuclear Steam Service Generators (NSSGs). This steam runs the turbo-generators that produce electrical power.

4.2.3   The comparison of energy produced by coal and nuclear fuel: ways of transfer of heat

4.2.3.1   *The comparison between the sizing of power plants*

One kg of uranium dioxide produces the same amount of heat as 25 tonnes of coal. The reason for this amount of power is produced due to the compactness of the fuel core and the reactor vessel. A large radiant heat-transfer area is required in the case of coal-fired power stations so that the radiant thermal energy of the burning fossil fuel is transferred to the light water in the tubes of the steam boiler with resultant steam generation. This steam runs the turbo-generators which generate electrical power.

The above description highlights the difference between nuclear fuel-fired and fossil fuel-fired stations.

4.2.3.2   *The comparison between nuclear fuel-fired and fossil fuel-fired stations regarding fuel storage, waste disposal and greenhouse-gas emission*

4.2.3.2.1   Fuel storage requirement and extent of Fuel feed for power-generation

4.2.3.2.1.1   Fossil fuel-fired plants

Elaborate arrangements like the provision of fuel storage yards are required to ensure maximum working days for the plants. Fuel transportation handling and feed to the boilers in terms of thousands of tonnes per day are also required.

4.2.3.2.1.2   Nuclear fuel-fired plants (HWPTRs)

Automated fuelling by way of double-ended fuelling machines is required to ensure positive reactivity to ensure the generation of set reactor power capacity. About 15 kgs of nuclear fuel U238 is fed into the already-formed nuclear fuel core in the pressure channel tubes to achieve the set power capacity. The spent 15 kgs of fuel evicted from the pressure tubes are automatically sent to the spent-fuel bay. It is pertinent to point out that 1 kg of U238 produces the same thermal heat energy as 25 tonnes of coal.

4.2.3.2.2   Burnt fuel handling and disposal

An elaborate ash handling system is required in the case of coal-fired stations. The burnt nuclear fuel is stored temporarily in water, filled and cooled in storage bays and later re-processed for further storage as required.

The liquid waste in nuclear plants is led into a local evaporation facility. Residual wastes are enclosed in concrete and later buried in deep underground vaults. Further details about these processes are discussed in the upcoming sections.

4.2.3.2.3   Greenhouse-gas emissions

Sulphur dioxide and carbon dioxide, which constitute greenhouse-gas emissions, are created in fossil fuel-fired plants. On the other hand, there is no significant greenhouse-gas emission in Nuclear Power Plants.

CO_2 is one of the greenhouse-gases whose generation is as follows:

a)   Coal-based stations: 888 tonnes/GWhr

b)   Nuclear-based stations: 29 tonnes/GWhr

4.2.3.3   *The difference in the heat-transfer processes between fossil fuel-fired and nuclear fuel-powered plants*

There are quite some differences between the two types of stations. The transfer of heat in fossil fuel-powered stations is a single occurrence from the radiant heat of the burning fuel directly to the light water in the boiler tubes. Whereas, in a nuclear-powered station, there is a double transfer of heat: first from the fuel to the pressure coolant and then from this coolant to the light water in the tubes of the NSSGs. There is hence greater thermal efficiency in the case of high-pressure high-temperature fossil fuel-fired stations (some have even breached 40%) than in nuclear fuel-fired stations with an efficiency of 30%. While super-critical thermal power units fired by fossil fuels are possible, nuclear stations have to do with only saturated steam.

There are, however, immensely greater advantages in nuclear fuel-powered stations which have already been touched upon and will continue to be discussed in the following chapters.

4.2.4   Various buildings/facilities required for Nuclear Power Plants (NPPs)

4.2.4.1   *General buildings*

We will now look into more details about the nuclear power stations, with particular reference to the HWPTR.

Nuclear power stations generally consist of the following buildings.

4.2.4.1.1   The Reactor Building

Nuclear equipment like the Calandria, coolant headers, feeder pipes, fuelling machines, etc., can be found in this building.

4.2.4.1.2   The Service Building

The equipment and facilities required for day-to-day maintenance, monitoring of health physics, radiation safety of the operating and maintenance personnel, spent-fuel storage, waste-management facilities, locker/washrooms, etc., are located here.

4.2.4.1.3   The Turbine buildings

It houses the power-generating equipment (TGs), HV and LV switchgear 415 V Mccs/load centres, HV and LV motors, de-aerators, ventilation equipment, turbine auxiliaries, condenser, main control and instrumentation panel, control equipment panel, relay panels, channel temperature monitoring panel, fire alarm and indication panels, control distribution frame, etc.,

4.2.4.1.4   Buildings and facilities inside the NPPs

The area between the turbine and service building provides the space for the instrumentation, electrical and station chemistry laboratories.

Mechanical workshops and maintenance stores are also located in this area.

4.2.4.1.5   The Administration Building

It accommodates offices for all the senior maintenance staff of the station including the station director, chief superintendent, technical services superintendent, supporting staff of these officers, administration officer, accounts officer, etc.

The offices of the Industrial relations officer and Labour welfare officer are also located here.

All the required communication facilities, computer facilities and conference facilities are also made available from this building.

More details about these buildings and facilities will be discussed in the upcoming sections.

4.2.4.2   *The Reactor Building (RB)*

4.2.4.2.1   Dimensions and Containment Details of the Dousing System

The reactor building of a typical HWPTR unit of 220 MWe unit capacity would be circular with an internal diameter of about 42.56 metres; it will be about 36.6 metres high from the ground grade level. The RB of the Rajasthan Atomic Power Project (RAPP), in particular, is 45.75 metres high from the ground grade level. This additional height of the RB of RAPP was considered due to the provision of a vapour pressure suppression distribution tank above the boiler room. This system was done away with in later Nuclear Power Projects.

The distribution tank, in the above-described system of the RAPP, is called the dousing tank. It has a capacity of 7.3 million litres of water and was expected to come into play in case of any gross leakage of heavy-water (D_2O) vapour in the boiler room and consequent pressure build-up in the floor. The dousing was to be carried out through a system of distributor piping. In latter reactors, this dousing system was replaced by a pressure-suppression pool instead.

4.2.4.2.2   The containment dome

A pre-stressed concrete dome about 300 mm thick at the springing level (bottom of the dome), about 200 mm thick beyond 39.5 metres height from the grade level and up to the top of the dome would span the Reinforced Cement Concrete (RCC) perimeter wall of the RB which is 1.22 metre thick. The concrete for all the structures would be poured at a chilled temperature of 23 degrees C with the use of a chilling plant to reduce the heat of hydration and thus the strength of the resultant dioxide. The perimeter wall along with the dome constitutes the primary containment for the RB and its auxiliaries.

The pressure in the RB is maintained below the atmospheric pressure on the outside ambience to ensure that the leakage of ambient air if at all any, is to the inside of the building rather than the other way about. Thus, any leakage out of the containment system is prevented and the safety of the personnel and the public is assured. It can be noted that the RB has been designed to withstand an internal pressure of about 0.42 kg/cm sq. and to ensure a maximum leakage rate of 0.1% of the RB volume per hour through all sources like the ventilation ducts, cables and all other project materials and equipment.

It is to be noted that the density of normal concreting was 2.4 metric tonnes per cubic metre while the density of haematite (which has more iron content) concrete, used for the thermal shield, was 3.63 metric tonnes per cubic metre.

4.2.4.2.3 Reactor building optimisation

The size of power stations needs to be optimised due to the constant increase in land costs and the growing demand for land. This applies particularly to the reactor building wherein the design, construction, and engineering efforts are required to limit and contain the leakage of the building active air into the outside atmosphere. The limitation of the building size will result in the optimisation of building penetrations, which are the single biggest cause of leakage of active air from the building. Building penetrations could happen to an extent as seen in the following figures:

(1) 1000 mm x 200 mm openings: 600

(2) Radiation openings: 1000

4.2.4.2.4 Leakage rate: personnel and emergency exits

The leakage from the reactor building is limited to 0.1% of the building volume per hour at a test pressure of 0.42 kgf/cm^2. The engineering design features cater to this requirement. The building itself is designed to withstand an internal pressure of 0.44 kgf/cm^2 and is tested initially at 0.55 kgf/cm^2 for a specific leak rate.

The reactor building provides two kinds of service openings, one for the entry of normal personnel and materials while the other is for emergency entry and exit. The normal personnel entry provides space for entry and exit of equipment and materials required for maintenance. Each of these entries is provided with double steel doors—only one of which will open at the time of commands—thus isolating the reactor building from the service buildings and outside ambience at all times during reactor operation. The emergency openings provide for the entry and exit of personnel under emergency conditions.

4.2.4.3   *The Service building*

4.2.4.3.1   Size optimisation: a fuelling machine maintenance requirement

The service building would be designed to service the number of reactors in the station and also to be adequate for the unit capacity of the reactors. In the case of 2×220 MWe RAPP units, the building was of an approximate size of 1200 square metres in its main operating floor—wherein the fuelling machine maintenance area, heavy-water storage and monitoring area, washroom/locker facilities, fuel storage bay, etc., are housed. The other two operating floors had an area each of 750 square metres. The sizing could be optimised in the case of stations having more than two units.

The advantages of optimisation could be seen from the example of the Pickering Nuclear Power Station in Canada which has 4×500 MWe HWPTRs. The station has only one pair of fuelling machines in operation. These machines run on rails located in a basement carriageway under the fuelling machine rooms of all the reactor units and are raised onto the respective fuelling machine rooms of the reactor when the re-fuelling takes place. The trade-off between the capital and running costs of the additional three pairs of fuelling machines, on the one hand—which would otherwise be required—and the capital and running cost of the basement, rail tracks, carriageways and the connected civil, mechanical and instrumentation workshops on the other, is adequate from the engineering and economic points of view to justify this optimisation. Spare fuelling machine heads would, however, have to be provided as required by the number of reactors in the specific plants.

4.2.4.3.2   Other facilities in the service building

Other facilities include a storage bay for the spent-fuel, active rooms for laundry change, lockers and showers and absolute filters for the active exhaust gases of the station. Waste treatment plant facilities for monitoring radiation safety and Health Physics—both in respect of personnel and equipment—heavy-water addition/ management, various shops required for maintenance and the Chemistry laboratory of the station are all located in the service building.

4.2.4.4 The Turbine Building

This building houses the turbine generators, steam re-heater condensers, air circulation units, water-treatment plant, valve stations, the 3.3 KV/415 V switchgear, the 415 V Mccs/load centres, station auxiliaries, power-supply equipment, Class-III 415 V emergency diesel generating energy sets, 250 V/48 V DC batteries, Class-II motor generator sets and the unbreakable power-supply equipment.

A typical turbine building for a 2×220 MWe HWPTR station would be 140 metres long, 45 metres wide and 13 metres high.

4.2.4.5 The chimney stack and the hydraulic cooling tower

4.2.4.5.1 The chimney stack

A free-standing concrete chimney, or stack as it is called, about 100 metres high is provided to cater to the number of units in the station. The sizing and other relevant details would be optimised, taking into consideration the number of reactor units, location of the units, etc. The primary function of the stack is to exit active gases from the reactor/allied systems after these pass through absolute filters at a point high enough to ensure the safe dispersal of the gases. Necessary aviation warning lights are provided at the top of the stack at the desirable levels.

4.2.4.5.2 The Hydraulic cooling towers

Forced draft, natural draft (ND) and induced draft cooling towers are provided when called for. Further details about a typical ND hydraulic cooling tower are discussed in Chapter 7.

4.2.4.6 The EHV Switchyard

The Extra High Voltage (EHV) Switchyard could be located either outdoors or indoors to suit the local environmental conditions. The function of the switchyard is to evacuate the power generated at the station through EHV lines to the required transmission load centres.

4.2.4.7 *The Fuel Storage Bay*

This is required for the short-term storage of the spent-fuel. The spent-fuel bundles are stored in racks underwater at a depth of nine metres. The fuel is stored long enough for its residual heat to die down, and then it is transported through specially designed fuel flasks to fuel re-processing plants for the re-processing of the fuel. The re-processed and depleted fuel is used in HWPTRs initially and then sent to other reactors.

The depth of water in the storage bay is adequate to contain the radioactivity emitted by the spent-fuel. The spent-fuel has its own circulation and clean-up system consisting of filters and ion-exchangers.

4.2.4.8 *The Waste Management Facility (WMF)*

The WMF takes care of the solid and liquid wastes of the nuclear plant. Liquid wastes are of two forms namely low-level and high-level liquid waste.

4.2.4.8.1 Low-level Liquid waste

The low-level liquid waste is of a very low level of radioactivity which is first held up in 31.8 kilo-litre (kl) tanks and then in dispersal tanks of 182 kl capacity where the very low half-life radioactive products die down quickly. The low-level liquid is discharged into the mainstream station discharge system after due monitoring by monitors which can detect 5×10 multiplied by the power of minus 5 curie/ml of gamma rays and dilution in a 1600 kilolitre capacity tank of plant effluents. It is estimated that the total radioactivity that is discharged into the plant effluent system results in the average dose of 3×10 multiplied by the power of minus 6 curie/ml for a two-reactor system. This is 2% to 3% of the International Commission for Radio-activity Protection's (ICRP) permitted figure of 10 to the power of minus 6 curie/ml for drinking water when the radionuclides present are unidentified but Ra-266 is present.

4.2.4.8.2   High-level Liquid waste

The other liquid waste which has a comparatively higher level of radioactivity is pumped into the solar evaporation facility, where under the control of shallow pan storage and solar evaporation conditions the radioactivity is allowed to die down over a longer period. The facility is an intensively monitored one and the remnant products are treated as high-level solid waste and disposed of accordingly.

4.2.4.8.3   High-level solid waste

The solid waste disposal area for a typical 2×220 MWe station normally occupies two acres. This area is enough to store the estimated solid waste for 30 years. The waste may be buried directly in the ground or containers. The method of burial depends on the activity levels. The following methods of disposal are adopted:

1. Concrete or tiled trenches, 3 metres deep and 1 metre or larger in diameter, are dug out according to the volume of the stored material wastes. These trenches are used to dispose of highly active wastes and are usually capped with concrete slabs
2. This method follows the same procedure as the first except that the trenches are covered with earth after the burial
3. Concrete monolith trenches, normally 3 metres wide and 3 metres deep, are used to store high-level solid wastes from the evaporator or incinerator. These wastes are mixed with cement to form drums. These drums are stacked and lowered into the trench which is concreted to form a monolith. This method is different in places where waste immobilisation plants are available. In this case, the product is first immobilised in vitrified glass blocks which are then stored underground or in other suitable storage facilities for periods of up to 30 years. These blocks would finally be buried in deep salt mines or other deep geological formulations. These locations would strictly be monitored for radioactivity.

4.2.5   The containment of radioactivity in HWPTRs

4.2.5.1   *Exposure limits*

We have already discussed how the HWPTR plant buildings and facilities are designed in such a way as to protect the plant personnel and the public from any harmful effects due to radioactive explosions. The exposure limit has been specified by the ICRP which is constantly being monitored at the stations.

4.2.5.2   *The containment of radioactivity within the nuclear fuel*

It may be noted that 90% of the radioactivity released due to fission is contained within the fuel itself. Nuclear fuel has a very high physical integrity; its melting point is 2000°C. The fuel is in the form of 8.17cm diameter, 49.5 cm long cylindrical bundle elements for a 200-MWe HWPTR. Each element is made up of pressed and sintered natural uranium dioxide pellets contained in a zircaloy-2 sheath of 1.52 cm diameter. Each bundle contains nineteen such fuel elements. There are twelve such fuel bundles in each coolant channel.

4.2.5.3   *The primary high-pressure boundary (PHPB)*

The primary high-pressure boundaries are confined to the piping, vessels and equipment which are designed and constructed to withstand the pressure, temperature and levels of activity encountered in the system. The system is designed for high integrity and all the components are engineered to restrict gross leakage of the primary coolant and thus prevent the harmful release of radioactivity to the ambient atmosphere. The maximum credible accident (MCA) that may arise in the system is the loss-of-coolant accident (LOCA). This accident may be caused by a double-ended rupture in the pipeline of the largest diameter of the high-pressure boundary. The consequence is the complete loss of primary coolant to the reactor. Measures are taken in such an event to cool the coolant and bring the reactor to cold shut-down conditions. The probability of catastrophic failure of this system loading to fuel core melting is of the order of 7.4×10^{-5} per reactor-year. The probability of this failure of the system loading to release radioactivity to the public domain is nil. Measures are taken to limit the spread of radioactive contamination between as well as outside these zones.

4.2.6 Health Physics: Radiation Protection

The buildings and areas of activity of nuclear power stations are classified into four zones depending on the extent of radioactivity that may be expected in the said areas.

4.2.6.1 *Zone 1*

This zone will contain no radioactive equipment and will be kept free of radioactive contamination at all times. It includes the administration wing, lunch rooms, first-aid centre, personnel clothing and locker rooms.

4.2.6.2 *Zone 2*

This zone contains no radioactive equipment and is not expected to be contaminated by radioactivity. However, some radioactive contamination may get into this area with the movement of personnel into this zone from Zone 3. The contamination in Zone 2 will be cleared as soon as it is encountered. This zone includes the turbine building, control room, non-active laboratories/shops, washroom, water-treatment plant and general stores.

4.2.6.3 *Zone 3*

This zone includes the service areas for the active equipment and materials that are potential sources of contamination and are expected to be contaminated at times. The use of equipment and work procedures will be planned to keep whatever contamination that occurs localised; loose contamination will be cleaned up whenever and wherever it occurs. This zone comprises the active shops and laboratories, the spent-fuel storage bay, the active waste management area and part of the de-contamination centre.

4.2.6.4 *Zone 4*

This zone contains the sources of all contamination. It includes the entire reactor building and the cleaning rooms of the de-contamination centre. The use of equipment and work procedures to be followed will be planned to keep the contamination localised and under control at all times.

The control points at all passages between the different zones will be equipped with protective clothing like lab coats, rubber shoes, etc., that are to be worn by the personnel. Personnel monitors are also provided at these control points.

4.2.7   General Survey of Radiation

Regular surveys of the reactor and service building areas are carried out to determine the extent and intensity of radiation fields in them. Fixed radiation monitors are installed throughout these areas. These monitors provide a warning about any increase in the ambient radiation fields. Immediate steps are then taken to locate the candidate equipment/materials causing the radiation which are followed by remedial measures for the containing or reduction of this radiation.

4.2.7.1   *The use of dosimeters*

All the operation and maintenance personnel carry individual direct-reading dosimeters, in addition to integrating dosimeters and film badges when entering radioactive areas. A record is kept of all the radiation doses accumulated, and necessary Health-Physics measures are undertaken to deal with cases of dosages accumulated more than the prescribed limits.

4.2.8   Classes of Electrical power in HWPTRs and other Nuclear Power Stations

We will now take a look into the various classes of Electrical power to be provided in the HWPTRs and other types of reactors in nuclear power stations to sustain the station's safe and reliable operation.

4.2.8.2   *Class-I Power (normally 250-volt DC)*

This is the most reliable class of power in nuclear power stations. It provides power to control circuits of the Electric power systems, regulatory systems and those equipment to which power is required to be fed at all times. This is provided by the station's batteries, of which there are two sets. The batteries are normally charged by the rectifier of the rectifier–inverter UPS system. The rectifier also powers the Motor Generator (MG) sets in the RAPP units. The MG set caters to a firm load of 1000 amperes. In case a 415-V Class-III system experiences an outage, the 250-V battery sustains the Class-III UPS through the inverter of the rectifier–inverter system or the MG sets as in the RAPP. The battery system is designed to maintain an emergency supply of 1600 amps for 20 minutes without the end-voltage dipping below the stipulated voltage.

4.2.8.3   Class-II (UPS) Power (415 volts)

This power is provided by the uninterruptible rectifier–inverter system or the MG set. The constant alternating current voltage rectifiers (ACVRs) which derive their power from the 415-V Class-III system charge the station batteries and also provide the UPS through the inverter system. The 250-volt battery system runs the MG set in the RAPP on the failure of the Class-III supply.

Please refer to Chapter 6 to follow the flow of power from the rectifier to the Class-II and Class-I power systems of the station.

4.2.8.4   Class-III Power (415 volts)

This system supplies power to the equipment that is needed for the emergency operation of the plant. The power-supply may be interrupted for a brief period without any compromise of the plant's safety. If the 415-V Class-IV power-supply—which normally feeds the Class-III power-supply—also fails, then a 415-V Emergency Diesel Generator (DG) starts working automatically (within a minute of failure of the Class-IV power-supply) and picks up the loads connected to the Class-III system. Three 1250/1450 KW capacity DG sets are provided for each reactor unit of 220-MWe capacity.

4.2.8.5   Class-IV power-supply (415 V/3.3 KV/6.6 KV/11 KV)

This power system feeds the load power-supply which may be interrupted without affecting the safe operation of the station. It includes the safe shut-down of the station if found necessary after a reactor trip.

This power is derived from the grid—to which the station is connected whenever it is not generating any power—and from the Turbo-Generating sets of the station whenever the station is generating power.

Start-up transformers, located in the switchyard, provide power for starting up the station. These are normally 220 KV/3.3 KV/6.6 KV/11 KV, with the two secondary windings being connected to separate buses from the point-of-view of redundancy.

Class-IV power with 415 V, 3.3 KV, 6.6 KV and 11KV systems may be described as the least desirable of the four classes of power available in the Nuclear Power Plants. The interruption of this supply does not affect the safety functions of the station. There are generally two unit transformers, the secondaries of which are connected to separate 3.3 KV/06.6 KV/11 KV buses for purposes of redundancy.

415-V Class-IV power is derived from (415 V/3.3 KV/6.6 KV/11 KV) the station's auxiliary transformers which are generally six in number for each reactor.

4.2.9   Heavy-water upgrading plants

Heavy-water (D_2O) upgrading plants, of either distillation or the electrolytic type, are provided in all HWPTR plants to upgrade the heavy-water recovered from various D_2O equipment system components such as pump glands, pump seals, pump connections, etc. The upgraded D_2O later joins the main heavy-water system of the plant.

The Erection of a Typical HWPTR Nuclear Power Station: Sequencing and Scheduling

5.0 Introduction

In the introduction to this book, we have discussed the advent of commercial nuclear power in India. Chapter 1 dealt with the genesis of the development of Atomic power in India. Chapter 4 has furnished a general description of typical heavy-water moderated and cooled units. It also detailed the differences between fossil fuel-fired and nuclear fuel-powered stations and described the various buildings and facilities in the HWPTRs.

The general approach in this book is to acquaint even a layman how nuclear power reactors have been designed, constructed and operated. The aim is to de-mystify the subject of nuclear power. This effort is carried further in this chapter as more details will be furnished about the various stages involved in the erection of HWPTR stations.

These stages are common to all HWPTRs. While this is the case, the RAPP units 1 and 2 have been chosen specifically for this chapter. Having worked in the RAPP 1 and 2 sites for over a decade, the author is more familiar with the various levels and room details of these two units. This particular aspect does not alter the sequence and planning of the erection of HWPTRs as they are generally applicable to all such reactors.

In Chapter 7, we will be discussing in detail about RAPP 1 and how it was commissioned within seven-and-a-half years, from the turning of the first sod of excavation at the site to the first concreting. The TAPP units 3 and 4 were commissioned within five years from the first concreting. This schedule is likely to be followed in the construction of future reactors, subject to no major constraints arising in the individual projects.

The overlapping of work has also been taken into consideration in detailing the schedule. The details of this schedule, based on experience in the individual projects, are as follows:

1) The preliminary site work and excavation for Major structures : 12 to 15 months
2) Concreting and structural work : 15 to 30 months
3) Reactor plant work, Reactor auxiliaries plant and piping work : 28 to 48 months
4) Electrical, mechanical, control and instrumentation work : 30 to 60 months
5) Pre-commissioning and commissioning work : 60 to 72 months

All the periods mentioned above are reckoned from the start date of work at the site.

5.1 Various stages in the erection of nuclear power stations

We will now discuss the various stages in the civil work of nuclear power stations. Quite a few of these stages are common to fossil fuel-fired power stations as well.

5.1.1 Pre-Civil work

Before Civil work begins at a site, the following studies need to be undertaken:

1. Soil investigation
2. Geological/tectonic studies
3. The study of wind and climate conditions
4. The study of tidal, coastal and under-sea conditions
5. Soil conditions for shore-based stations

5.1.2   Excavation

This is essential for providing a sound foundation for the power station structures and buildings. We may notice that the reactor buildings of 220-MWe and 540-MWe units are circular. The central line of the building is marked up on site by the use of both horizontal axes. This marked-up central line is used for scaling all distances before the excavation. The depth of excavation depends on soil conditions. It depends, among other factors, on accessing a surface that would safely support the entire load of structures, equipment, etc. The safe bearing capacity of different types of soil is given below:

1)   Compact sound dash - 2.5 kg/1000 mm sq.
2)   Compact sand - 6.5 kg/1000 mm sq.
     (prevented from spreading)
3)   Laterite - 2.5 kg/1000 mm sq.
4)   Granite upwards of 27.5 kg/1000 mm sq.

Note: Please refer to para 9.3.2.1 of Chapter 9, wherein it has been mentioned that the soil at the Narora site was of alluvial sand and hence the excavation was carried out to a depth of 17 metres below grade level. In para 7.37.1 of Chapter 7, the method of excavation at the RAPP site, which is of weather sandstone, has been described in detail.

5.1.3   The Foundation

The foundation work is taken up after the excavation to the required depth, as per soil conditions and level required for the plant, is completed. The foundation at the Narora Atomic Power Project (NAPP) site, which is of alluvial sand, was a mat of plain concrete 4.6 metres thick. The soil at the Madras Atomic Power Project (MAPP) site is similar to that of the NAPP site, with the difference that the former consists of coarse sand instead of alluvial sand. The difference at the MAPP site is that the drifting of sand was recorded at a certain depth in the reactor site which turned out to be about 1 million tonnes per hour. This called for the strengthening of the perimeter wall of re-inforced concrete by way of drilling deep steel anchor rods into the surrounding soil at various depths and locations. This was done at an inclination to the vertical face to add pressure to the concrete.

5.1.3.1 *The concreting method at RAPP units 1 and 2*

In Chapter 7, we will be discussing the wagon drilling method that was adopted for the excavation of deep rock formations in RAPP units 1 and 2. Once the required depth of excavation was reached, the surface was cleared of loose rock and debris. Interstice gaps in the rock surface were filled up and pressure pumping of cement was carried out—a monolithic surface was thus ensured for the commencement of building the foundation of the reactor raft.

5.1.3.2 *Centring for the Reactor Building*

The three-axes centring of the reactor building was first established before the commencement of the structural work in the building. These benchmarks will be used for all measurements and will periodically be cross-checked and re-validated as the building structure is raised further.

5.1.3.3 *The foundation of the connected buildings*

The foundation work for the connected buildings such as the turbine building, the service building, the pump houses, the cooling towers (wherever necessary), the administrative building and other structures commenced according to the scheduled date of completion of the plant. Pile foundations were resorted to as well, the details of which were dependent on the site conditions. Para 9.3.2.2 may be referred to better understand the foundation for the turbine building and service building of NAPP.

5.1.3.4 *Further structural work: the perimeter wall of the reactor building*

This is the very first auxiliary structure of the reactor building to commence work. The perimeter wall is made up of RCC.

5.1.3.4.1 Structural steelwork

Steel structures are erected to support the equipment, intermediate slabs, piping, etc. These structures require footing foundations—made up of RCC—and laid out on the foundation mud-mat or various floors as the case may be. Grout bolts to support the base plates of these structures are suitably embedded as well.

5.1.4   Embedded parts

The erection of embedded parts is one of the most important aspects of reactor construction. It is also one of the earliest to be taken up in terms of the stages of erection. These parts comprise a wide variety such as metal (GI/conduct) pipes; steel parts; steel pipes to pass and support other large, medium and small diameter pipes, ducts, overhead travelling cranes' cable and wire ways; wall/floor penetration supports, etc. These parts are numbered and detailed drawings indicating the method of fabrication/manufacture are furnished to the field forces well in time. The most important precaution, wherever embedded parts penetrate the reactor building, is sealing as per specification at these penetrations to guard against leakage of radiation. Gas-proof leakage requirement is also specified wherever required, particularly for power and cable penetrations.

5.1.5   Main and intermediate concrete slabs

5.1.5.1   Reactor building

Along with the rising perimeter wall, the main and intermediate floor slabs of the road buildings are laid. Some of the important floor levels of the RAPP 1 and RAPP 2 units are listed below:

1)   355.5 metres : Labyrinths on either side of the reactor
2)   357.3 metres : Moderator Room
3)   362.20 metres : F/M vault floor
     (access to the RB is through this floor and via the air-locks)
4)   365.80 metres
5)   369.50 metres
6)   373.47 metres at RB

The Boiler room which houses the pressurizing pumps, PHT pumps, boilers (heat-exchangers), driers, housing tank with its distribution piping, etc. is located on this floor in RAPP units 1 and 2. This tank was dispensed with in later projects.

7)   385 metres : The springing of the dome is located on this floor
8)   400.3 metres

This is the top of the dome and the highest level of the RB. Lightning protection is provided at this level outside the top of the dome.

Note: All levels are 'AMSL' (above mean sea level). The base ground level of the RAPP units 1 and 2 is 357.3 metres.

5.1.5.2 *The Turbine Building*

1) 351.2 metres

    The mechanical workshop, waste-management facility and operation and maintenance stores are located here.

2) 357.3 metres

    This is the base-ground level which corresponds to the road level. Right outside the Turbine buildings 1 and 2 (TB-1 and TB-2) is the loading bay for all the major heavy equipment, particularly the conventional equipment. The 150/50-tonne over-head travelling crane in the TB-1 and TB-2 main level is used for hosting the above-mentioned equipment on the respective floors.

    The ventilation equipment, Diesel-Generating (DG) equipment, Class-III de-mineralised water storage tank and water purification equipment are all located on this floor.

3) 360.3 metres

    This is the operating floor of the Turbine-generating sets and their auxiliary equipment. The TG set control and supervisory panels, 3.3 KV switchgear, 415-V Class-IV switchgear, etc. are all located on this floor.

4) 353.0 metres

    Crucial equipment such as the 21-KV isolated phase bus duct (metal clad), neutral grounding transformer, steam condenser, steam re-heaters, etc. are located below the TG sets at this level.

5) 362.20 metres

    The plant control room and control equipment room, relay panels, channel temperature control and monitoring panel control distribution frame are located on this floor. Boiler feed pumps and miscellaneous equipment are located below this level.

6) 369.50 metres

   415-volt Class-II and Class-III switchgears, 415-volt mccs, 250-volt power batteries, 48-volt control batteries, 250-volt/48-volt DC switchgear, 250-volt rectifier and rectified control panel are located on this level.

7) 373.47 metres (TB)

   Feed-water heaters and the de-aerator are located on this floor.

8) Outside terrace area on top of the 373.47-metre floor.

   Coding water/air equipment for the air-conditioning system of the control room and control equipment room are located on this floor.

5.1.5.3 *The Service Building*

5.1.5.3.1 Other higher levels: 362.20 metres

Washrooms and locker rooms, instrumentation laboratory, electrical laboratory, fuelling machine service bay, spent-fuel storage bay and connected equipment are located on this level. Active and inactive laundry, waste-management facility, Physics facility and other facilities are located in this building.

5.1.5.3.2 The levels above 362.20 metres

Absolute filters for active exhaust air, active and inactive laundries and the Chemistry laboratory are located above 362.0 metres floor level.

5.1.6 The pre-stressed concrete dome

Once the civil construction reaches 385 metres in elevation, the pre-stressed concrete dome is cast. All critical and heavy-reactor equipment—requiring hoisting through heavy-duty cranes—would be in position before the dome is cast.

5.2 Permanent erection of the reactor building

5.2.1 Stainless steel liner and thermal/biological shield work

The stainless steel liner work on the walls of the fuelling machine (FM) vault and thermal/biological shield steelwork in the reactor building is essential from the point-of-view of providing effective shielding of radiation. They are thus carried out at appropriate stages as part of the reactor plant work.

5.2.2 Grounding and cable pan work

So far, we have seen in paras 5.1.3.2 to 5.1.4 about the stages of erection of the plant. We have also seen how the work of embedded parts plays a crucial part in the direction of the main plant. The other two important works of erection that have to be started at the earliest stages of construction are grounding and cable pans/cableways.

The work of grounding is crucial from the point-of-view of the safety of the plant and personnel. Therefore, it is to be taken so that this work is carried out strictly as per the specifications. The work of erection of the cable pans and the cableways precedes that of laying control and power cables. The cable-bridge and pipe-bridge between the RB and TB provide the means of laying control and power cables and large-sized types between RB and TB. These types including erection work form an extensive part of the plant work.

5.2.3 The reactor plant: The erection of conventional plant equipment

The erection of equipment is taken up to align with the completion of floors in their ascending form. The erection of moderator pumps, motor units, moderator heat-exchangers and Ion exchangers is taken up as soon as the moderator room is ready in all respects for the erection of equipment.

The next important stage for the erection of major reactor plants is when the FM vaults become available for the erection of equipment. The erection of the Calandria and end-shields, heavy-water headers, heavy-water feeder pipes, coolant channels, Calandria tubes, end-fittings shield tanks, dump tanks, atmosphere coolers, breakout panels, etc. is taken up at this stage. The erection of ventilation systems for both active and inactive ambient air is also taken up.

In case a particular floor or area is not becoming available for any reason—for eg. some civil work is not yet completed, relevant equipment to be installed in these areas has not yet been received at the site, etc.—then the erection work is taken up in other areas/floors that are available. The equipment at these sites is stepped up accordingly to make up for the overall progress. In para 16.9.4.1 of Chapter 16, a unique way of carrying out Civil work without delaying other work has been described in detail.

Some of the pre-erection work related to the Calandria, dump tanks, pressure tubes, Calandria tubes, etc., were carried out at a component workshop set up for the RAPP units 1 and 2. (paras 7.15 and 7.16 of Chapter 7 can be referred to in this regard)

Giant crawler cranes were utilised from KAPP units 1 and 2 for lifting heavy equipment above the reactor building and further lowering and positioning them at their assigned locations.

5.2.4   Instrumentation and Control Equipment Systems

This is one of the most extensive of all systems associated with the reactor and its auxiliaries; so also are the reactor regulating, monitoring and control systems. Pressure, temperature and flow controllers are located extensively in the reactor plant. These controllers are wired to field marshalling boxes which are further wired to the Control Distribution Frame (CDF) located in the control and instrumentation section of the turbine building. The control distribution frame serves as the interface between connections from the marshalling boxes to those at the main control panel located in the plant control room in the turbine building. Approximately, 2 lakh connections are involved at the control distribution frame.

5.2.5   Power and control cabling

Power and control cabling, running several hundred kilometres in length, is taken up at a fairly early stage to ensure that cable termination and connections may be completed to line up with the erection of the equipment.

5.2.6   Piping work

Piping work in the reactor building comprises mostly of the nuclear system. The diameters of nuclear piping mostly vary from 25 mm to 250 mm and larger measures in some cases. The types of piping used are cast black, mild steel, cast steel, stainless steel and carbon steel. Piping connection types are welded, screwed, and tight-fitted and rolled, among others. The major non-nuclear piping work is the main steam lines which are 400 mm in diameter. Those in the turbine building are the circulating water system lines for the condenser which are 1200 mm to 1600 mm in diameter.

5.2.7 The 220-KV switchyard

The station switchyard, being a separate and independent system by way of planning and sequencing of plant activities, is taken up at a fairly early stage to line up with the time-wise requirement of the 220-KV system power for start-up commissioning activities.

5.2.8 Crucial Pre-commissioning Work

5.2.8.1 *The flushing of the moderator system*

The moderation system contains ultra-high-purity heavy water (99.9%). It is hence necessary to ensure that the system is ready to take in this heavy water. It is also necessary to ensure that the system is free from any leaks before the D_2O is filled in. The system is thus flushed with light water and the leakages are plugged. Hot air is then blown into the system at various points to ensure that the system is fully dry. The flushing also ensures that all foreign matter is removed from the system. The system is then filled with heavy water.

5.2.8.2 *Hot-conditioning of the PHT system*

This is the process by which the system is conditioned in a light-water environment under the supervised norms of temperature, pressure and water chemistry. The objectives of hot-conditioning are as listed below:-

1) The formation of an adherent, protective, impervious and uniform magnetite coating on carbon steel and other structural materials. This coating reduces corrosion.
2) The cleaning of the system under high pressure and temperature conditions to free the system from loose crud. This also ensures that crud and other unwanted infiltration matter are reduced.
3) Ensuring the integrity of the system and plugging leakage spots.

The temperature is set at a value slightly lower than the under-operating conditions. The pressure is set a little lower than the set value of pressure relief valves.

5.2.8.3 *Running in the TG set at low speed*

The auxiliary steam supply is availed of to fill all the glands and also to run in the set at low speed. This process helps in sealing off the leakage location. It also helps in the checking of all bearings, thus ensuring the thermal equilibrium of the set before nuclear steam becomes available for full-scale commissioning of the set.

5.2.9 Maintenance and Other Plant Support Services

Most of the maintenance and support services of the plant are required to be completed and ready for plant operation. More details on the same can be found in paras 6.2.6.1 to 6.2.6.6 of Chapter 6.

5.2.10 The coordination between the construction, quality surveillance and commissioning groups

The pre-commissioning and commissioning activities require 6 to 12 months. The pre-commissioning activities largely consist of the taking over of the systems from the construction group. The quality surveillance group also comes into the picture at this critical stage; their concurrence in the handing over of documents from the construction group to the commissioning group is mandatory from the point-of-view of conformance to quality requirements. This part of sequencing for commissioning is vital and has to be built into the overall programming for the commissioning of every nuclear plant.

The following chart provides details about the organisational setup.

Chart Nos. 5.2.11.1 (t1) and 5.2.11.2 (t2) elaborate on the positioning personnel.

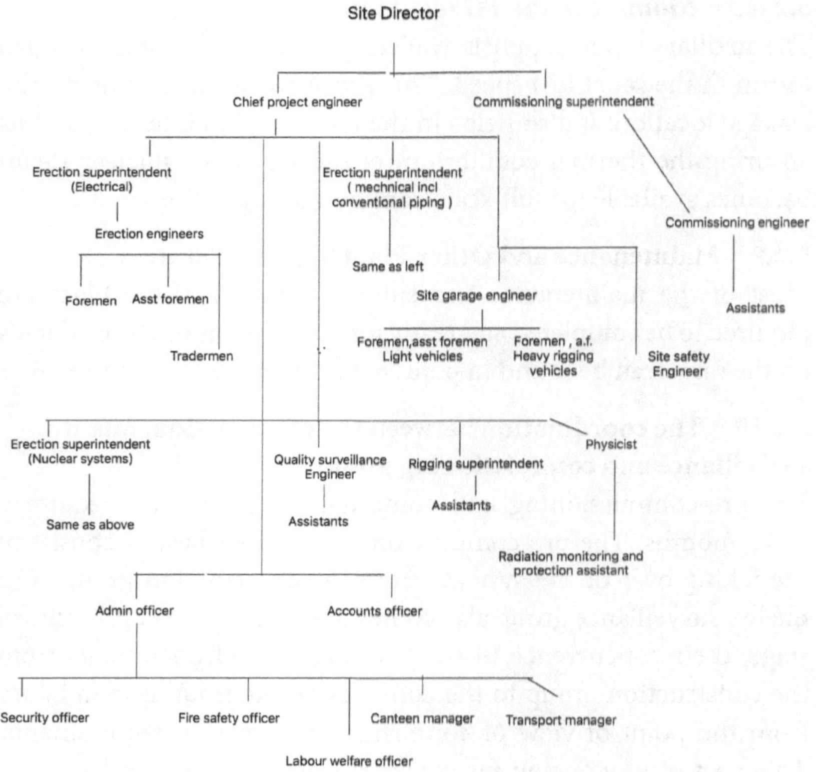

Chart no. 5.2.11.1 (t1) indicates the positioning personnel for plant erection and commissioning

OPERATION and MAINTENANCE

ORGANISATIONAL CHART FOR TYPICAL TWIN UNIT

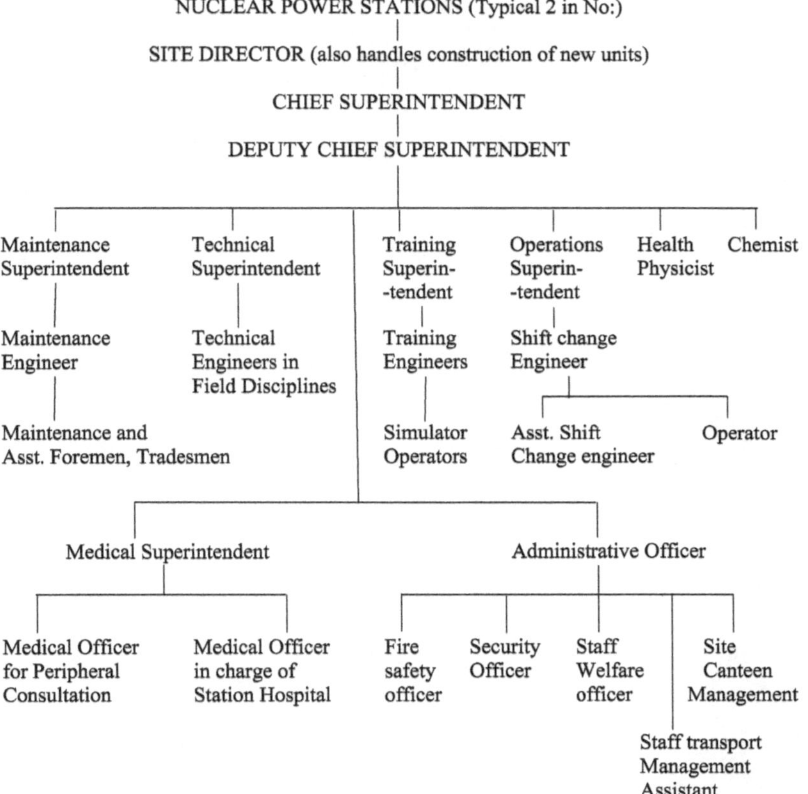

Chart no. 5.2.11.2 (t2) indicates the positioning of staff for operation and maintenance

Further Description of a Typical HWPTR with Particular Reference to RAPP Units 1 and 2

6.1 Broad principles of the HWPTRs: further details about the construction and commissioning of Nuclear Power units in India

In Chapter 5, we have seen the broad principles and plant requirements of the HWPTRs. We will now go into more details about RAPP 1 and RAPP 2 which would be the country's first Nuclear Power units to be constructed and commissioned in India. These units would go on to fulfil the agreement between the Indian and Canadian Governments. Complete details about the erection of these units are dealt with in Chapter 7 of this book.

6.2 Various systems of the plant

There are several systems present in the plant that house the RAPP units 1 and 2. They are listed below:

6.2.1 The Reactor plant system

6.2.2 The Reactor plant auxiliaries system

6.2.3 The Reactor control, protection and regulatory systems

6.2.4 The Reactor equipment system

6.2.5 The Conventional Equipment System

6.2.6 The Reactor and conventional maintenance services

6.2.7 Other plant support service systems

Note: The reactor plant system and its auxiliary systems are housed in a containment building called the reactor building. The details of these buildings have been furnished in section number 4.2.4.2 of Chapter 4.

We will now elaborate on the reactor plant system.

6.2.1 The Reactor plant system

6.2.1.1 *The Reactor building access system*

Air-locks provide a barrier between the low and high-radiation areas of the building. Access to the building is controlled by two air-locks namely:

(a) The main air-lock for personnel and equipment entry

(b) The emergency air-lock intended for use when the main air-lock is not available for normal operational purposes

Each of the above-mentioned air-locks has two doors which are operated pneumatically in a push-button pattern—only one door will open or close at one time, thus ensuring the isolation of the main reactor buildings from the rest of the reactor complex at all times.

6.2.1.2 *The Reactor vessel or Calandria*

The Calandria is a horizontal pressure tube-type vessel. More details about the vessel can be found in para 7.9.5 of Chapter 7.

6.2.1.3 *The Coolant assembly system*

Three hundred and six thick-walled coolant tubes, also called pressure tubes, are placed inside the Calandria tubes from end-to-end of the pressure vessel; they are extended by end-fittings through the seals at both ends. These fittings are made of type-403 ferritic stainless steel. The end-fittings are rolled onto the Calandria tubes at both ends. The above-mentioned 403-ferritic SS material was chosen for its strength, resistance to corrosion, irradiation coefficient and its low coefficient of thermal expansion. All of these are very much essential for the making of a sound-expanded joint with the zircaloy-2 Calandria tubes; the pressure coolant tube along with the two end-fittings is called the coolant assembly.

The coolant assembly houses the nuclear fuel and the heavy-water coolant. The coolant tubes are made up of seamlessly drawn zircaloy-2 tubes with a maximum outer diameter (OD) of 9.08 cm and a wall thickness of 0.39 cm. The Calandria tube, made up of nickel-free zircaloy-2, has an OD of 11.02 cm and an inner diameter (ID) of 10.76 cm. The coolant tube is supported on garter springs to ensure uniform annular air space between the Calandria tube and the pressure tube.

6.2.1.4   Support assembly arrangement for Calandria

The Calandria is hung from the Calandria vault roof concrete by four pairs of 'Invar' (Nickel–Iron alloy) rods of 9.52 cm diameter. The rods are secured to adjustable supports located in a pit on the 373.47-metre elevation of the boiler room above the Calandria for the vertical positioning of the Calandria during installation. Provision for the adjustment of supports

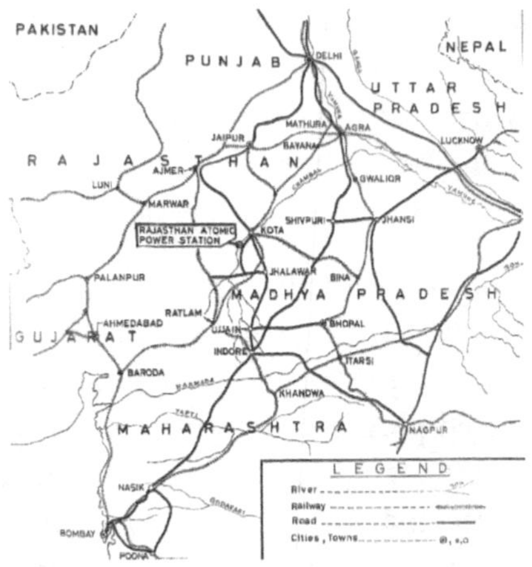

General Location of Atomic Power Station

to compensate for the long-term changes in the supporting structures is available. The load is shared equally by the four pairs of hanger rods. The rods are designed for low thermal expansion and low strain. The rods are cooled by heavy water; the ultimate tensile strength of the rods is 4569 kgs per square centimetre with a factor of safety of 8.6. The diameter of the rods is 8.25 cm and the length is 6.4 metres. More details about the hanger rods of the reactor are described in para 7.26.1 of Chapter 7.

6.2.1.5   The End-Shields

The reactor end-shields are steel and water-sealed. As the name implies, the end-shields shield both the fuelling machine vaults (each at either end of the end-shield) from the effect of radiation due to nuclear fission. The end-shield assembly consists of six semi-circular slabs contained within a steel shell; cooling water flows within and around the slabs. The slabs and the outer face of the shell are made up of carbon steel; the inside face of the slab and the end of the shield hole liners are made up of low-transition temperature ASTM A203 grade-D steel. The rate of flow of cooling water is adjusted so that the temperature of the steel does not fall below 93 degrees C during operation and 66 degrees C when shut down. The dimensions of the end-shield are furnished in Table 15.4(t3) of Chapter 15.

6.2.1.6 Support arrangement for the end-shields

The end-shields, like the Calandria, are supported by hanger rods which form part of the support structure located in a pit on the boiler floor above the Calandria vault roof. Unlike the support rods for the Calandria, there is no cooling arrangement for these rods.

6.2.1.7   End-fittings assembly

The end-fittings make transition joints between the pressure tubes and the primary circuit piping. The end-fittings accommodate a sealed plug and a high-pressure closure which are operated by the fuelling machine during re-fuelling operations. The end-fittings are fabricated from type-403 ferritic stainless steel. This type of steel was chosen for its strength, hardness and resistance to irradiation damage. The other advantage of this material is its low coefficient of thermal expansion which is necessary to make a sound expansion joint with zircaloy-2. The design allows for some movement of the coolant assembly relative to the end-shield because of the differences in thermal expansion. Adjustable stops on the end-fittings set a limit to this expansion.

The end-fitting assembly dimensions are as stated below:

End-fittings: length 2.14 metres, outer diameter 15.24 cm

Shield plug: length 0.99 metres, outer diameter 8.25 cm

6.2.1.8   The end-shield cooling system

A heat quantum of 939 KW is generated in the two end-shells, and the shield-cooling heat-exchangers are provided to remove this heat. The cooling water, which is de-mineralised and de-aerated, flows through the annuli around the lattice positions to the outlets at the top of the end-shells. The coolant enters the end-shield at the bottom through multiple inlets. The inlets can be adjusted to ensure the flow distribution required to obtain an even temperature across the end-shields.

The temperature and other parameters are as stated below:
1)   Temperature inlet to shields : 72 degrees C
2)   Temperature outlet from shields : 75 degrees C
3)   Coolant flow rate : 9.1 klits/minute
4)   Shield temperature : 93 degrees C
5)   Temperature at the inlet of heat-exchangers : 22 degrees C
6)   Temperature at the outlet of heat-exchangers : 38 degrees C

There are three heat-exchangers and three cooling pumps connected in parallel. Two of these are capable of meeting the load while the third is on stand-by. The loss of coolant to either end-shield will result in a reactor trip.

6.2.1.9   *The moderator system*

This is a low-pressure low-temperature heavy-water system that comprises helium cover gas. The heavy water in the system is held in the Calandria by helium gas pressure in the dump tank beneath the Calandria vessel. The gas pressure acts on one leg of each U-shaped dump port at the bottom of the vessel. The helium gas pressure in the dump tank and the dump port is maintained by water-jet exhausters in the circuit which act to compress the gas. The moderator level in the Calandria is controlled by regulating the flow of helium through the control valves between the dump tank and the Calandria. The pressure of helium in the space above the Calandria is 1.20 kg/cm sq. The maximum permissible outlet temperature from the Calandria is 63 degrees C and the minimum permissible inlet temperature is 21 degrees C. The heat generated in the moderator during normal operation is 40 MWe which is removed by two heat-exchangers with a rate of flow of D2O of 27.3 klits/min, of which 25 klits/min is from the Calandria and 2.3 klits/min is from the dump tank.

The moderator heavy water is circulated through heat-exchangers from a point above the mid-level of the Calandria to a point at the bottom of the vessel; the rate of flow of the heavy water is 16.4 klits/min. A side stream with a rate of flow of 6.82 klits/min is sprayed into the top of the Calandria, to cool that part of the Calandria and the array of Calandria tubes above the operating heavy-water level of the moderator. There are other side circuits in the system, which are described as follows:

1) The ion-exchange circuit with a rate of flow of 228 klits/min
2) The pump-up circuit with a rate of flow of 15 klits/min
3) The adjuster rods cooling circuit with a rate of flow of 3.41 klits/min
4) The water-jet exhauster circuit for normal helium circulation with a rate of flow of 1.55 klits/min

The ion-exchange columns contain a deuterated resin to remove poisonous Boron and other impurities from the moderator heavy water. It takes five hours to completely remove the poisonous Boron from the moderator heavy-water system. There are five 415-V 150-HP moderator heavy-water circulating pumps, three of which are fed from the 41-V Class-II system while two are fed from the 41-V Class-III system.

The moderator can be rapidly dumped by opening the dump port valves in the helium line between the dump tank and the top of the Calandria, thus equalising the pressure below them. This permits the moderator to be dumped into the dump tank by gravity. Such a method is adopted for the tripping of the reactor. Helium lines are provided to purge Deuterium and Oxygen that are produced as a result of dissociation of the moderator heavy water.

6.2.1.10   *The Dump Tank System*

This system is one of the most important links necessary for maintaining the level of moderator heavy water in the Calandria to suit operating conditions, and also for tripping the reactor in case of operational necessity. The dump tank is located below the Calandria and is connected to the latter by a transition section and an expansion bellow. The dump tank is made up of stainless steel. It is designed to accommodate an internal pressure of 0.8 kg/cm sq. and the full hydro-static head and impact loads during dumping. It is nearly empty during normal operation but contains helium at a pressure of 0.64 kg/cm sq. to maintain and hold up the level of the moderator in the Calandria. The helium circuit contains devices to prevent over-pressure in the dump tank. The pressure in the tank is always maintained at 169 kg/cm sq. When the heavy water is dumped from the Calandria to the dump tank, the dump tank is filled to a level a few centimetres from the top of the tank. The dump tank along with the transition and expansion bellows will hold 169 tonnes of heavy water. The tank is designed to hold heavy water from the moderator and the primary circuit. This leaves a maximum of 38 tonnes of heavy water at the bottom of the Calandria. This retained quantity is much less than the minimum heavy water required to make the reactor critical.

The dimensions of the tank are as stated below:
Length : 8 metres
Width : 5.08 metres
Depth : 3.185 metres
Wall thickness : 3.8 cm

The Calandria end-shield dump tank assembly may be stated to be one of the most critical assemblies in a HWPT reactor plant. This is the assembly wherein nuclear fission takes place. The radiation due to fission is contained (mostly within the nuclear fuel itself) in the dump tank in which heavy water may be dumped to shut down the reactor in case of any abnormal occurrences within the reactor system.

6.2.2   The Reactor auxiliary system

6.2.2.1   *The biological and thermal shield cooling system*

6.2.2.1.1   General
It would be useful to have a look at the ways of shielding various sub-areas of the reactor containment from the intense radiation produced in the Calandria as a result of fission.
1)   The two circular ends of the Calandria are provided with water and steel shields and end-fittings which act as the extension of the pressure tube that is made complete with a shield plug.

These effectively shield the fuelling machine vaults from the effects of intense radiation.

2)   The walls of the Calandria vault are constructed with heavy concrete i.e. haematite concrete (increased iron content) with granite metal that is added to the concrete mix (density – 3630 kgs/cu. metres). This helps in withstanding the high temperature caused by radiation. The vault concrete is also cooled by water flowing through pipes embedded in it.

3)   The walls of the Calandria vault are lined with a 3.2-mm thick stainless steel liner to contain the vault atmosphere of air and any heavy-water vapour that may be present.

6.2.2.1.2   The thermal shield system

The side wall of the Calandria vault is provided with a steel plate which serves as a thermal and radiation shield. A quantity of approximately 600 KW of heat is dissipated in the thermal shield. This heat is removed by two blower fans. A constant purge flow enters the system at the blower plenum chambers and is exhausted through the filters to the stack.

The details of the cooling system are furnished below:
1)   Coolant : air
2)   Coolant temperature
    a)   At inlet : 27 degrees C
    b)   At outlet : 44 degrees C to 49 degrees C
    c)   Coolant flow rate : 1698 klits/min
    d)   Coolant purge rate : 33.96 klits/min

The radioactive discharge from the stack on account of the coolant outflow is 2.2 curies per hour as Argon-4L.

6.2.2.1.3   The Cooling of the biological shields, end-shield rings and thermal shields

The coolant is pumped through various parallel circuits in the system and cools the biological shield, end-shield rings, the upper and lower labyrinth thermal shields and the top-hatch beams of the Calandria.

The details of the cooling system are as stated below:
a)   Coolant : De-mineralised water
b)   Coolant temperature at the inlet : 29.4 degrees C
c)   Coolant temperature at the biological shield outlet : 32.7 degrees C
d)   Coolant temperature at the end-shield ring outlet : 34.9 degrees C

6.2.2.1.4   The shield tank and its cooling system

The shield tank, made up of steel, contains de-mineralised water and provides a biological shield for the fourth side (east side) of the Calandria vault. The shield tank water is provided with a cooling system.

The details of the system are as stated below:
a)   Coolant : de-mineralised water
b)   Shield tank water temperature at the inlet of heat-exchangers : 33.3 degrees C
c)   Shield tank water temperature at the outlet of heat-exchangers : 29.4 degrees C
d)   Coolant flow rate : 374 lits/min

6.2.2.1.5   Air-flow in the fuelling machine vault and the Calandria vault

The walls of the fuelling machine (FM) vault are lined with 1.5-mm-thick stainless steel plates to provide thermal and radiation protection to the walls and hence exterior of the vaults. About 1500 klits of air per minute are circulated through the coolers in each FM vault. The temperature in the vaults is maintained at 36 degrees C. Air is also circulated in the Calandria vault in a closed system to cool the vault and to condense heavy-water vapour arising from leaks in the primary circuit. Side air circuits with a flow of 45 klits/min are circulated through a dryer and scrub system to maintain the dew point in the vaults at 0 degrees F and remove any nitric acid that may be found due to irradiation. The heavy water is also recovered for further upgradation in the D2O upgrading plant. The total air-flow through the vaults of the Calandria and fuelling machines vaults is about 148.5 klits/min, of this the total flow from both the FM vaults is 56.6 klits/min.

6.2.2.1.6 Fuel for the reactor

We have seen in para 4.2.2 of Chapter 4 about the type of fuel used in HWPTRs and how the fuel causes fission and nuclear heat which leads to the generation of nuclear steam that runs the TG set. Each reactor is filled with 41.60 metric tonnes of natural uranium in the form of uranium dioxide pellets. The pellets are sheathed in zircaloy-2 tubes to form cylindrical elements of 1.522 cm outer diameter and 49.5 cm length. Nineteen of these elements are assembled to form a fuel bundle which is held together by end supports and spaced by wire helixes on the six elements of the inner ring and alternate elements of the outer ring. All joints in the fuel bundle are resistance welded. The details of the bundle are indicated in the sketch provided in para 7.33.6 of Chapter 7.

Some of the important parameters of the bundle are furnished below:

1) Density : 10.4 to 10.79 kgs/cm^3
2) Cladding : zircaloy-2
3) Movement diameter of the bundle : 81.69 mm
4) Number of fuel bundles per channel : 12
5) Heat-transfer area per channel : 4500 cm sq.
6) Total heat-transfer area of all fuel bundles : 1390 metre sq.
7) Maximum fuel pressure : 105 kgf/cm sq.
8) Pressure drop across the fuel channel of 10 bundles : 7.38 kgf/cm sq.
9) Inlet temperature of the fuel channel : 249 degrees C
10) Outlet temperature of the fuel channel : 293 degrees C
11) Maximum mass flow of D2O per channel : 4.54 × 104 kgs/hour
12) Maximum fuel temperature : 1930 degrees C
13) Maximum power per unit length of the bundle : 8.71 kW/cm
14) Average fuel life in the reactor : 1.65 years
15) Maximum fuel life in the reactor (nominal) : 2.20 years
16) Burn-up rate : 7000 MWd/tone + (10%)

More details on the fuel elements are provided in the form of photographs in para 7.33.6 of Chapter 7

6.2.2.1.7　The fuelling machine system

Once an HWPTR is in operation, re-fuelling daily is required to maintain positive reactivity in the reactor to enable the unit to produce rated thermal power. This re-fuelling counteracts the negative activity caused by the evolution of xenon as a by-product of the fission. It may be mentioned that the positive reactivity change from 0 to full power is 4.4 milliK.

The system takes care of re-fuelling and the removal of spent-fuel. The reactor is double-ended, hence two fuelling machines are in use at the same time. While the machine at one end loads fresh fuel in a channel, the other machine at the opposite end retrieves the spent-fuel from the same channel. The two machines are operated remotely from the control room. The machine carriages travel horizontally across the reactor faces; the operating head assembly, which refuels at one end and retrieves spent-fuel at the opposite end, moves vertically and is remotely visible at any lattice position. The head assembly comprises a mechanical hydraulic ram, a notary magazine containing fresh fuel and a snout. The head assembly seals with the end-fitting of the Calandria coolant assembly. This sealing has to be positive and complete before the commencement of the re-fuelling operation.

The rotary magazine has twelve positions which are enough to load or unload one complete channel at a time. It carries two channel-seal plugs, one channel-shield plug, the snout-seal plug, one ram-adaptor for specific operations, one guide-tube to guide the fuel to and from the magazine and twelve fuel bundles. The FM normally refuels six bundles daily. The fuel in adjacent channels moves through the channel in the opposite direction to maintain fuel symmetry. Thus, when a spent-fuel bundle is at the end of its travel through one channel, a new fuel bundle is simultaneously placed at the start of its travel through an adjacent channel assembly.

6.2.2.1.8　Spent-fuel transfer and storage system

Spent-fuel bundles are first transferred from the fuelling machine to the fuel-transfer room, and then they go through an underground tube to the spent-fuel bay in the service building. The bay is capable of storing, cooling and shielding the spent-fuel through 20 years of operation of the two reactor units. It is filled with water to a depth of 7.17 metres. The bundles are stored in racks in the storage bay. A minimum cover of 4.12 metres of water over stored fuel is ensured. A cover of 3.66 metres is adequate for fuel after five years of storage.

An inspection bay, adjacent to the storage bay, is equipped for underwater inspection of failed fuel leaks, testing and canning of failed or suspect fuel.

A shipping flask is available to ship the spent-fuel bundles to the re-processing plants for the recovery of plutonium.

6.2.2.1.9 The Primary Heat Transport (PHT) system

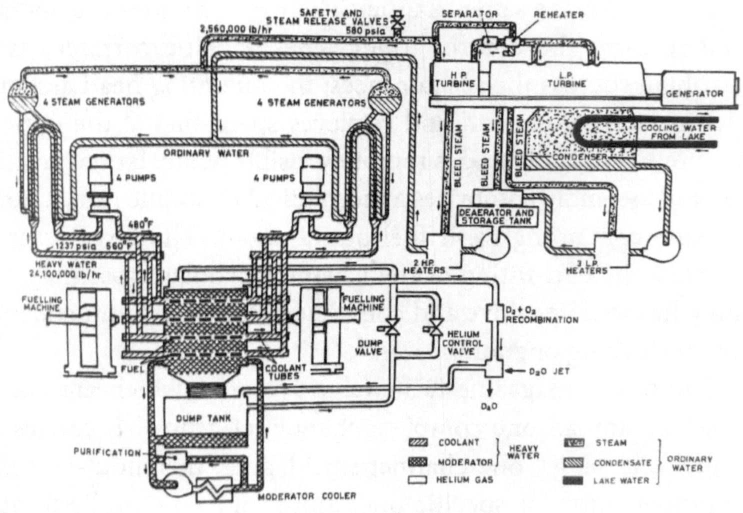

Simplified Flow Diagram

6.2.2.1.9.1 General

In the previous sections, we have seen some of the essential parts of the reactor plant and reactor auxiliary plant systems. We will now deal with the PHT system which, along with the moderator system, constitutes the very heart of the reactor system of a HWPTR. As the name implies, this system constitutes the primary system of the reactor plant which serves to transfer or transport the nuclear heat from the fuel to the high-pressure high-temperature heavy water circulating in the coolant tube. The adjacent coolant tubes have the heavy water flowing in opposite directions. This is to ensure uniform heat transfer across the fuel core and to keep the differential temperature between the channel ends within limits. This hot heavy water in turn transfers the nuclear heat to the secondary system which comprises the feed water, from the feed-water system, that was pumped by the boiler feed pump to the shell-side of the nuclear steam service generators (NSSGs) in which steam is raised.

6.2.2.1.9.2 More details about the PHT system

There are two heavy-water feed pumps, each driven by 3.3-KV 350-HP pumps, which serve to maintain the PHT system press at 91 kg/cm sq. in the reactor outlet headers using valves which feed to or bleed from the heavy water in the system. The feed pumps draw heavy water from the heavy-water storage tank located in the boiler room. A surge cooler/receiver, located in the boiler room, receives surge heavy water from the PHT circuit. The line from the surge cooler feeds the ion exchanger which removes fission by-products from the PHT circuit. The bleed cooler draws heavy water from the inlet header and cools the heavy water that is bled for the purposes of maintaining the PHT system pressure within the limits. There are two auxiliary feed pumps in the system as well.

6.2.2.1.9.3 The PHT pumps

There are eight heavy-water coolant circulating pumps, each driven by 3.3-KV 1170-HP motors divided into two banks of four each on either side of the boiler room pit. Eight corresponding steam boilers are similarly situated. Each pump circulates 25.8 klits/min of heavy water at an operating head of 14 kgf/cm sq. Each of the two outlet headers (one in each FM vault) feeds one set of the four boilers on the boiler floor.

The PHT pump shafts are sealed by throttle bushings that are backed by low-pressure mechanical seals to restrict the leakage of heavy water. The seal chambers are fed with 73 litres of cool heavy water from an external supply at a pressure of 3.5 kgf/cm sq. which is higher than the pump inlet pressure. This is done to prevent leakage through the seals and to cool them.

The main parameters of the main circuit of the PHT system are as given below:

1) Total coolant flow rate : 10.7×10 to the power of 7 litres per hour
2) Reactor inlet temperature : 249 degrees C
3) Reactor outlet temperature : 293 degrees C
4) Mean heavy-water temperature : 269 degrees C
5) Mean D2O density : 0.8469 kg/cu. cm
6) Piping material : Carbon Steel
7) D2O temperature low power : 262 degrees C

6.2.2.1.9.4   Heavy-water headers/feeder pipes

6.2.2.1.9.4.1 Headers

There are two inlet headers, one in each FM vault. Similarly, there are two outlet headers, one in each FM vault. Each header is of 40.6 cm outer diameter, 9.75 metres in length and 3.82 cm wall thickness. The headers are made up of ASTM-A 106 grade-B carbon steel and are designed for a maximum tensile stress of 1055 kgf/cm sq.

There are relief valves in the outlet headers which are set at a nominal pressure of 91 kgf/cm sq. The minimum acceptable pressure in the outlet headers is 84.7 kgf/cm sq. Triplicated instruments in the system will shut down the reactor if the pressure in the outlet headers is outside of the set limits. All the headers are supported by springs or other variable supports; they are provided with insulated enclosures to restrict heat loss to the vault atmosphere.

6.2.2.1.9.4.2 Feeder pipes

Feeder pipes take off from the end-fittings of the coolant channel assembly and connect to the headers at the other end. The feeder types are of four diameter sizes, namely 3.18 cm, 3.81 cm, 5.08 cm and 6.35 cm. The sizing caters to the varying coolant flow requirements arising from different fuel flux densities in the coolant lattice positions. The feeder pipes are made up of schedule-80 carbon steel of ASTM-A 106 grade-B. Each outlet feeder pipe is provided with a throttling device for coolant flow control, two temperature sensing elements and a sample tube for activity monitoring. A total of 5% of the coolant channels are provided with flow-sensing elements.

The feeder pipes are supported on springs or other variable supports and are provided with an insulated enclosure to prevent heat loss to the vault atmosphere.

6.2.2.1.10   Nuclear Steam Service Generators (NSSGs)

There are eight NSSGs, a bank of four boilers on each side of the pit on the boiler floor. More details about the NSSGs are furnished in para 7.9.6 of Chapter 7.

6.2.3   The Reactor control, protection and regulatory system (RCPRS)

6.2.3.1   The adjuster rod operation

The reactor generates rated power with the moderator at the full level of the Calandria. Small fluctuations in reactivity or radial flux distribution are corrected by the East and West adjuster rods. All six regulating rods are normally inserted fully into the Calandria. These rods are in the East, West, South-East, South-West, North-East and North-West positions of the Calandria. There are two elements in each regulating rod. The East and West position rods are called adjuster rods. They are so located that they control radial flux distortion or tilt. They control the reactivity variations caused by daily on-load re-fuelling. The four elements of the East and West rods are worth +3.75 mk. The maximum operating speed of the adjuster rods, when used for flux tilt control, is 20% of that provided for the overall reactor control.

The adjuster rods in the South-East, South-West, North-East and North-West positions are used primarily for booster functions. The worth of the eight elements on these four rods is +7.25 mk. These rods, which are normally inserted fully into the core, are removed from the core to overcome the temporary loss of reactivity caused by Xenon-135 build-up above the equilibrium value after a reactor trip or a shut-down. The reactivity worth of these eight elements—as well as that of the four elements of the East and West rods—permits the reactor start-up without poisoning out within 25 minutes of a trip. The reactivity of these rods contributes to the continued operation of the reactor even after reactor power reduction from 100% to 70%.

6.2.3.2 *The range of power for reactor regulation: signal generation for the adjuster operation (RPRR)*

The reactor regulating system is designed to provide automatic control of the reactor from 10 raised to the power minus 8 of full power to 110% of full power. The high time constant of the measuring circuit (450 seconds) in the range of power—10 raised to the power of minus 8 to 10 raised to the power of minus 5 range of power—makes the system stable. The manual operation of the regulating system is hence resorted to while the reactor power is in this range.

Six uncompensated ion chambers, one each for the triplicated reactor regulating channels and one each for the triplicated reactor trip channels, are provided. These channels are housed in a lead plug located 2.13 metres below the horizontal central longitudinal plane of the Calandria. The output from each ion chamber is used to generate linear and log signals of the rate of change of neutron power. Twenty-four signals in all are generated by the ion chamber and transmitted to the regulating and trip systems (twelve each for the two systems).

Several groups of triplicated Resistance Temperature Detectors (RTDs) transmit ΔT and outlet header temperature signals to the reactor regulator system. The ΔT signal is added to the log N signal and the combined signal is used as a control signal for power changes. The log N signal is dominant at low power while the ΔT signal is dominant above 20% reactor power. ΔT and log N signals have equal effect at 20% reactor power. The set point for the combined ΔT and log N signals is driven through a servo motor with a maximum speed corresponding to a ΔT of 0.9 degrees C per second or 2% of full power per second. Thus the slow rates of change of power are taken care of.

6.2.3.3 *The Low coolant flow alarm*

Sixteen coolant channels are monitored for coolant flow using differential pressure measurement across venturies. The signal pressure is transmitted to an indicator alarm for low-flow annunciation in the main control panel.

The measurement, as discussed earlier, is used to monitor the heat generation in the core and to observe the pattern of coolant flow through the core.

6.2.3.4   *The Reactor Trip and set-back signal (RTSS)*

6.2.3.4.1   The Trip signals

The following signals cause a reactor trip:

1) When the primary coolant outlet header pressure is less than 84.7 kgf/cm sq. or above 91 kgf/cm sq.
2) The neutron power is more than 110% of the normal full-power value
3) The reactor Calandria vault pressure is 0.071 kgf/cm sq. above the atmospheric pressure
4) The reactor building pressure is 0.036 kgf/cm sq. above the atmospheric pressure
5) The gross temperature rise between hot and cold headers is greater than 105% of the normal full-power temperature rise
6) When less than three primary heavy-water circulating pumps on one side of the boiler room pit (normally done by four) take the load
7) Both the coolant channel temperature and the zonal channel outlet temperature simultaneously exceed 299 degrees C and 297.9 degrees C, respectively
8) There is a low water level in the NSSGs (heat-exchangers)

6.2.3.4.2   The reactor set-back signals

The following events cause a set-back of power in the reactor:

1) The coolant channel outlet temperature and the zonal coolant outlet temperature simultaneously exceed 297.9 degrees C and 295.7 degrees C, respectively
2) Low vacuum at the steam turbine exhaust
3) When the steam discharge valves are open
4) Moderator level is less than 80% of the fuel level
5) End-shield cooling of the water flow is low and the shield temperature is high

6.2.3.4.3   A triplicated channel system for the reactor trip

6.2.3.4.3.1   Ion chamber signals

We have seen in para 6.2.3.2 that twelve signals are transmitted by the ion chambers for the triplicated trip channels. The reactor is tripped if two out of these three channels receive the trip signals.

6.2.3.4.3.2   The coolant channel mean-temperature variation signals to detect fuel failure

A duplicated analogue instrumentation monitors the fuel channel mean-temperature. The fuel core is divided into eight zones. These zones consist of four quadrants and comprise one inner and one outer zone per quadrant. One coolant channel is selected for each zone and is provided with a deviation indicator circuit having two high-alarm signals. The zone average and set point for deviation trips will be read on one set of common metres per zone. The entire installation is duplicated and two-out-of-two deviation high signals will cause a reactor trip. The high deviation system is backed up by a mean-temperature system that is also duplicated. Two-out-of-two signals based on this measurement cause a reactor trip. A low-temperature measurement system that is again duplicated causes a reactor set-back. This set-back also requires a two-out-of-two setback signal. The set-back reduces the reactor power at the rate of 1% of full power per second to bring down the power to 2% of the normal power. Thus, any further set-back is automatically prevented.

6.2.3.4.3.3   Fast pump-up after a reactor trip

A reactor trip shuts down the reactor at once and is caused by the dumping of the moderator from the Calandria to the dump tank. The dumping is controlled by two parallel sets of springs to open and close the air valves in the helium gas dump lines. Dumping reduces the reactor power from 100% to 50% in 7 seconds, 50% to 20% in 13 seconds and 20% to 10% in 10 seconds.

Fast start-up pumping of the moderator heavy water into the Calandria at a rate of 0.6% of full load per second is used to help overcome xenon poisoning after a short shut-down. The fast start-up is automatically prevented above 90% of full power. The fast pump-up must be started within 13 minutes of a trip to prevent xenon poisoning. A heavy-water flow of 15 klits/min is fed to the helium pump tank to ensure a proper high rate of flow of heavy water (D2O) into the Calandria. The pump circuit is drained of D2O when not in use.

If the reactor gets poisoned out, the poison-out will last for 35 hours. The residential heat is adequate to keep the reactor and boiler at the required stand-by temperature. The start-up from this condition will require about 90 minutes to come up to full-power operation after the attainment of criticality.

The start-up from the cold dumped state will require two-hours-and-a-quarter on a 167 degrees C per hour warm-up rate.

6.2.3.4.3.4   The dousing system

This system is designed to limit the pressure rise in the reactor building, particularly in the boiler room and in the fuelling machine vaults, by condensing the system escaping from the PHT system due to any break in the system. The system, for the area in the RB outside of the fuelling machine vault, utilises ordinary water from the 1820-kilo-litre storage tank—located on the 385.06-metre elevation floor—through a pipe distribution system located beneath the storage tank. Any break in the PHT system results in the massive release of heavy-water steam and hence increases the pressure in the boiler room and the surrounding areas. This pressure rise is sensed by a sensor and a resulting signal opens a butterfly valve in the deluge distribution system. The deluge spray, with an initial rate of 3460 klits/minute, condenses the steam and thus brings down the pressure in the building to a safe value.

The dousing system for the FM vaults utilises heavy water from the moderator system and condenses any heavy-water leak through the spray from nozzles. Nearly 6.84 klits of heavy water are routed to each FM vault. The system takes care of the heavy-water leak of the order of 1.36 to 22.7 kgs per second.

6.2.4 The Reactor Services System

6.2.4.0 General

This system provides the various services required for the operation of the reactor plant and its allied systems. This system is crucial for the satisfactory operation of the plant. One may legitimately ask: how the facilities for satisfactory operation are dealt with at the early design state? A modern 'state-of-the-art' design concept takes into account all the systems necessary, important and crucial for the seamless operation of a plant, much more for a multi-disciplinary plant as a Nuclear Power Plant. The reactor services system consists broadly of the following sub-systems:

6.2.4.1   Storage and management system for handling heavy water

6.2.4.2   The Waste Management System

6.2.4.3   The Radiation control, monitoring and surveillance system: Health Physics system

6.2.4.4   Active and Inactive Laundry System

6.2.4.5   Personnel change/lockers/shower rooms

We will now explore the above-mentioned systems in detail.

6.2.4.1   Storage and management system for handling heavy water

This is one of the most critical reactor service systems that impinge directly on the economy of operations and hence the profitability of the reactor station. About 140 metric tonnes of heavy water is contained in the moderator system and about 40 metric tonnes are stored in the PHT system. This is a large quantity of water. It takes about Rs. 10,000 to manufacture 1 kg of heavy water. The need for containing the loss of heavy water during operations needs no emphasis. The following are the sources of loss of $D2O$:

1)   Shaft seals

2)   Pump glands

3)   Joints in pipelines

The following measures are adopted for minimising the loss and preventing the down-grading of $D2O$:

1)   The primary containment boundary is made as continuous as possible

2) Connections are welded and integral seals such as bellows are used

3) Double seals are used in locations such as the shaft seal

4) All pump glands and valve glands are provided with double seals

Para 6.2.2.1.5 may be referred to better understand the recovery of heavy water from the FM vault atmosphere. All the D2O that is mopped up and recovered from the drier and scrubber are upgraded in the heavy-water upgrading plant. A daily account is maintained of the mopped-up heavy water and the throughput from the D2O.

6.2.4.2   *The Waste Management System*

This system forms an integral part of the station's corporations. Full details about this system are furnished in para 4.2.4.8 of Chapter 4 of this book.

6.2.4.3   *Radiation control, monitoring and surveillance system: Health Physics system*

The International Committee on Radiation Protection (ICRP) has recommended the maximum limits to the quantity of radiation permissible for the general public and radiation workers. The maximum exposure for the general public to air-borne radiation is 0.5 rem/year. The maximum limit for radiation workers is 10 raised to the power of minus 6 mc/ml (micro-curie per millilitre) in drinking water. The radiation level in working areas is limited to 0.25 mrem/ hour.

Access to areas where high radiation levels are possible is restricted. Among other measures for coping with ambient radiation is the use of proper protective clothing and the adoption of safe operating procedures. Film badges that store radiation quantum and direct reading dosimeters are used to monitor radiation levels. Areas are divided into radiation zones to access control and monitoring radiation.

6.2.4.4   *Active and Inactive Laundry Systems*

Clothing such as overalls and other personnel apparel like gloves, shoes etc. are segregated, depending on the radiation dosage and intensity of the field in the areas of usage. The clothing used in active areas is cleaned up in the active laundry while the others go into the inactive laundry. Both of these laundries are located in the service building. A dumb waiter has been provided in it for the transport of laundry goods.

6.2.4.5 *Personnel change/locker/shower rooms*

All the personnel deputed for work in the active areas of RB and SB are provided with lockers wherein personal belongings and clothing of the personnel are kept. The personnel are provided with their work clothes which are discarded for laundering purposes at the end of the day's work. The personnel must have their shower in the shower rooms before they change into their street clothes. All these areas are closely monitored for levels of ambient radioactivity.

6.2.5 The Conventional Systems

6.2.5.0 *General*

The conventional systems may broadly be described as systems other than the reactor plant system, reactor auxiliaries systems, reactor control, protection and regulatory system and the reactor services system. Some of the conventional systems that follow may apply to the reactor plant and its connected systems. It may be noted that the reactor plant and the connected systems are housed in the reactor and service buildings while the conventional systems are mainly housed in the turbine and connected buildings.

6.2.5.1 *Different types of conventional systems*

They are comprised of the following:

6.2.5.1.1 The Electrical system

6.2.5.1.2 The 220-MWe TG Set and TG auxiliaries system

6.2.5.1.3 The Feed-water System

6.2.5.1.4 The Steam Re-heater System

6.2.5.1.5 The Condensing system

6.2.5.1.6 The Water supplies system

6.2.5.1.7 The Ventilation System

6.2.5.1.8 The Fire protection system

6.2.5.1.9 The Communication system

6.2.5.1.10 The Drainage system

6.2.5.1.11 Control and Instrumentation System
Let us look into the above-mentioned systems in detail.

6.2.5.1.1   *The Electrical System*

6.2.5.1.1.1   The Generator output system: output step-up transformer systems

The Turbo-generator output voltage is 21 KV which is delivered to the main output transformer (412.G1) of 250-MVA capacity. The unit is delta star connected (DYNII) with a 21 KV/220 KV rating with four off-load taps on the HV side. The HV side neutral is solidly grounded. The generator output is delivered to the main output step-up transformer and the station's two step-down transformer units through a 21-KV 10-kA metal-clad phase-isolated bus duct.

The measures of protection undertaken for the main output transformer are as follows:
1)   Magnetic differential
2)   Gas trip
3)   Phase over-current
4)   Ground back-up

The transformer is of the oil-forced water-cooled (OFW) type. Station-type lightning arresters (LA) are provided on the 220-KV HV sides of the transformer. These lightning arresters are designed to protect the output transformer, unit station service transformers, isolated phase bus ducts and the turbine generator from travelling voltage waves and peaking voltages caused by system disturbances on the 220-KV spur line—connecting the HV terminals of the output transformer to the 220-KV switchyard of the station.

6.2.5.1.1.2   The Unit Station Service Transformers (USST)

There are two USSTs of unit capacity 9 MVA, 21 KV/3.3 KV delta star (DYNI) for providing power to the 3.3 KV bus. The transformers are oil natural cooled type and have four off-load taps. The 3.3 KV neutral terminal is grounded through a 1.90-ohm 1000-ampere capacity resistor. The protection for the USSTs is more or less identical to the main output transformer.

6.2.5.1.1.3   The Main Bus Duct

6.2.5.1.1.3.1 The 21-KV isolated phase bus duct (21-KV IPBD)
The bus duct is air-insulated and naturally air-cooled. Each phase conductor has its enclosure. Both the phase conductor and the enclosure are made up of an aluminium alloy. The bus conductor is octagonal, 500 mm across the corners. There are two gaps in the conductor, one at the top and one at the bottom. These gaps are 100 mm wide and closed at suitable intervals by plasma welding, thus constituting an integral continuous bus conductor. The bus conductor is formed by a 16-mm thick obtuse angle and flat aluminium sections which are again welded by plasma welding. The bus conductor is supported by three insulators, one each at the bottom and two sides of the bus conductor, which are fixed onto the container closure. The enclosure is also made up of aluminium which is 6.35 mm thick and 1000 mm in diameter.

6.2.5.1.1.3.2 Indoor and outdoor sections of the main bus duct: wall frame and seal-off bushings
There are indoor and outdoor sections in the run of the bus duct from the generator terminals to the LV terminals of the main output transformer. It is necessary to prevent the interchange of ambient air between these sections to obviate condensation—due to the difference in the temperature between the two ambient atmospheric air spaces—and also to prevent the ingress of dust, moisture and other undesirable matter from outside of the TB to the inside of the building. The wall frame-work, which is made up of Aluminium angles, sheets etc., is embedded in the TB outside wall and wall bushings to isolate the inside section of the bus from the outside.

Arrangements for draining water that is condensed in the enclosure are provided near the TB wall, both inside and outside of the building. A breather with silica gel is also provided on either side to remove moisture in the enclosure that may be formed due to changes in the ambient temperature and load changes in the bus conductor.

All the three-phase enclosures are grounded by a common aluminium structural member. Their grounding point is connected to the nearest plant grounding pad.

6.2.5.1.1.4   The tap-off 21-KV bus ducts

There are tap-off bus ducts from the main bus duct as stated below:

1) 300-amp-rated bus ducts are connected to the two 9 MVA 21/3.3 KV unit station service transformers. The bus duct comprises of a channel section Aluminium bus conductor supported on a single insulator at the bottom and fixed onto the enclosure. The enclosure is again circular and made up of Aluminium that is 6.35 mm thick and 600 mm in diameter.

2) There are ducts connected to the Potential Transformers and generator surge protection cubicle.

There is also a link from the shortened six neutral terminals of the generator to the neutral grounding cubicle. It can be noted that the generator has two parallel stator windings and six neutral terminals.

6.2.5.1.1.5   Different classes of power

The different classes of power have been discussed in general in para 4.2.8 of Chapter 4. We will now look into this topic in detail with particular reference to the RAPP units.

6.2.5.1.1.5.1 The 3.3-KV Class-IV and 415-V Power Systems

6.2.5.1.1.5.1.1 The 3.3-KV Class-IV Power System

The two 21 KV/3.3 KV unit service transformers of the station service are 522 T-1 and 522 T-2, while the 20 MVA 220/3.3 KV three-winding service transformer 521 T-1 feeds the four different 3.3-KV Class-IV buses namely 5241-BUC, 5241-BUD, 5241-BUE and 5241-BUF. The buses BUC and BUD are linked by the bus coupler breaker 5241 CB-2 while buses BUE and BUF are linked by the bus coupler breaker CB-6. These two breakers are normally open to enable the above four buses to feed the loads connected to them independently. The separate buses may, however, be coupled in an emergency to enable the continuity of supply to redundant 3.3-KV Class-IV loads. The critical 3.3-KV redundant loads of various systems are connected to different buses to enable the continuity of supply to the systems. Redundant loads are normally 2x100% or 3x50%. The failure of a single source results only in the partial loss of the essential loads. The power flow of the above-mentioned system and more details on the bus links can be explored in the key one-line diagram shown at the end of this chapter.

6.2.5.1.1.5.1.2 The 3.3-KV Metal-Clad Switchgear

6.2.5.1.1.5.1.2.1 Construction

The switchgear is metal-clad with 2-mm-thick cold-rolled grain-oriented treated steel on the load-bearing side and with 1.5-mm-thick metal on the non-load-bearing side. The breakers are of the air-break type (it would be more apt to describe the breakers as of the air-assisted type) with chutes for leading the ionised air, formed by the fault current, out of the breaker with the magnetic field, formed by the fault current—assisting in the onward passage of the arc product. The breaker is housed in a truck which can be pushed into and out of the switchgear panel, thus making the maintenance of the breaker possible.

There are three positions for the breaker:

1) The operating position where, under both the power and control, the contacts are fully in operating condition.

2) The test position where, under the power, the contacts are disengaged and the control contacts are in position; the testing of the circuit breaker control functions is possible in this position.

3) Completely disengaged position wherein both power and control contacts are completely disengaged from the switchgear panel. A routine check-up and minor repairs to the switchgear are possible in this position.

Insulated barriers between the bus bar and the contacts of the breakers fall into place in the portal before the breaker can be moved into the 'test' physician. The barriers are withdrawn automatically as the breakers move from the 'test' to the 'connected' positions;

the screened ventilation louvres are provided for the flow of cooling air to the switchgear. Space heaters are provided to prevent condensation in the switchgear panels.

6.2.5.1.1.5.1.2.2 The control voltages for the switchgear panels: relay operation

The impulses for the breaker opening and closing functions are provided by 48-volt DC supervisory relay contacts. These functions are carried out by 250-volt DC relays receiving impulses from the 48-volt DC supervisory relays. The 48-volt relays also carry out annunciation and monitoring functions of the breakers. The breakers are remotely operable from the main control panel in the control room.

6.2.5.1.1.5.1.3 The 415-V Class-IV, Class-III and Class-II systems

There are six 3.3 KV/415 V/1250 KVA station auxiliary transformers to feed the station auxiliary loads through nine 415 V buses (four from Class-IV, each rated for 2000 Amps; 1 from Class-IV rated for 2000 Amps to enable transfer of loads two from Class-III, each rated for 3000 Amps and two from Class-II, each rated for 1000 Amps).

The flow of power, under normal operating conditions, is from Class-IV to Class-III and then from Class-III to Class-II. Of the six station-auxiliary transformers, two feed the Class-III and Class-II stand-by power systems. When the grid power-supply is lost, the stand-by power system is supplied power from the four emergency power sources. These sources are the 1.5-MVA DG sets through buses 5231-BUL and 5231-BUM and the two no-break 300-KW unit capacity motor generator sets through buses 5231-BUN and 5231-BUP. The two MG sets are driven by the two-set 250-V batteries through buses 5232-BUA and 5232-BUB to provide UPS for the loads while the two DG sets are being started. The two DG sets supply all the loads to which a short power interruption of 1 minute for the DG set to automatically start and pick-up load is permissible.

The emergency power system consists of two physically separate sub-systems, each consisting of a Class-III bus, a diesel-generator unit and a DC–AC 'no–break' motor generator unit. The two sub-systems are independent with their own Class-IV and emergency power supplies. The two main supply sources are not to be run in parallel at the 3.3 KV or 415 V levels in order not to exceed equipment interruption ratings. All important loads are duplicated and independently supplied from the two sub-systems. The sub-systems may be interconnected by a Class-III or Class-II tie-breaker in any emergent situation. Following partial or total Class-IV power failure, in the case of total Class-III loads on the two buses exceeding the available emergency power-supply, the loads not essential for the safety of the plant are switched off automatically to bring down the loads to the available power-supply. These loads are sequentially restored manually once the Class-IV power-supply is resumed.

6.2.5.1.1.5.1.4 The 415-V Switchgear and Motor Control Centres

6.2.5.1.1.5.1.4.1 Design and Construction

The 415-V switchgear is of metal-clad construction, fabricated from cold-rolled grain-oriented steel, of modular type, vermin proof and provided with screened mesh for purposes of ventilation. Several modules, as required, constitute a vertical section of the switchgear. Several such vertical sections, as required for meeting the required number of loads, constitute the switchgear. The breakers are of the draw-out type and move-on rails.

There are three positions for the breakers which are listed as follows:
1) Connected
2) Test
3) Drawn-out

The breakers are of a tilting type to enable the inspection of contacts, minor repairs, etc. The power and control contacts are fully engaged and in the connected position. The power contacts are out, but the control contacts are in the 'test' position. All tests on the breaker are possible in this position. Both the power and control contacts are out in the 'drawn-out' position. Inspection, minor repairs, etc., on the breaker are possible in this position.

The insulated barriers between the bus-bars and the breaker contacts fall into position in the portals before the breaker is drawn out. Similarly, the barriers are drawn up before the breakers are moved to the connected position.

The 415-volt breakers supply individual motor loads of capacity 60 HP to 225 HP. All motor loads of capacity 225 HP and above are at the rated voltage of 3.3 KV.

All the breakers, supplying critical loads, are remotely operable from the station control room. The 48-volt rated relays carry out supervisory and monitoring functions and provide the closing and opening impulses for the 250-V DC-rated relays. The opening operation, in the case of manually operated breakers, charges the closing spring which completes the closing operation.

The 415-volt Motor Control Centres (MCCs) and load centres are provided for controlling local loads through contactors or switches. These gears are provided with space heaters for the prevention of condensation.

6.2.5.1.1.5.1.4.2 Ratings of the 415-V breakers

Breakers are rated for continuous current capacities of 600 amps, 1 kA, 2 kA and 3 kA. The minimum interrupting capacity of the breakers is 43.6 kA, asymmetrical. Each of the phase and neutral buses has a momentary current rating of 60 kA asymmetrical and 600 amps continuous ratings.

The maximum voltage withstand value of the breakers is 2.2 KV at 50 cycles frequency standard-wave form. The inter-turn test on coils is carried out with induced high frequency for 15 seconds.

There are two no-break motor generator sets: the AC motor has a rating of 300 KW at 0.8 power factors, and the DC generator of 400 HP rating at 250 volts maximum rating and a 200 volts minimum rating at a rated speed of 1000 rpm. The sets are rated for 20 minutes of operation on the failure of 415-V Class-III/Class-IV power.

6.2.5.1.1.5.1.4.3 Protection for miscellaneous service motors and generators: RTDs for the equipment

The breakers of all current ratings of equipment, except that of generator vehicles, are provided with direct-series over-current dual-selective adjustable trip settings on each phase. The 48-volt DC annunciators are provided for the alarm contacts to convey signal or trip conditions of the main control room. Generator breakers are tripped on fault through relays. Supply breakers, tie breakers and MCC feeder breakers are provided with long and short-time delay trip devices. All other breakers also have long-time and short-time delay trip devices.

High-capacity motors are provided with embedded-resistance temperature detectors and thermocouples to monitor the motor-winding temperature and enable corrective steps if the temperature exceeds the set values.

6.2.5.1.1.5.1.4.4 The Auxiliary Transformers

Six 1250-KVA unit auxiliary transformers feed the station with 415-volt loads. The transformers are of the 3.3 KV/415 V ratio, the vector group being DYII with the neutral solidly grounded. The main protection for the transformer is the 'Bucholz' relay with the two mercury-in-oil contacts—the lower one providing an alarm in case of gradual pressure rise in the transformer, and the upper one protecting the breaker trip in case of abnormal pressure rise caused by internal fault.

6.2.5.1.1.5.1.4.5 The 415-V Diesel-Generating sets

There are two 1500-KVA 0.8-PF 1000-rpm 50-cycles 415-V sets with the neutral grounded through a 0.48 ohm 500 amps-rated resistor.

6.2.5.1.1.5.1.4.6 Tests and protection for 3.3-KV rated motors

These motors have 225 HP and higher capacities, hence they should withstand a voltage of 7.6 KV for 1 minute after the mounting of all wheels and before coupling. The motors should also withstand a voltage of 7.6 KV for 15 seconds after coupling and before pre-heating. The squirrel-cage rotor motors should undergo a locked rotor test at reduced voltages. It may be noted that slip-ring motors are used whenever high starting torque is required.

6.2.5.1.1.5.1.4.7 Main output system protection, emergency transfer, synchronisation, metering and billing system panels

The relay panels, for all of the systems discussed so far, are grouped and installed in the control equipment room of the turbine building. These panels cater to the various generator output transformer unit stations/service, breaker synchronisation, emergency transfer and interface logic requirements. The billing requirements for the station's generation are also catered to. These panels constitute major critical ones.

6.2.5.1.1.5.1.5 The cabling system with station HV and LV

High voltage (HV) and low voltage (LV) cabling are one of the most important of the conventional systems. They provide feeds for all the electrical equipment and inputs for all the power systems. The system feeds safety-related loads and hence forms part of the safety systems of the plant.

The cabling consists of the following:
1) Power cabling
2) Control and instrumentation cabling

6.2.5.1.1.5.1.5.1 Power Cabling

This consists of HV and LV cabling.

6.2.5.1.1.5.1.5.2 HV Cabling

3.3-KV-rated HV cabling consists of paper-insulated lead-covered (PILC) and polyvinyl chloride (PVC) cables. Let us now look into the specifications of these cables.

6.2.5.1.1.5.1.5.3 The PILC cables

The PILC cables are 3.3-KV 3-core paper-insulated lead-covered cables made up of a multi-stranded aluminium conductor. The cables are non-draining screened and sheathed with lead alloy 'B' of sizes 75 mm sq. and 300 mm sq. used for feeding the PHT primary feed and circulating pumps, respectively.

6.2.5.1.1.5.1.5.4 The 3.3-KV multi-core PVC cables

The 3.3-KV 3-core PVC cables are made up of a stranded aluminium conductor that is shielded with semi-conducting carbonised cotton tape, PVC-insulated aluminium tape, a laid-up inner sheath of extruded PVC and armoured with galvanised steel single-wire. The PVC cables of sizes ranging from 100 mm sq. to 500 mm sq. are used for feeding motors ranging from 525 HP to 2500 HP.

6.2.5.1.1.5.1.5.5 The 3.3-KV single-core cable

The 3.3-KV single-core paper-insulated lead-covered cables are made up of a multi-stranded aluminium conductor of 800 mm sq. size. It is mass-impregnated and non-draining type screened; it is sheathed with a load alloy 'B' and armoured with steel wire. It is overall PVC sheathed for feeding 3.3-KV buses from 3.3-KV units and station service transformer secondary terminals (two runs per phase).

6.2.5.1.1.5.1.5.6 The 3.3-KV cabling for PHT pump motors

The PILC cables are used for part of the cabling till the cables cross the perimeter wall; PVC cables are then used for the rest of the cabling. PILC–PVC cable joints are carried out at the cable pans after entry into the RB. The 3.3-KV cable couplers are used for entry into the PHT pump motors. The movable portion of the couplers is screwed onto the fixed portion of the motors to secure positive engagement and contact at the motor end. The PHT pump motors are the most critical equipment in the PHT system and thus the immediate availability of the motor is essential for maintenance purposes; this system of quick coupling and de-coupling of the feeder cable was decided upon.

6.2.5.1.1.5.1.5.7 Single-core/multi-core aluminium conductor cables

These are low-voltage cables of 1100 V-grade whose conductor is made up of single-core/multi-core stranded aluminium, PVC/HR PVC insulated and armoured with single-round galvanised wire/ strips. They are overall HR PVC/ PVC sheathed as well. Such types of cables with other varying specifications are listed below:

(1) Single-core cables of 800 mm sq. to 1000 mm sq. cross-section

(2) Single-core cables of 150 mm sq. cross-section but with a copper conductor

(3) 2-core cables of 6 mm sq. to 35 mm sq. cross-section

(4) 3-core cables of cross-section 6 mm sq. to 400mm sq.

(5) Three-and-a-half core cables of cross-section 16 mm sq. to 400 mm sq.

(6) 4-core cables of 6 mm sq. cross-section

(7) LV cabling of 1100 V-grade single-core flexible cables with a stranded copper conductor, heat-resistant (HR), PVC-insulated unarmoured, PVC outer sheathed with a cross-section of 625 mm sq. for the turbo-generator excitation system.

(8) 650/1100 volt grade multicore cables made up of a stranded aluminium conductor that is silicon rubber-insulated, armoured with a galvanized steel wire strip, and overall sheathed with a size ranging from 4 mm sq. to 300 mm sq.

(9) Single-core PVC-insulated 650/1100 V-grade cable made up of stranded aluminium conductors of sizes 4 mm sq., 6 mm sq. and 10 mm sq. that are used for feeding lighting fixtures and 3-pin receptacles; 2-core 2.5 mm sq. copper-conductor cables and 2-core 4 mm. sq. aluminium cables are also used for this purpose.

6.2.5.1.1.5.1.5.8 Cabling length: number of terminations and connections

The total length of the HV cabling is 13 km and that of the LV cabling is 182.5 km. The total number of HV cabling connections is 258: of which 128 are of single core and 130 are of multi-core. The number of LV multi-core cable terminations is 5,852 and that of the single core is 435. All cable connections are through crimp-type lugs. The lugs are made up of copper—for connections of copper cables to copper terminals—and of aluminium—for connections of aluminium cables to aluminium terminals—and of bi-metallic crimp-type lugs—for connections of aluminium cables to copper terminals.

About 80% of the total cabling is run on cable trays and 20% through conduits, tunnels, trenches on-ground and ducts. The GI cable markers are installed on all cabling at 30-metre intervals;

aluminium tags with cable numbers are embossed at 10-metre intervals.

The power and control cables are run on separate cable trays. When the cable trays are run one above the other, the HV cables are run on the top-most tray, the LV Cables on the following tray below and the control cables on the lowest tray.

6.2.5.1.1.5.1.5.9 The cable terminations

All cable terminations in the reactor building, about 500 in number, are carried out by using heat-shrinkable sleeves. The details of the terminations in other places, cable category-wise are given below:

(1) MV PVC & NV HR PVC cables : Cable glands
(2) HV PVC cables : Cable glands and epoxy seal kits
(3) HV PILC cables : Compound-filled/resin-filled pot heads and cable sealing kits
(4) LPE cables : Heat-shrinkable type termination kits
(5) Mineral-insulated cables : Gland sealing with PVC sealing compound, lock-nuts, etc.

6.2.5.1.1.5.1.5.10 Cable sealing at the RB cable penetrations

6.2.5.1.1.5.1.5.10.1 General

Suitable locations are provided for the steel boxes/steel enclosures with glands to be welded onto the gland plate of the RB openings. These ducts are embedded in the perimeter wall. In the case of plants with double containment walls (RAPP has only one containment wall), these boxes and enclosures are embedded in both walls. There are 1542 such penetrations in RB 1.

6.2.5.1.1.5.1.5.10.2 Power cables with stranded conductors

6.2.5.1.1.5.1.5.10.2.1 The inside of the perimeter wall

Core sealing is provided with type-A moulds. The outer sheath and the conductor insulation

are first removed. The exposed conductor is covered with copper ferrule and crimped. The type-A mould is then placed in position and the epoxy resin is poured into the mould. The casting is then pulled into the gland gaskets placed at both ends, and the gland nuts are tightened to ensure gland tightness.

6.2.5.1.1.5.1.5.10.2.2 The outside of the perimeter wall
Composite sealing of the cable is carried out with the use of a type-B mould.

6.2.5.1.1.5.1.5.10.2.3 Power cables with a solid conductor
The outer sheath only is removed. The cores are carefully separated and the space in between the course is sealed with a type-B mould, both on the inside and outside of the perimeter wall. The other procedures are as discussed before.

6.2.5.1.1.5.1.5.10.2.4 The DC walls
In the case of plants with double containment walls, the sealing is done on the outside of the ICW and outside of the OCW. The sealing process is the same on both walls. The armouring of cables is separately earthed on both sides.

6.2.5.1.1.5.1.5.10.2.5 Control Cabling
Control cabling is used with the following specifications:

(1) 650V-grade annealed, tinned, stranded copper conductor; PVC-insulated; cores laid up; PVC inner sheathed; armoured with galvanized steel wire and overall PVC sheathed.

2-core, 4-core, 7-core, 10-core and 24-core cables; each core is of 2.5 mm sq. cross-section while the 2-core is of 10 mm sq.

(2) 650V-grade, PVC-insulated cable with copper conductor, overall shielded and PVC-jacketed as stated below:

1 pair 1.51 mm sq. to 10 pairs 1.51mm sq.

3-core 1.5 mm sq. to 10 pairs 1.5 mm sq.

25 pairs 19 AWG

50 pairs 1/0.912 mm

2-core 2.5-mm sq. to 10-core 2.5 mm sq.

(3) Core 6mm sq.

(4) Thermocouple cables, PV-insulated, overall shielded and PVC-sheathed

1 pair 20 AWG; 4 pairs 12 AWG; 12 pairs 20 AWG

1 pair 16 AWG; 4 pairs 20 AWG

(5) Same as that of (4) but with a copper Constantine

24 pairs 20 AWG

(6) Same as that of (4) but with an Iron Constantine

12 pairs 20 AWG and 16 pairs 20 AWG.

(7) Cable with copper conductor, PVC insulated, and armoured with galvanized steel wire and PVC-jacketed.

2-core, 3-core, 4-core and 7-core with 15 mm sq.2

(8) Flexible rubber-insulated, ATC wire fibre glass braided and neoprene jacketed cable

2-pair 26/0.16 mm

25-pair 26/0.16 mm

4-core 65/0.16 mm

50-core 65/0.16 mm

(9) 300-volt cable with copper conductor, PVC-insulated, overall shielded and PVC-jacketed

1 pair, 3 conductors; 5 pairs, 12 AWG

6.2.5.1.1.5.1.5.10.2.5.1 Total length of cabling: number of terminations and connections

The total length of control and instrumentation cabling is approximately 288 kilometres. The total number of control cabling terminations is 15,860.

The total number of connections in the controlled distribution frame is, approximately, 200,000.

6.2.5.1.1.5.1.5.10.2.5.2 The marking of cables

Separate and distinctive marking of all safety-related power and control cables has been carried out.

6.2.5.1.1.6   The cable pans

6.2.5.1.1.6.1 The general description details of the construction

An all-extensive system of cable pans, to accommodate cabling runs, has been used. About 80% of all cabling in the station is run on cable pans.

6.2.5.1.1.6.1.1 Ladder-Type pans

The pans are fabricated from galvanized steel plates of (GS) 1.6-mm thickness. The sides of the pans are bent to form shallow pans. The rungs of the same GS material are welded to the bottom of the pan at suitable intervals to form the pans. The standard lengths of the pans are joined together by coupler plates.

6.2.5.1.1.6.1.2 Perforated cable trays/pans

The 1.6-mm-thick galvanised steel plates, of the required width, are folded using a power press to form trays. The bottom of the trays is then perforated to form perforated trays. The standard lengths are then joined by coupler plates.

6.2.5.1.1.6.1.3 Cable tray locations and tray supports

The trays are used in all the plant buildings—on the cable bridge between the RB and TB, in the pump house, in the trenches between the TB and the 220-KV switchyard and the annular area between RB and SB. The pans are supported on vertical galvanized steel angles of size 50 mm×50 mm×6 mm, and also horizontally on the bottom of the trays. Only horizontal supports are provided wherever the site conditions warrant.

The trays vary in widths of 150 mm, 300 mm, 400 mm and 600 mm. They are arranged in multiple tiers both horizontally and vertically.

6.2.5.1.1.6.1.4 The arrangement of cables in trays: quantity of pans

The HV, LV and control cables are run on separate tiers as follows:
1) HV cables on the top-most lower tier
2) LV cables on the next lower tier
3) Control cables on the next lower tier

The minimum separation distance between safety-related cables and non-safety-related cables should be 500 mm for horizontal separation and 1000 mm for vertical separation. These distances should be 1000 mm and 1500 mm, respectively, in the case of redundant safety-related cable pans.

Wherever it is not possible to maintain the separation distances, the safety-related cable pans should be provided with solid galvanized steel covers, fabricated from a 16mm-thick sheet.

The quantity of cable pans of RAPP units 1 and 2 is approximately as stated below:
(1) Ladder-type cable pans of width 150 mm, 300 mm, 400 mm and 600 mm : 21.97 km

(2) Perforated-type cable pans of similar sizes as mentioned above : 3.5 km

(3) Cable pans with solid covers : 6.4 km

6.2.5.1.1.7 The Lighting system

6.2.5.1.1.7.1 System objectives: forms of lighting

It is essential to define some of the terms that will be used to describe the system. Lux is the lumens per square metre area of rooms/surfaces. Lumen is the lighting output per watt of power used in the lighting fixtures which varies from 30 to 120 lumens per watt, depending on the kind of lighting fixtures used.

The system forms another very important part of the conventional systems. The objective of the system is to provide a work-comfortable ambience that is dependent on the kind of work that is carried out in an area. Thus, a system is designed to provide varying illumination levels in the range of 750-1000 lux in areas like the instrumentation laboratory, wherein miniature components are handled; 200 lux in areas like the galleries, trenches and the like, where only periodical checks are carried out.

6.2.5.1.1.7.2 Different kinds of fixtures: areas catered to

6.2.5.1.1.7.2.1 High-bay lighting

This is required for the lighting of areas like the turbine operating floor, loading bay and other connected areas. These areas are huge and have a very high ceiling. High-pressure Mercury-vapour colour-corrected lamps of 250/300 W are used for providing an intensity of 300 lux.

6.2.5.1.1.7.2.2 Low-bay lighting

Industrial-type, 100-W incandescent lighting fixtures with a heat-resistant, clear-glass housing reflector and wire-guard protector are used for low-bay lighting in the reactor building. The fixtures are designed to operate satisfactorily in areas having a maximum pressure of 1.25 kg/cm sq. in a 100% steam atmosphere and an ambient temperature of 112 degrees C. The illumination levels are 200-300 lux.

6.2.5.1.1.7.2.3 Lighting in other areas: emergency lighting

The low-bay lighting fixtures, as explained earlier, but with 300-watt lamps are used in other areas of the reactor building, particularly as emergency lighting fixtures. These fixtures will provide lighting in case of Class-IV power failure.

6.2.5.1.1.7.2.4 Lighting large rooms: laboratories and other areas

Indoor industrial-type fluorescent lighting fixtures of 2×40 W with a vitreous, enamelled, slotted reflector, and complete components, are used for lighting the general area of large rooms, laboratories, etc. The fixtures are installed in continuous rows with Mild Steel (MS) support members welded to the inserts on the ceiling or other suitable fixing arrangements. The intensity of lighting of 750-1000 lux for the control and instrumentation laboratory and mechanical workshop, and about 300-500 lux for other laboratories like electrical, chemical, etc., is envisaged. Spot-lighting near large lathes and machinery equipment, preferably, with adjustable incandescent lighting fixtures in the mechanical workshop is also provided. This kind of spot-lighting will also serve to counteract the stroboscopic effect of fluorescent lighting on high-speed rotating parts of mechanical equipment.

6.2.5.1.1.7.2.5 The fuelling machine vaults

The vaults are areas of intense activity, whether during the operation or shut-down mode. A very high level of illumination (2500-3000 lux) is required for catering to this kind of work. The area is covered by closed-circuit television (CCTV) when the station is in operation to follow the operations of the fuelling machines. The 4×65 W fluorescent fixtures are fixed on the SS liners of the walls of the vaults which are used to provide the intensity of illumination that is required.

6.2.5.1.1.7.2.6 The fuel storage and inspection bay

The 400-watt linear-type high-pressure sodium-vapour lamps are used in the above-mentioned area. The bay contains de-mineralised water to a depth of nearly seven metres. The lamps are designed to proof against water pressure and tightness against leakage of water into the fixture. The fixtures are hung from stainless steel rails, provided on the sides of the bay using tough nylon trailing ropes, and are lowered or raised using the nylon rope through an aluminium pulley block that is supported by a hook and handle.

The components of the fixtures such as starters, chokes, fuses, etc., are mounted on a separate lighting control panel near the bay. The wiring for the fixtures is run from this panel to receptacles fixed on the walls of the bay above the highest water level. The water-proof fixture trailing wires terminate in a plug which is hooked onto the receptacles mentioned above. It is possible to hang about 20 such fixtures to provide sufficient illumination levels underwater for the remote inspection of the fuel bundles. Assuming normal attenuation, illumination levels of 1000 lux on the horizontal face and 600 lux on the vertical face of the bay water will be available. This should suffice for the detailed remote inspection of the stored spent-fuel.

6.2.5.1.1.7.2.7 The control room: luminous ceiling

A luminous ceiling is provided for the control room to ensure a general illumination of 600 lux. The ceiling is formed by polystyrene square grids which are supported on an aluminium angle frame. This frame is further supported by a chain suspended from the ceiling. The lighting is provided by twin 40-W fluorescent fixtures, supported from the ceiling and positioned just above the luminous ceiling. The fixtures line up with the trough which in turn follows the contour and lay of the main control panel to ensure uniform illumination on the horizontal and vertical faces of the panel. The fixtures in the other areas of the control room follow the layout of the room.

6.2.5.1.1.7.2.8 Illumination in other areas of the plant

The levels of illumination in these areas are as stated below:

a)   Switchgear/DG set rooms : 200 lux
b)   Piping/cable pan galleries and trenches : 100 lux
c)   Water-treatment area and pump houses : 200 lux
d)   Air compressor area : 200 lux
e)   General areas of service building : 100 to 300 lux
f)   General areas of reactor building : 100 to 300 lux

6.2.5.1.1.7.2.9 Street lighting

The traffic on the streets of the nuclear plant is usually very light. So an illumination level of 1 to 2 lux at street level is adequate. Hence 125-watt Mercury-vapour fixtures and 80/125-watt post-top lantern-type fixtures are used for illumination.

6.2.5.1.1.7.2.10 Flood lighting/security lighting

In strategic areas of the plant, 250-watt mercury-vapour lamps and heavy-duty outdoor-type 400-watt flood-light fixtures are used.

6.2.5.1.1.7.2.11 Total number of lighting fixtures

Nearly 4,250 lighting fixtures have been used in the RAPP units 1 and 2.

6.2.5.1.1.8   The grounding of non-current carrying metal-clad equipment: related equipment and personnel protection systems

6.2.5.1.1.8.1 General

These systems are of utmost importance from the point-of-view of the safety of the personnel and equipment of the station. An efficient earthing system enables the limitation of rise-of-ground electrical potential even while passing very high earth-fault currents. It protects the personnel by letting stray voltage potential dissipate by way of current passage to the ground. The equipment protection system depends, among others, on the earth-fault relay which senses spill current or unbalanced current passing to the earth. An efficient grounding system is a must for the operation of such relays.

The grounding system of large plants consists of the following:
6.2.5.1.1.8.1.2 Equipment earthing
6.2.5.1.1.8.1.3 Lightning protection
6.2.5.1.1.8.1.4 Earth screen for the EHV switchyard
All of these forms of direct and indirect earthing are essential for the protection of equipment and personnel.
We will now look into the above-mentioned system in detail.

6.2.5.1.1.8.1.1 Equipment earthing

The locations for which earthing is to be provided are listed below:

6.2.5.1.1.8..1.1.1 Non-current carrying metallic parts of electrical equipment such as motors, switchgear, motor control centres (MCCs), load control centres (LCCs), etc.

6.2.5.1.1.8.1.1.2 Cable pans

6.2.5.1.1.8.1.1.3 Metallic conduits carrying electrical cables

6.2.5.1.1.8.1.1.4 Power receptacles

6.2.5.1.1.8.1.1.5 Lighting fixtures

We will now look into the same in detail.

6.2.5.1.1.8.1.1.1 Switchgear, MCCs, LCCs, etc.

6.2.5.1.1.8.1.1.1.1 Earthing

An earth bar of rectangular cross-section made up of bare copper, and fixed on a suitable support insulator, is provided usually at the bottom of the equipment such as the switchgear, MCCs, and LCCs. Duplicate earth wires from the metallic body of each piece of equipment, such as breakers, contactors, load switches, etc., are run and connected through the 'Dowell' type crimped connectors to the earth bus. The earth bus is in turn connected to the nearest copper alloy grounding pad or the continuous 2/0 grounding conductor on the nearest cable pan through clamps by duplicate grounding conductors.

6.2.5.1.1.8.1.1.1.2 The HV/LV Motors

The motors are fed by 3-core or 3½/4-core cables. In the case of the latter, the fourth core is connected to one of the two grounding studs/terminals, normally provided on all motors. The other stud is connected by a tinned stranded copper conductor—bare or insulated as dictated by the site conditions of 10 mm sq. and 35 mm sq. size—to the continuous 2/0 grounding conductor on the nearest cable pan or the nearest copper alloy grounding pad. The connections are done through the 'Dowells' type crimped ferrules or the copper clamps.

Wherever three core cables are used, both the earthing studs and the terminals on the motor are connected by the above-mentioned cables of the green colour conductor (the other two being of white and blue colours) to the 2/0 grounding conductor on the cable pans or the nearest copper alloy earthing pad.

6.2.5.1.1.8.1.1.2 Cable pans

A 2/0-stranded copper grounding conductor is run continuously on the cable pans and fixed to the pans by copper saddles. The conductor is connected to sections of the cable pans using parallel groove clamps or similar suitable glands. Wherever multi-tier cable pans are run, the conductor is run on one cable pan and is connected to the other cable pans using short lengths of 2/0-stranded copper conductor and earthing clamps at various sections of the multi-tier pans.

6.2.5.1.1.8.1.1.3 Metallic conduits carrying electrical cables

The ends of the metallic conduits are cleared of enamel in the case enamelled conduits are used, and copper or GI clamps are fixed at the ends. The GI clamps are preferred for GI conduits. Fourteen SWG tinned copper is connected to the screwed ends of the clamps. The other end of the conductor is connected to the earthing stud on the metallic box or terminal box, on which the conduit is terminated or the nearest point of the earthing grid as the case may be.

6.2.5.1.1.8.1.1.4 Power receptacles

Three-phase receptacles are provided with double earthing studs, and single-phase receptacles receive one stud. A continuous 10 mm sq. tinned stranded copper conductor is used for earthing. One end of the conductor is connected to the stud and the other to the nearest point of the earth in the grid.

6.2.5.1.1.8.1.1.5 The lighting fixtures

A continuous 16/18 SWG copper conductor is run along with the wiring for the fixtures and is connected to the earthing points using studs provided on the fixtures.

6.2.5.1.1.8.1.1.1.1.7 Earthing and armouring of power cables

The armouring of the 3-core/3½-core/4-core cables is bent back at the point of termination of the cables. Tinned copper or G I earthing clamps are fixed onto the armouring and are connected by copper G I conductors to the earthing studs on the motors. The armouring of the above-mentioned cable is earthed both at the SWGr end and the motor ends. The armouring of single-core cables is earthed only at the SWGr end.

6.2.5.1.1.8.1.2 Lightning protection

High structures like the reactor building, turbine building and service building are provided with lightning protection.

6.2.5.1.1.8.1.2.1 The reactor building

The building is provided with a mast of heavy-duty copper pipe with a 3.5-metre length, 90 mm outer diameter and 75 mm inner diameter at the top of the dome. The pipe is anchored onto the dome at its base plate and is provided with copper spikes at the top. Fabricated tinned copper alloy plates of 12 mm thickness are laid on the dome and anchored to the surface of the dome to suit the profile of the top of the dome, starting from the base plate and extending to a length of 1.562 metres. These plates are connected to the base plate of the copper pipe by 2.5 mm×5 mm×140/210 mm-long copper links or by equivalent positive electrical bonding. A 4/0 tinned stranded copper ring is laid around the dome at the springing level at the bottom of the dome. Two lengths of 4/0 tinned copper are laid from the lightning protection copper pipe to the above ring. The conductors are connected to the copper pipe by tinned copper alloy clamps and by parallel grooves or similar clamps to the ring. Two lengths of 4/0 down-comers from the ring are directly connected to two separate station earths by brazing. There are no joints in these down-comers which are of single length. A sketch of the arrangement is enclosed below:

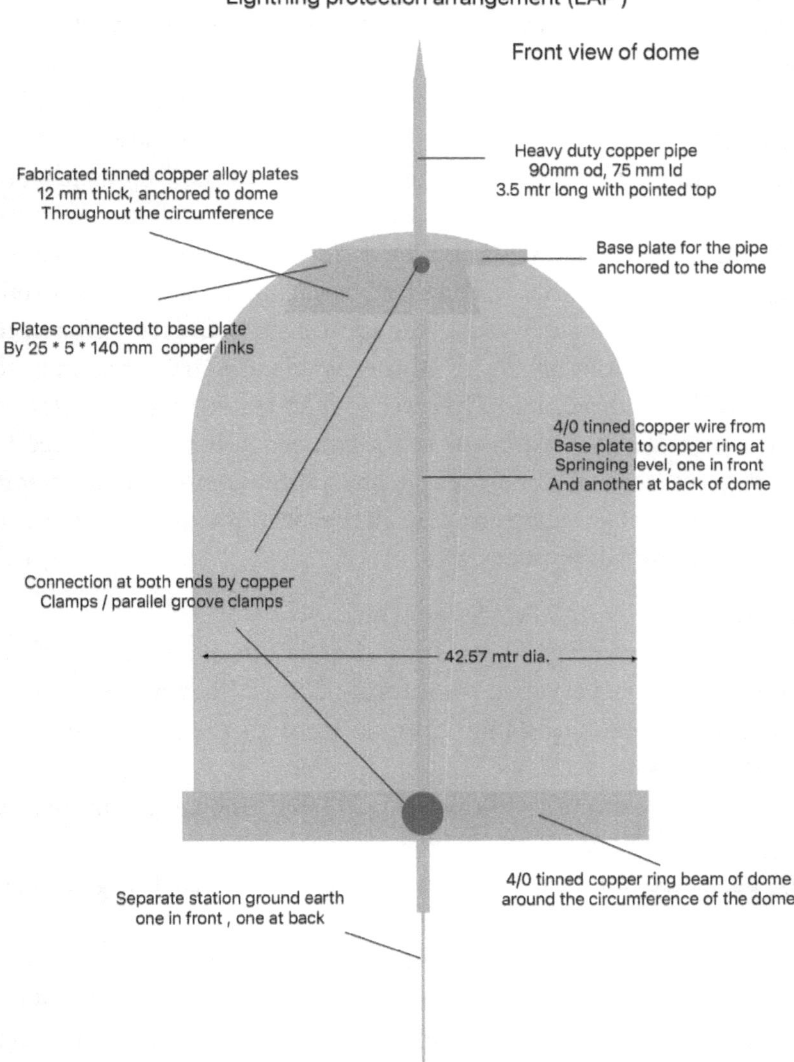

6.2.5.1.1.8.1.2.2 The Turbine and service buildings (TB and SB)

The turbine and service buildings are each provided with four vertical air terminations of 800 mm-long 62-mm-diameter 3-mm-thick copper pipes with spikes at the top. The terminations are each connected by a 4/0 AWG tinned copper conductor of a single length without any joints to the station earth by brazing.

6.2.5.1.1.8.1.4 The Earth screen for the switchyard

The switchyard is of the outdoor type. It is essential to protect the switchyard equipment from direct strokes of lightning. All the outgoing transmission lines are provided with line-type lightning arresters. The system service transformers (SSTs) are provided with station-type lightning arresters. The rest of the switchyard equipment such as bus breakers, current and potential transformers, wave traps, etc., also need to be provided with protection against lightning strokes. This is done by providing an earth-screen full-stop earth conductor, usually made up of standard galvanised steel that is strung on top of all the switchyard structures in both perpendicular directions. The screen is connected to the switchyard earth grid. The screen protects equipment within a 15-degree to 30-degrees vertical cone area at any point on the earth screen. Given the extensive earth screen, all the switchyard equipment will be covered by full protection.

6.2.5.1.1.9   The 220-KV Switchyard

The 220-KV switchyard is of the double bus type and has four bays. There is a provision for a future 220-KV bay as well. Power evacuation from the station is done through the following EHV transmission lines:

a)   A double-circuit, 49-km-long, 220-KV line to the Industrial Estate in Kota

b)   A single-circuit, 220-KV line to the Kota 220-KV grid sub-station about 50 km in length

c)   A single-circuit 190-km-long 220-KV line to Udaipur

d)   A 220-KV spur line to the Heavy Water Plant Kota (HWPK)

Power Line Carrier Communication (PLCC) through wave traps is provided.

The 220-KV short spur lines are strung from the output terminals of the 225/250 MVA output transformers to gantries in the switchyard to feed the output system of the station. The bus bars are of tubular-type and designed to carry 1.3 kA of current. The 220-KV breakers are of the air-blast type. A compressor house in the switchyard houses the air compressors along with a control panel. The compressors feed the air required for the breaker operation.

6.2.5.1.1.9.1 The 220-KV Protection System

The 220-KV outgoing feeders are provided with line-type lightning arresters. Six sets of 220-KV Current Transformers (CTs), each set consisting of three single-phase CTs of double taps at 800 amps and 1200 amps are provided for the protection and metering. Two sets of bus Potential Transformers (PTs)—each of the primary 230 KV divided by root 3, 115 volts divided by root 3—and double secondary PTs are also provided for protection and metering. One single-phase line PT 230/√3 KV/ 110V/√3 V is provided on each outgoing 220-KV feeder for synchronising purposes. The system start-up transformers (SUTs) are provided with station-type lightning arresters. All the non-current-carrying metallic parts of the switchyard equipment and all the structures of the switchyard are connected by duplicate earthing wires to the grounding grid of the switchyard system. This grid is in turn connected by separate duplicate 2/0 stranded tinned copper conductors without any joint to each of the two separate earths of the switchyard.

6.2.5.1.1.9.2 The switchyard protection system

The following switchyard electrical protection system has been provided:

(1) 220-KV bus protection
(2) 220-KV line protection

6.2.5.1.1.9.2.1 The 220-KV bus protection

Differential current protection has been provided.

6.2.5.1.1.9.2.2 The 220-KV line protection

The line protection is provided as follows:

a) Directional distance phase fault detector
b) Directional distance ground fault detector
c) Directional over-current ground fault detector
d) Selective over-current local breaker back-up detector
e) MHO face and directional over-current ground relay back-up detector
f) Transfer trip detector

6.2.5.1.1.9.3 The System Service Transformers: start-up transformers

For the start-up power of the station, two 12/20 MVA ONAN/ONAF three-winding transformers of 220 KV/3.3 KV ratio and YNYO/YNODI type—with two separate 3.3 KV windings earthed through 1.9 ohm 1000 amps neutral earth resistors and the mid-point of the open 3.3 KV delta tertiary winding (provided for reducing harmonics within the transformer) solidly earthed—are provided.

6.2.5.1.2 The 220-MWe TG set and auxiliaries

6.2.5.1.2.1 The 220-MWe TG sets: critical speed and low-power operation

The details of the set have been furnished in paras 7.9.1.1 and 7.9.1.2 of Chapter 7.

Further relevant details are furnished below:

a) Electro-magnetic seismic detectors are mounted on bearing pedestals. A relay has been provided for alarm indication on the local TG control panel in case of excessive vibration in the set.

b) The first critical speed for the TG set is 1600 rpm (revolutions per minute); the second is at 4200 rpm. The figure for the LIP shaft is less than 2,400 rpm.

c) Low-power operation of the set should not be less than 15 MW. At this level, automatic sprays would come into effect at a route blade temperature of 82 degrees C to keep the temperature below the alarm setting of 149 degrees C.

6.2.5.1.2.2 The electrical turning gear

The TG set is run by an electrically operated turning gear for some time before the set is started up and also before it is shut down after a trip or for maintenance purposes. The aim is to ensure the freeness of the bearings and the thermal equilibrium of the set. The gear will disengage itself automatically when the shaft speed exceeds the rated speed of the turning gear. The turning gear will start only when the turbine lube oil supply is on and also when the jacking oil pump is on. The jacking oil pump will be put off once the full speed of the gear is reached, and will automatically shut down if the lube oil supply fails.

6.2.5.1.2.3 The lubricating oil system

The main lubricating oil pump circulates the oil from the storage tank through the bearings before the TG set is started up. Once the set is started up, the shaft-driven pump takes care of the oil circulation and the main lubricating oil pump is shut down. Part of the lubricating oil is circulated continuously through a side circuit consisting of a centrifuge-type oil purifier which removes all foreign matter from the oil; the purified oil is returned to the main lube oil storage tank.

6.2.5.1.2.4 The TG set electrical protection system

The following measures are undertaken to ensure protection:
a)   Differential current protection
b)   Loss of excitation to the exciter
c)   Phase-current balance protection
d)   Stator ground fault protection
e)   Field ground protection alarm
f)   Supplementary set starting protection

6.2.5.1.3 The Feed-water system

There are five feed-water heaters which heat the condensate water before it enters the boiler feed pump (BFP). Two 100% 1,725 KW BFPs are provided. The pumps deliver the feed water at a temperature of 171 degrees C to the NSSGs. There are two 415-V Class-III stand-by feed pumps. These pumps sustain the feed water when there is the failure of 415-V Class-IV power. There are three high-pressure heaters, two high-pressure heaters and one direct contact de-aerator. The de-aerator limits the oxygen content of the feed water to 0.005 em3 per litre of water.

6.2.5.1.4 Steam separators and re-heaters

There are two steam separators in parallel positions. Steam from the last stage of the HP turbine passes through the moisture separator where it is dried to 99%, and thence to the re-heaters where it is heated by bled steam from the turbine before it enters the LP turbine. The drain from the re-heaters is pumped into the steam drums of the NSSGs.

6.2.5.1.5 The Condensing system

One surface condenser, which is a single-shell twin-nest tube two-pass type with a surface area of 20,465 sq. metres, condenses the steam from the LP turbine. The cooling water is on the tube side and the steam is on the shell side. It is of box-type with a steel body of adequate thickness that is re-inforced with stiffeners. The vessel is welded onto the flanges of the turbine casing. The condensate enters the feed-water system. The condenser back-pressure is Mercury 24 mm vacuum (38/mm mercury absolute pressure). The circulating water flow is 626 klits per minute, with a dynamic head of 70.31 metres, which goes through four circulatory cooler water pumps.

The condenser is designed to condense 762.8 m tonnes of steam per hour when supplied with circulatory cooling water (CCW) at 21 degrees C. The temperature rise of CCD is 8.8 degrees C. The condenser is designed to store about 2 minutes of full condensate flow. A high-level reserve feed-water tank at 372.6 metres elevation serves to make up for the feed-water requirement.

6.2.5.1.6 The plant water-supply system

6.2.5.1.6.1 General

The plant uses 1680 klits of water per minute from the Rana Pratap Sagar Lake (RPS). About 40% of this requirement is for the Closed Cooling Water system (CCW). Two other systems utilise the supplies and they are described in the upcoming sections.

6.2.5.1.6.2 The process water system

The process water system is fed by two 100% duty pumps from 415-V Class-IV power generators. Each pump is of 105.80 klits/minute capacity at 31.2 metres and has to meet at least one of the following cooling requirements:

1) Moderator heat-exchangers
2) Shield-tank cooling heat-exchangers
3) Ventilation cooling

The pumps also meet the domestic use and make-up water for the closed circuits.

The domestic water-supply is filtered and chlorinated. The make-up water is purified in a water-treatment plant which consists of a coagulator, filters, a cation exchanger and a mixed bed de-mineraliser. The mixed bed de-mineraliser is used to polish the water-treatment plant effluent and serves as a stand-by for the anion and cation exchangers. The make-up water caters to the following:

1) Boiler feed-water system
2) Spent-feed cooling water system
3) Generator stator cooling system
4) Reactor and end-shield tank cooling system
5) Biological cooling system

6.2.5.1.6.3   Stand-by water and fire-fighting system

This is supplied by three 50% duty pumps from 415-V Class-III power generators. Each pump has a capacity of 8.1 klits/minute at a head of 90 metres. The system is used for fire-fighting, essentially, and also supplies cooling water to essential auxiliary heat-exchangers if the 415-V Class-IV power-supply fails. These heat-exchangers are also connected to the dousing tank so that the continuity of cooling water is maintained by gravity during the time required for starting up the Class-III pump.

6.2.5.1.7   The plant ventilation system (VS)

6.2.5.1.7.1   General

This system is designed to exhaust the plant air and provide necessary air changes required for the plant area.

6.2.5.1.7.2   The reactor building VS (RBVS)

The system is designed to provide three to four air changes per hour in all accessible areas. The air is exhausted from all inaccessible areas and blown through absolute filters located in the service building to the nearly 100-metre-high stack. There are two independently operated exhaust fans, one of which is required for normal operation. The ventilation system provides 425 kl/ minute of fresh air during reactor operation. Air-supply units contain both process and chilled water coils without supply fans. The normal operation does not provide air supply to the boiler room. Conditioned air is supplied to the boiler room during purge operation when under maintenance or prolonged shut-down with both exhaust fans operating. Purge is required during maintenance and prolonged stay of the plant personnel in the area; should an emergency arise when the primary-circuit heavy water escapes as steam with consequent pressure build up in the building, gasketed dampers—designed to hold 0.42 kgf/cm sq. pressure differential across them—will automatically close. These dampers are by-passed by reverse pressure-relief dampers to prevent the pressure in the building from falling below one atmospheric pressure due to steam condensation. Maximum temperatures in the various areas of the reactor building vary from 27 degrees C to 43 degrees C.

6.2.5.1.7.3   The Service Building VS

This building is ventilated by an air supply of 1020 kl/minute which caters to four air changes per hour in small or medium-sized rooms, and a half to one in large rooms. The cleaning hoods in the de-contamination area and the zone-3 Chemical laboratories are supplied with additional outside air without any cooling which is called hood air. All the air from the building is exhausted through filters to the stack.

6.2.5.1.7.4   The Turbine Building VS

This building is ventilated by air drawn in through wall louvres and in-take filters. The hot air is exhausted through the turbine-roof extractor fans.

6.2.5.1.7.5   The Administration Building VS

This building has its own independent air supply and ventilation system.

6.2.5.1.8   The Fire protection system

6.2.5.1.8.1   General

A fire station with necessary fire tenders, hoses and protection clothing for personnel was established at RAPP units 1 and 2 with a crew of one fire officer, an assistant fire officer and fire assistants for fighting external and station fires.

6.2.5.1.8.1.1 Fire protection for oil-filled transformers

All the high-capacity transformers (9 MVA capacity and higher) are provided with foam-type protection, complete with piping and foam supply arrangements for fighting oil fires.

6.2.5.1.8.1.2 Plant buildings: the fire protection system (FPS)

The buildings have been provided with the following FPS:

(1) Fire water-protection

(2) Fire warning-system

6.2.5.1.8.1.2.1 Fire water-protection

Fire hydrants are provided at various points with water, pumped using pumps powered by a 415-V Class-III power-supply, to fight any fires that may arise.

6.2.5.1.8.1.2.2 Fire warning-system

Smoke and fire detectors are provided at various points in the plant buildings. These detectors work on the infra-red protection principle and are capable of detecting ordinary chemical and plastic fires. The detectors are wired to zonal fire-indication panels, suitably located in different parts of the buildings. These panels indicate the location of the fire to enable fire officers to undertake suitable fire-protection measures.

6.2.5.1.9   The Communication System (CS)

A private automatic exchange for telephone communication, both within the station and with the various township colonies, has been established. Several telephone lines for senior officers have been provided for outside communication as well. FAX and wireless communication with other nuclear power stations and the NPCIL HQ at Mumbai are available. A personnel address system for paging of employees within the station premises is available. Box and horn-type loudspeakers have been installed at vantage points as part of station communication and paging of personnel. With increasing online computerized functioning, intensive and extensive online inter-station computerised functioning is quite on the cards.

6.2.5.1.10 The Drainage system

The sewage draining system is a conventional one. Drainage from potentially active areas is dealt with separately in the liquid waste-management system of the waste-management facility which was already discussed in detail in paras 4.2.4.8.1 to 4.2.4.8.3 of Chapter 4.

6.2.5.1.10.1 Valve actuators

Globe-type and butterfly-type valves are two of the most important types of valves used for controlling fluid flow. Single-port and double-port types are used in globe valves, while butterfly valves are of the following types:

a) Air pressure-assisted
b) Pressure-balanced piston
c) Electro-hydraulic
d) Mechanical lever type

These types of valves are used for pipes which have a large diameter.

6.2.5.1.11 The control and instrumentation system: instrumentation laboratory

6.2.5.1.11.1 General

The control system takes care of all the control needs of the station. The instrumentation system looks after all the measuring devices and also relays the measured outputs required for control, monitoring and protection purposes. The laboratory provides the necessary inputs for the system.

6.2.5.1.11.2 The instrumentation system: marshalling boxes and local indication

All the instrumentation cables, emanating from the equipment and monitoring devices, are terminated at the local marshalling boxes. Measuring and indicating devices are located on these boxes wherever necessary and called for. The cables are laid further from the marshalling boxes to the control distribution frame (CDF) located in the control equipment room of the turbine building. The cables are terminated at the verticals of the CDF and are connected using horizontal connections to the control cables from and to the relay control panels and main control panel for initiating the various logic, protection and supervisory functions.

6.2.5.1.11.3   The Measuring devices

The instrumentation system pertains largely to the relaying of measurements and the integrating of outputs of the various sub-systems to meet the operating and maintenance requirements of the plant. The measurements are explained in detail in the upcoming sections.

6.2.5.1.11.4   Pressure and flow measurement

Pressure transmitters and flow-measuring devices are used to measure and relay the outputs to the control regulating and protection equipment. These are field devices and are thus located accordingly.

6.2.5.1.11.5   Temperature measurement

Resistance temperature detectors (RTDs) and thermocouples are largely used to measure and relay the outputs. A detailed description of the same with regards to the reactor control and regulatory systems, which are based on temperature outputs, has been provided in para 6.2.3.4.3.2 of this chapter.

6.2.5.1.11.6   Energy transducers

These are installed on the metering and billing panels, which are located in the control equipment room, to record the energy generation and evacuation parameters of the station. They are also used to enable the necessary bills to be raised and charged to the consuming State electricity board.

6.2.5.1.11.7   The main plant control system

6.2.5.1.11.7.1 General

The most essential of all plant control and protection systems is the main plant control panel. The panel consists of different control modules, such as the fuelling machine and spent-fuel transfer control, TG and auxiliaries' control, reactor and auxiliaries' regulation and protection control, switchyard system control, main power output system control, unit and station power system control, station auxiliaries' system control, DG system control, control power system, other conventional systems control, etc.,

Thus, the entire station is operable from the control room.

6.2.5.1.11.7.2 The fuelling machine control module

This module controls the six hundred sequential operations required for the remote re-fuelling and spent-fuel transfer system. An aide to the remote operation is the closed circuit television system (CCTV) installed in the FM vaults. The CCTV system indicates, on the monitor installed in the control room, the entire operation on an online basis. The system enables the operators to closely follow the operations.

6.2.5.1.11.7.3 The system status: visual and audio annunciation

The main control panel displays the status of all operating equipment and systems. The 48 V annunciation window panel displays the operating status of all equipment along with audio indication, in case of tripping or malfunctioning of all the operating equipment and systems. The operators are, thus, enabled to take corrective action as called for.

6.2.5.1.11.7.4 The date logger and acquisition system

The date logger and acquisition system records, on a continuous real-time basis, the operating parameters of all equipment and systems. The operator is also able to call for the operating parameters of any particular system when required. This system is one of the most important systems in the control room. The system helps the operator to check the entire sequential occurrences of the plant, particularly in the case of reactor tripping and or other malfunctions of any system in the plant.

6.2.5.1.11.7.5 Mimic diagram: electrical system status

The luminous mimic diagram shows the status of the entire electrical system with indications for 'on', 'off' and 'tripped'. It also displays the power flow of the system prominently. The status enables the initiation of action on the part of the operating personnel to ensure the continued and satisfactory operation of the plant.

6.2.6   Conventional Equipment Maintenance Services

6.2.6.1   General

As mentioned at the beginning of this chapter, all services required for the satisfactory operation of the station have to be decided upon in the initial stages of the station project design. These services are discussed in detail in the upcoming sections.

6.2.6.2   Instrumentation and Control Laboratory

This laboratory takes care of the calibration, testing and repairs of all the instrumentation equipment. It also validates on-going renovation, repair and maintenance of the equipment.

6.2.6.3   Electrical Equipment Laboratory

This laboratory takes care of all requirements, as discussed above, but for the electrical equipment.

6.2.6.4   The Mechanical Workshop

This workshop undertakes the fabrication of minor parts required for the maintenance of all mechanical equipment.

6.2.6.5   Chemistry Laboratory

The Chemistry laboratory of the station takes care of prescribing, monitoring and testing all the additives of the station fluids like feed water, moderator and PHT system heavy water, process water, etc. The laboratory routinely tests the samples from these fluids to ensure the conformance of chemical fluids to the prescribed standards.

6.2.6.6   The Compressed Air System

A reliable compressed air system is a must for modern power plants. The system provides compressed air of requisite purity and quality for the following purposes:

1) Fire protection as per the system requirement.
2) Pneumatic requirement for the air-assisted valves.
3) Instrument air to serve the instrumentation requirements.

6.2.7 Other plant support services

6.2.7.1 General

These services are vital for the plant operations and are dealt with detail in the upcoming sections.

6.2.7.2 The Administrative Services

6.2.7.2.1 Personnel-oriented services

The services offered are listed below:

a) Salary disbursement and deduction of tax at source by the accounts section; the overseeing of staff service interests, pensionary benefits, medical benefits and labour relations.

b) An industrial relations officer looks after labour interests and settles labour/administration disputes if any arise.

6.2.7.2.2 Stores procurement and inventory

This can be said to be the heart of the maintenance of the plant. The set-up has the vital task of maintaining the requisite inventory of vitally needed spares and equipment. A consistent and meaningful liaison with the operating and maintenance personnel is a 'must' for building and maintaining this inventory.

6.2.7.3 Canteen services

Catering to the food needs of the operating and maintenance personnel is another important part of the station administration. A canteen committee, headed by the labour-welfare officer, looks after these needs.

6.2.7.4 Security services

These services are provided by the Central Industrial Security Force (CISF). Details of such services are outside of the scope of this book, however, a few observations are in place. The security has to commence from the 1½-km exclusion zone around the station location. Near and close proximity security of the plant, plant equipment and plant systems are a must. Biometric identification of the personnel and the visitors is a requirement nowadays, and hence the constant state-of-the-art practices on this head are very much essential.

6.2.7.5 *Human development region: special services for staff welfare*

More details on this regard can be found in paras 7.41.7.1 to 7.41.7.5 of Chapter 7.

6.2.7.6 *Station roads*

Station roads are essential, even at the beginning of construction, to cater to the transportation of equipment and materials. These internal roads are suitably augmented to improve the access to various buildings and structures of the station.

6.2.7.7 *Landscaping*

This is a very important part of the overall design of any nuclear power station, the emphasis being made on the provision of an environmentally benign and attractive view of the station ambience. This scaping should include locally available flora and herbal plants if possible, thus making way for a green cover over the entire station and its environment.

6.2.7.8 *The concept and description of the Electrical system*

So far, we have described a typical 200-MWe nuclear project, with particular reference to the RAPP units 1 and 2. Such a project has wide ramifications, particularly in the electrical system; the electrical system is a large one, with user voltages ranging from 250 V AC to 220 KV AC for power and 48 volts to 250 volts DC for control systems. The motor/equipment power and facilities range from 25-50 HPs to 2500 HPs in unit capacity. The transformer capacities range from 250 KVA to 250 MVA. The schematics for such a large-sized nuclear plant would call for a jumbo-sized print. It is thus not possible for such a large schematic to become part of this book. The schematics were instead broken down into a typical sub-system format to suit this book.

The schematics have been so structured that they follow each other in a sequential sub-system manner, maintaining the continuity of the sub-systems to follow the power flow in a top–down narration.

Thus, the schematics start from the 220-KV switchyard systems which provide the start-up power for the station. Once the station is in operation, the power flow is reversed with the station's turbo-generator, thus providing power to the station auxiliaries for transmitting power through the outgoing 220-KV power evacuation lines.

Seven pages of schematics are attached below for better understanding. The legend and scheme efficiently explain the electrical systems of the whole station.

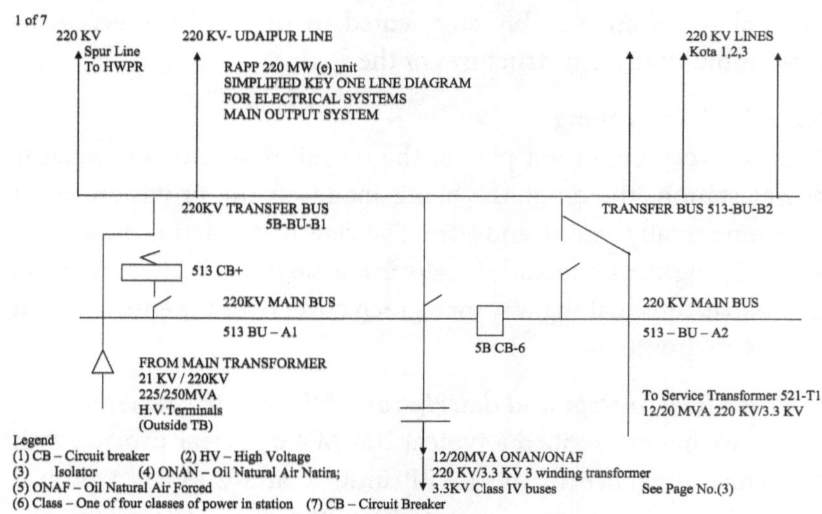

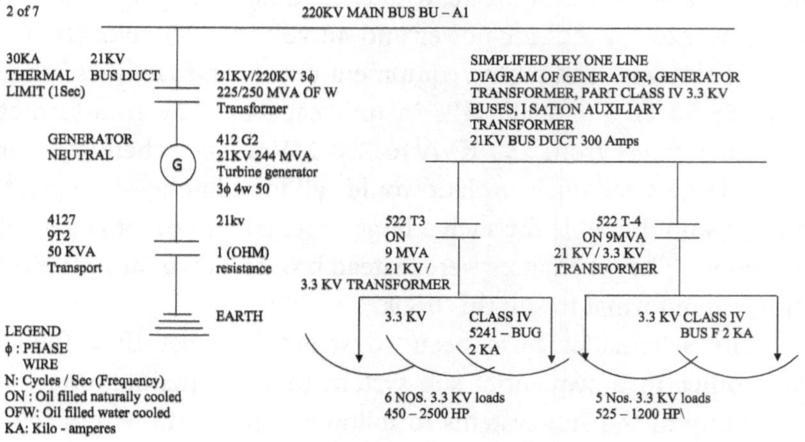

3 of 7

SIMPLIFIED KEY ONE - LINE DIAGRAM
OF AUXILIARY STATION SERVICE TRANSFORMERS

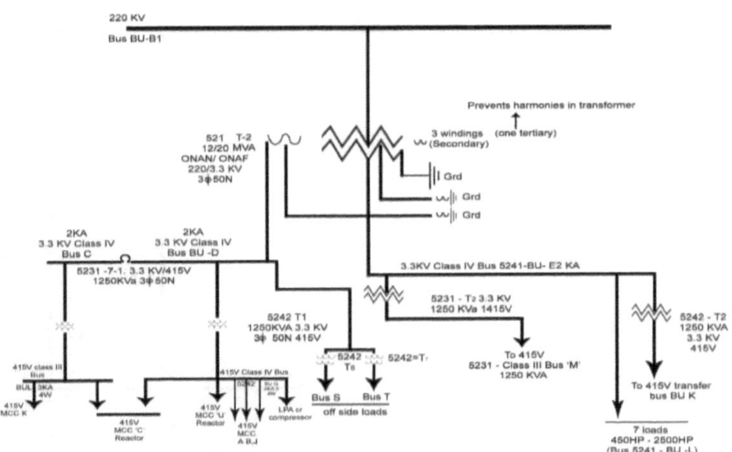

Legend

ONAN - Oil Natural Air Neutral cooled

ONAF - Oil Natural Air Forced Cooling

KV - Kilo Volts

Class - one of 4 classes of power in station

HP - Horse Power

W - Wire

4 of 7

3.3 KV CLASS IV BUS, 5241 - BUF , AUXILIARY
3.3 KV/ 415V TRANSFORMERS,
415V CLASS III BUSES 5231 - BU L AND 5231 - BU M

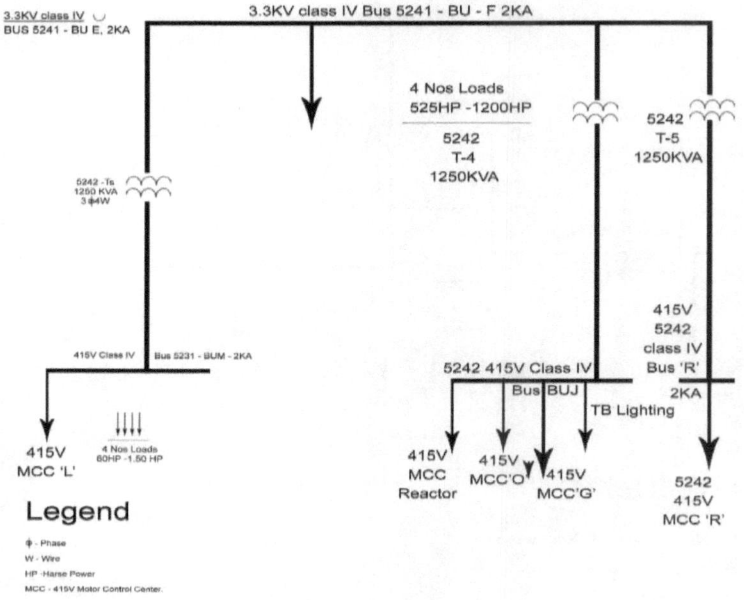

Legend

φ - Phase
W - Wire
HP - Horse Power
MCC - 415V Motor Control Center.

5 of 7

SIMPLIFIED KEY DIAGRAM OF CLASS IV 415V TRANSFER BUS K, BUS H, DIESEL GENERATOR, MG SET

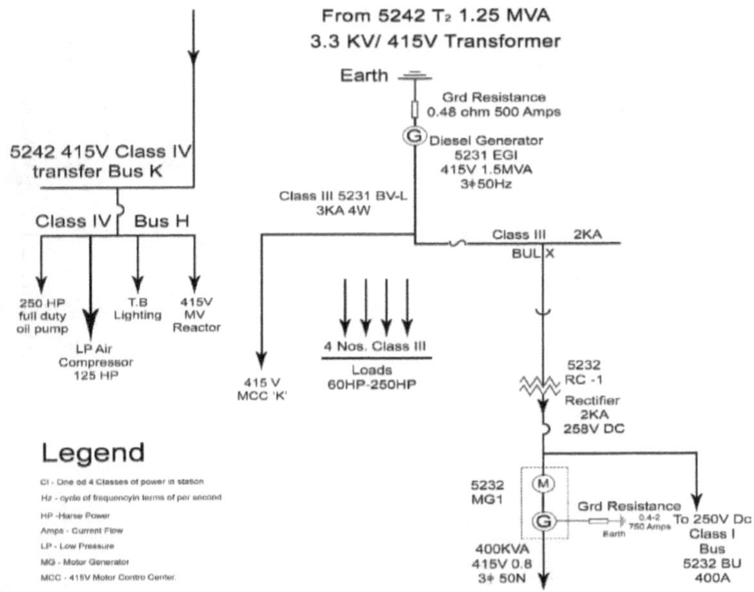

Legend

CI - One od 4 Classes of power in station
Hz - cycle of frequency in terms of per second
HP - Horse Power
Amps - Current Flow
LP - Low Pressure
MG - Motor Generator
MCC - 415V Motor Contro Center.

6 of 7

SIMPLIFIED KEY ONE -LINE DIAGRAM FOR CLASS II 415V AND CLASS III BUSES

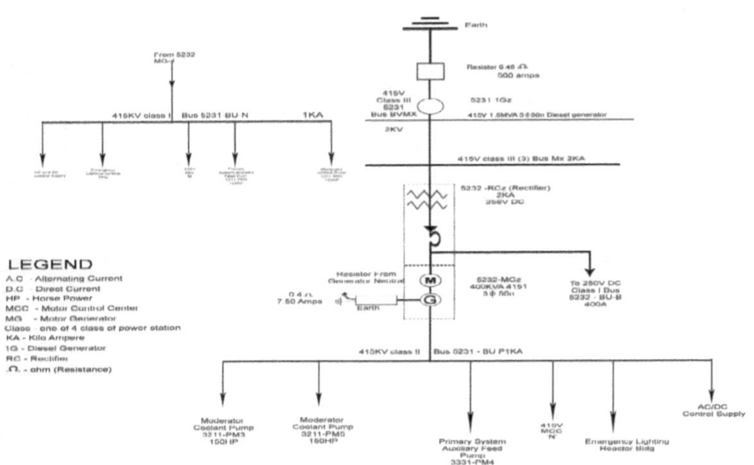

LEGEND

A.C - Alternating Current
D.C - Direct Current
HP - Horse Power
MCC - Motor Control Center
MG - Motor Generator
Class - one of 4 class of power station
KA - Kilo Ampere
1G - Diesel Generator
RC - Rectifier
Ω - ohm (Resistance)

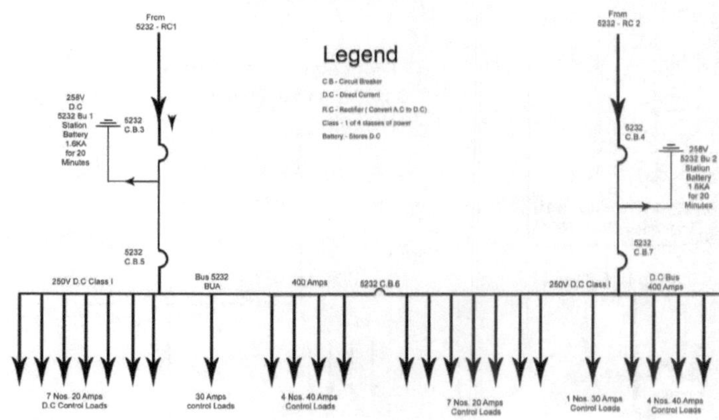

6.2.7.9. Flow diagram for a typical HWPTR is enclosed in P-13(b)

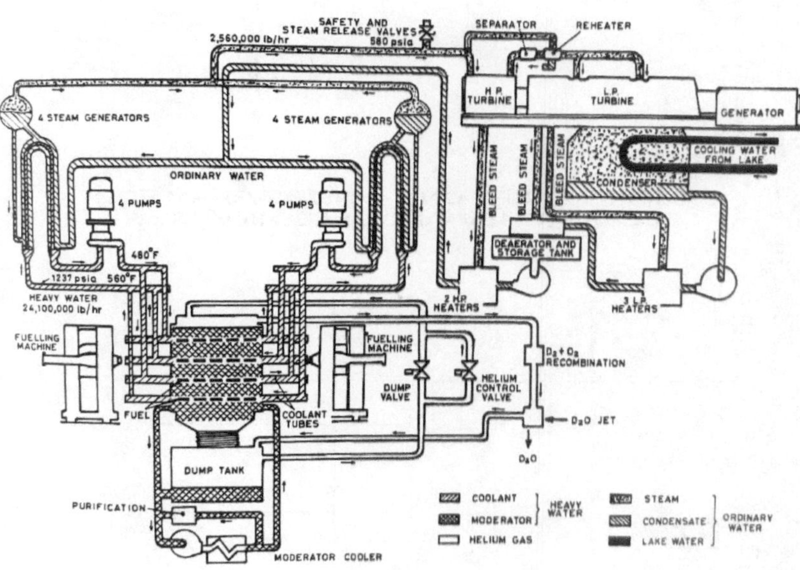

Simplified Flow Diagram

Rajasthan Atomic Power Project (RAPP) Units 1 and 2

7.1 Setting up of the Nuclear Power Stations: RAPP units 1 and 2

7.1.1 General

A Site Selection Committee (SSC) was set up by the Department of Atomic Energy (DAE) during the late 1950s to look into the suitability of various sites for the setting up of Nuclear Power Projects in India. In Chapter 4, we have seen the criteria for the selection of sites. Environmental impact studies have, over the years, gained importance in the selection of sites. The obsession towards the safety of the public and operating personnel, and the preservation of the environment and bio-diversity has always governed the study of any plans for the setting up of nuclear facilities in India. The plan usually begins with the SSC making its recommendation for a nuclear project commissioned by India.

The SSC made its recommendations in the early 1960s. The very first site that was accepted, for the purpose of commencement of construction, was a site about 13 km from the Rawatbhata village in Chhattisgarh district, Rajasthan. The site was located at 25.1° North Latitude and 75.5° East Longitude. The nearest city, Kota, was located 63 km from the chosen site. The project would be constructed with technical and financial assistance from the Canadian Govt. with whom the Govt. of India had earlier started negotiations for the project. Photo P-1 enclosed shows the site location.

7.1.2 Rajasthan State: A Brief History

Let us now take a brief look into the State of Rajasthan where India's first Nuclear Power Project was commissioned and built.

The State of Rajasthan, as we know it today, was once famous for the number of its Princely States before independence. These princely states were later integrated into India after the attainment of independence. Before independence, Rajasthan contained nearly 80% of the total number of Princely States whose rulers lost their sovereignty when Great Britain handed over power to the then Dominion of India. The sovereignty involved 300, or so, out of 400-odd Princely States. It looked as if the Indian Union Govt. was grounded even before it took off. Given this historical harshness, it took all of the indomitable courage and willpower of the late Sardar of India (Sardar Vallabh Bhai Patel) to ensure the integration of all these States into the union of India.

For a long period, Rajasthan was looked upon as a feudal state because of the nature of the rule of its Princes before the integration of states. Although the governance in a large majority of the Princely States might have been feudal, there was benevolent rule in some of the States like Bikaner, Jaipur, Jodhpur, Jhalawar, Kota, etc., which had resulted in the advancement of education and local self-governance, among others. However, the large majority of the population of Rajasthan was illiterate at the time of independence. Its colourful people had their hands full coping with the harsh conditions of the semi-arid areas like the districts of Jodhpur, Jaisalmer, Ganganagar, Churu and Bikaner. It was a fight with nature to eke out with such precarious living. Many of the animal herders of Rajasthan used to migrate to neighbouring States in search of pasture for their animals to graze in, especially during the harsh summers of the State.

In the following sections, we will see how such conditions have been changing in the State. We will also learn about how the first Indian nuclear project has been instrumental in changing the economic face of the State of Rajasthan, among other development projects taken up after independence.

This was also the State of Rani Padmini, the wife of Prithvi Raj Chauhan, whose legendary beauty was such that the Mughal Emperor who had invaded Chittorgarh was content to have a look at her image reflected by a mirror glass. Rajasthan is a State whose princesses thought nothing about jumping into fires to safeguard their honour from marauding invaders. It is also the State of Rana Pratap, whose valorous exploits are known to every Indian even today. The State also boasts the most extensive irrigation system ever constructed, namely the Indira Gandhi Canal, which brings 'golden water' to its parched fields and, in its wake, sound agricultural development to as many as 13,91,000 hectares of the previous arid land. The face of Rajasthan is destined to change with the developmental activities initiated since its independence in 1947. In spite of all these changes, Rajasthan is still being considered a backward State as one of the BIMARU (Bihar, Madhya Pradesh, Rajasthan and Uttar Pradesh).

7.1.3 Details of the Project Site: RPS dam power-house for start-up power

The site is located down-stream of the Rana Pratap Sagar (RPS) dam at Rawatbhata. The dam compounds the water of the Chambal River. Further up the river is the Gandhi Sagar dam which is the power-house in Madhya Pradesh with a capacity of 140 MW. The RPS dam was in an advanced stage of construction and was expected to provide about 170 MWe of hydro-power when completed.

The project site consisted mostly of semi-arid forest areas. Land acquisition for the proposed nuclear project would hence not present any major problem. The RPS dam would provide the required circulating, processing and domestic water requirements of the nuclear plant and the township. Hence the start-up power would also be provided by the RPS power station.

7.1.4 Installed power capacity of the State of Rajasthan: the effect on RAPP 1 and RAPP 2

The installed power capacity of the State, as at the commencement of RAPP units 1 and 2, was about 250 MWe only. There was quite some doubt in the informed power sector about whether the

Rajasthan power grid would be able to absorb the proposed nuclear power (400 MWe), that would be generated by RAPP 1 and RAPP 2, without causing grid problems. There were, in fact, quite some problems in the early stages of the operation of the RAPP 1 project. It is of relevance to point out here, that the installed power capacity of the Rajasthan State Electricity Board (RSEB) went up twice to 500 MWe by the time that RAPP 1 was commissioned. This was taken into account and considerably strengthened the stability of the RSEB on the commissioning of the RAPP project.

7.2 Financial Sanction

Financial and administrative sanctions for the setting up of a 1×220 MWe heavy-water cooled and moderated pressure-tube type reactor unit (HWPTR) was accorded by the Government of India during 1963-1964 for Rs. 37 crores.

7.3 The concept of design and construction

Rajasthan Atomic Power Station (RAPS): The Forerunner of Country's PHWR Programme

7.3.1 General

In Chapter 5, we have described the design and construction features of typical 220 MWe HWPTRs in a general manner. The differences between fossil fuel-fired and nuclear fuel-fired stations and the similarities between them were also described with specific details regarding RAPP units 1 and 2. These details will further be described in the upcoming chapters. Similar details concerning newer HWPTRs, like the changes periodically introduced in them by way of improvements in design and construction, will also be described in the respective chapters.

7.3.2 Preliminary work: the Agreement with Montreal Engineering Co. (MECo) of Canada

7.3.2.1 *The agreement between the Indian and Canadian Governments*

The two Governments started negotiations in 1963, for the design, commissioning and construction of a nuclear project in India to be aided by Canada. The negotiations would also cover Canadian financial aid for the project. The negotiations ended in 1966, when the Canadian Government agreed to render assistance, as stated above, for the building of two HWPTR 220-MWe units. Montreal Engineering Co. (MECo) of Canada would be the executing agency for the projects. MECo, apart from providing the basic design and procuring necessary imported equipment and materials, would offer on-site consultation for the engineering and construction work as well. The construction responsibility would lie entirely with the Indian engineers while Ontario Hydro of Canada would provide the commissioning 'know-how'. The Atomic Energy of Canada Ltd. (AECL) would provide the design for the reactor systems.

A photo showing how the station would appear when completed

7.4   Civil work: contractors for the plant work

Preliminary work at the site involving the construction of roads, site offices, mechanical workshops, warehouses, etc., began in 1963. Hindustan Construction Co. (HCC) was awarded the civil work for the RAPP 1 unit and they began their work by turning the first sod at RB-1 on 28.12.1964.

7.4.1   Seismic force

Before we proceed further with the progress, it is to be noted that the RAPP site falls under 'seismic zone-2' as per Indian Standards (IS) 1893-2002. All safety-related buildings, structures and systems have thus been designed with a zero-period ground acceleration of 0.05 g.

7.4.2 Further HCC work at the site: executive personnel of the construction company

Sri. K. Mokashi was designated by the HCC as the Construction Manager and A.S. Godbole as the Project Manager for the construction work at the RAPP site. Sri. M.R. Sampath, a senior HCC manager who was then based at HCC's Bhilai steel plant site in Chattisgarh, was deputed to the RAPP site whenever required to oversee the mechanical and electrical requirements, in particular the air compressor equipment and other related equipment that are required for wagon drilling at the reactor-1 site. These were critical construction equipment and their uninterrupted running was a pre-requisite for keeping to the construction schedule. Hence this required the expert attention of Sri. M.R. Sampath.

The HCC's scope of work for the RAPP 1 site was the entire civil work for the reactor building, turbine building and chimney stack. It also included the work involved in the erection of a departmental-free supply of embedded parts, water stops, structural steel and other related parts. The work on the pre-stressed concrete dome was entrusted to Gammon India Pvt. Ltd. The technical details of the dome work are listed as follows:

a) Design pressure : 0.42 kgf/cm sq.
b) Test pressure : 0.55 kgf/cm sq.
c) Grade of concrete : M-21
d) Pre-stressing of wires : 'Fressinet' type with twelve wires of 7 mm diameter each
e) Pre-stressed wire capacity : 74 metric tonnes.

A large area and an additional area of about 1 km from the plant site was made available to the HCC for housing their labour force, setting up a bazaar and other day-to-day amenities. These amenities were also availed of by the departmental residents of the phase-1 colony located one-and-a-half km from the site.

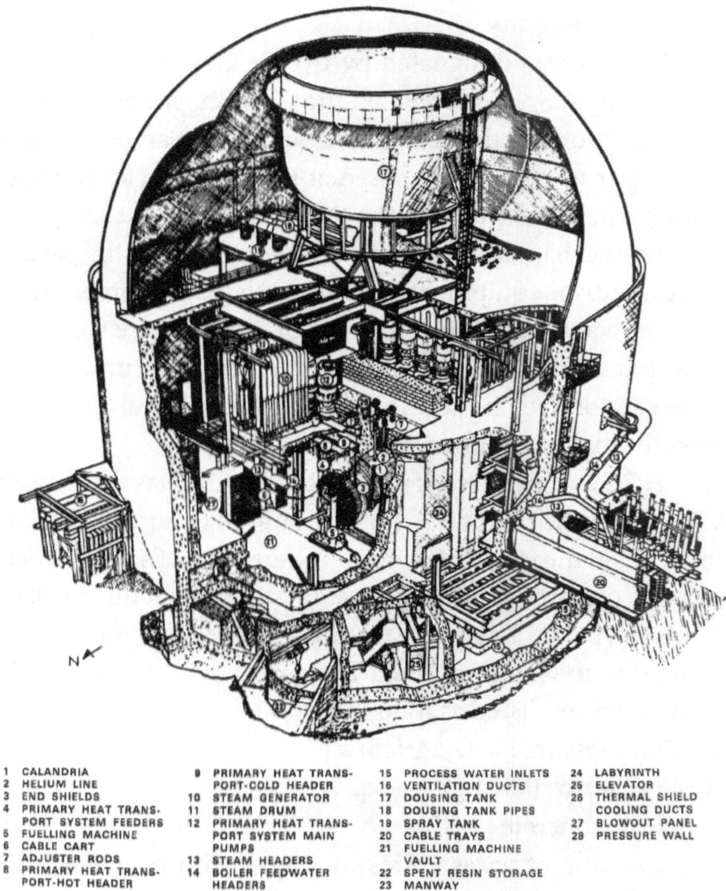

1	CALANDRIA	9	PRIMARY HEAT TRANS-PORT-COLD HEADER	15	PROCESS WATER INLETS	24 LABYRINTH
2	HELIUM LINE			16	VENTILATION DUCTS	25 ELEVATOR
3	END SHIELDS	10	STEAM GENERATOR	17	DOUSING TANK	26 THERMAL SHIELD COOLING DUCTS
4	PRIMARY HEAT TRANS-PORT SYSTEM FEEDERS	11	STEAM DRUM	18	DOUSING TANK PIPES	27 BLOWOUT PANEL
		12	PRIMARY HEAT TRANS-PORT SYSTEM MAIN PUMPS	19	SPRAY TANK	28 PRESSURE WALL
5	FUELLING MACHINE			20	CABLE TRAYS	
6	CABLE CART			21	FUELLING MACHINE VAULT	
7	ADJUSTER RODS	13	STEAM HEADERS	22	SPENT RESIN STORAGE	
8	PRIMARY HEAT TRANS-PORT-HOT HEADER	14	BOILER FEEDWATER HEADERS	23	MANWAY	

Reactor building view looking south-east

Photo P-2 shows the appearance of the station as and when it was completed, while photo P-3 shows the aerial view of the reactor building.

7.5 The overseeing of the project work by MECo

Mr. John. L. Tasker was the very first resident engineer from the Montreal Engineering Company (MECo) to be present at the site. Tasker was a Civil Engineer by profession. His long stint in MECo helped him keep an overview of the initial mechanical and instrumentation work at the site.

Tasker stayed at Kota city, which is about 63 km from the RAPP site, during the initial months of commencement of the RAPP work. It was during the third quarter of 1965 that Tasker moved to the Senior Officers' quarters at the phase-I colony.

At the site, the work picked up tempo when more MECo engineers arrived at the site to oversee the various disciplines of work.

7.6 Amenities provided by the Government of India for the MECo engineers

The engineers were provided with free accommodation in the phase-I colony initially, and then later moved to the permanent township when the quarters were ready for stay. They were entitled to living allowances for their family, consisting of a couple and a maximum of three children living with their parents at the site.

They would be entitled to bring their vehicles, free of import duty, for use at the site with the condition that they could sell their vehicles to private parties in India at the end of their tenure at the site. If the vehicles were not required to be taken back home, they would have to be turned over to the Canadian Embassy in New Delhi or the State Trading Corporation of India (STC). The engineers would be entitled to reimbursement of charges for the transportation of their families as well as personal effects both to and from the site. No Indian income tax would be liable to them for all of their allowances.

7.7 The Deputation of Indian Engineers to Canada

It was in 1964 that the Indian Engineers were deputed to Canada for training in nuclear facilities under the aegis of MECo and Ontario Hydro. These Engineers would later return from their training course, which lasted up to 18 to 24 months, to take up suitable positions at the site in the construction and commissioning fields.

7.8 Change in the MECo nomenclature

Sometime around 1969, MECo set up Montreal Engineering Co. International Limited (MEIL) to look after all the international operations that were previously under MECo. MEIL would thus look after the operations of the RAPP units 1 and 2.

7.9   The ordering of equipment for the RAPP units 1 2

The ordering of all the major nuclear and conventional equipment for RAPP 1 and RAPP 2 was completed by the year 1964. The details of the same are explored in the following sections.

7.9.1   Two 220-MWe Turbo-Generators (TGs)

These generators were of the General Electric (GE) Canada make. The purchase order placed by MECo on John Inglis, suppliers on behalf of the DAE of the Govt. of India, stipulated that a minimum value of 60% of the equipment cost would be sourced from Canada. The cost of the TG set was 6 million dollars for RAPP 1 and 6.6 million dollars for RAPP 2. The above-mentioned stipulation for the Canadian content of the TG equipment was made by the Canadian Government while signing the Inter-Government agreement for aid to the RAPP 1 and RAPP 2 projects.

7.9.1.1   Technical parameters of the TG sets

The sets are of the 21 KV 244 MVA 220-MWe 2-pole non-salient type, with 3000 rpm (revolutions per minute) hydrogen-cooled at 38 psi (per square inch) with a 0.6 short-circuit ratio and two parallel stator windings, with the neutral of the windings brought out and earthed through a 50-KVA single-phase transformer. The stator windings are formed of hollow copper conductors, de-mineralised water cooled through heat-exchangers. Excitation for the set is provided by a 1000-KW shaft-driven pilot and the main exciters are of the amplidyne controlled magnetic amplifier-type. The rotor is cooled by two 100% hydrogen coolers in parallel. The set is provided with local control and supervisory panels. The hydrogen coolers have a built-in fouling factor between 15% and 20%.

7.9.1.2   The 220-MWe turbine

The turbine is a tandem compound cylinder with five stages of steam expansion. The incoming steam supply at 565 psig (pressure per square inch gauge) and 251 degrees C is via two 400-mm-diameter pipes at opposite sides of the TG cylinder. With the above parameters of the main steam supply, the steam is to be reheated to 221 degrees C at a pressure of 3.8 kgf/cm sq. at admission to the LP stage through steam re-heaters. The TG is provided with local supervisory panels, servo-oil governing system, combined interrupter cum emergency steam valves (ESV), throttle valves, re-heaters, inter-connected piping, main lube oil pump, emergency lube oil pump, lube oil circulating system and a lube oil purification system. Photo P-12 shows the overall view of the TG set.

A view of the turbine - generator.

7.9.1.3 Feed-water temperature, heat rate and thermal efficiency

The final feed-water temperature is 172 degrees C. The heat rate for the system is 277 kcal/KWHr while the thermal efficiency for the plant is 29.2 degrees C.

7.9.2 Turbine following the reactor operating system

The reactor turbine interface has been designed on the turbine following the reactor principle of operation. The reactor generates thermal power based on the pre-set thermal power value. Minor deviations from this value are to be catered to in the reactor turbine inter-facing. These changes in the reactor thermal power-supply are sensed by the turbine control and supervisory system which changes the steam flow to suit the reactor power. Any large variation in the reactor-set power will activate the reactor regulating system. The system steps in and brings the reactor thermal power to the set value.

7.9.3 Primary heat transport pumps

The pumps (eight each for RAPP 1 and RAPP 2) have been imported. The pump units were of 'Bingham' make while the Motor units were of the 'Tamper' make.

7.9.4 Boiler Feed Pumps (BFPs)

These units were also imported.

7.9.5 Calandria (The Reactor Pressure Vessel)

Two Calandrias for RAPP 1 and RAPP 2 were ordered from Larsen & Toubro (L&T). The Calandria is a 6.5-metre-diameter cylindrical vessel made out of austenitic stainless steel, designed to house the coolant high-pressure tubes along with fuel and the Calandria low-pressure tubes. Austenitic stainless steel has the desirable quality of toughness at very low temperatures and strength at high temperatures. This quality improves the irradiation strength. The second desirable quality is its resistance to corrosion in different types of atmospheres at high temperatures.

Heavy water circulates in the pressure tubes located in the Calandria on a 22.86 cm × 22.86 cm square pitch. The Calandria tube sheet is penetrated by 306 nickel-free zircaloy-2 tubes with an inner diameter of 10.78 cm and a wall thickness of 1.24 mm. These tubes are attached to the tube sheet by specially designed roller joints. All other joints in the Calandria structure are full-penetration welded to the shell body to ensure additional stiffness and strength. The reactor vessel is designed for a maximum stress of 1,195 kgs/cm sq.

The stainless steel plates for the Calandria were imported. The cost of the two Calandrias was Rs. 60 lakhs. Photos showing the Calandria and the coolant channel assembly are enclosed as P-6 and P-8.

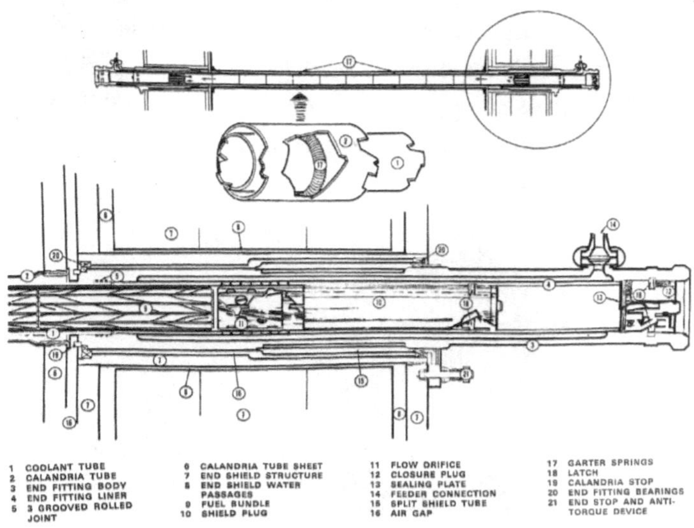

1 COOLANT TUBE
2 CALANDRIA TUBE
3 END FITTING BODY
4 END FITTING LINER
5 3 GROOVED ROLLED JOINT
6 CALANDRIA TUBE SHEET
7 END SHIELD STRUCTURE
8 END SHIELD WATER PASSAGES
9 FUEL BUNDLE
10 SHIELD PLUG
11 FLOW ORIFICE
12 CLOSURE PLUG
13 SEALING PLATE
14 FEEDER CONNECTION
15 SPLIT SHIELD TUBE
16 AIR GAP
17 GARTER SPRINGS
18 LATCH
19 CALANDRIA STOP
20 END FITTING BEARINGS
21 END STOP AND ANTI-TORQUE DEVICE

Coolant Channel Assembly

7.9.6   Nuclear Steam Service Generators (NSSGs)

Sixteen heavy-water-to-light-water heat-exchangers were ordered from L&T. Each hair-pin type heat-exchanger consists of 195 Monel (SB 193) inverted U-tubes of 12.5 mm outer diameter and 1.25 mm wall thickness. Each NSSG would generate 1.17×10 raised to the power of 6 kgs of saturated steam per hour at a pressure of 39.7 kgf/cm sq. and a temperature of 251 degrees C. The steam would be 99.75% dry. Each NSSG would comprise ten inverted U-type exchangers with heavy water on the tube side and light water on the shell side. Heavy water enters the tube side at the bottom of the boiling leg and returns to the pre-heater leg of the reactor inlet headers located in the RB. The light-water feed from the boiler circuit enters the shell side of each heat-exchanger at the bottom of the pre-heater leg and is heated to saturation temperature in the reduced diameter part of the leg. It boils in the upper part of the leg and enters the steam drum.

The minimum temperature difference between the heavy water and the light water in the NSSGs is 52 degrees C. The design pressure for the shell side is 51kgf/cm sq. and that for the tube side is 95.8 kgf/cm sq. The design temperature for the shell side is 264 degrees C and that for the tube side is 299 degrees C. The pre-heating leg area is 1027 metres sq. and the boiling leg area is 6,920 metres sq. Photo P-11 shows the primary coolant circuit equipment.

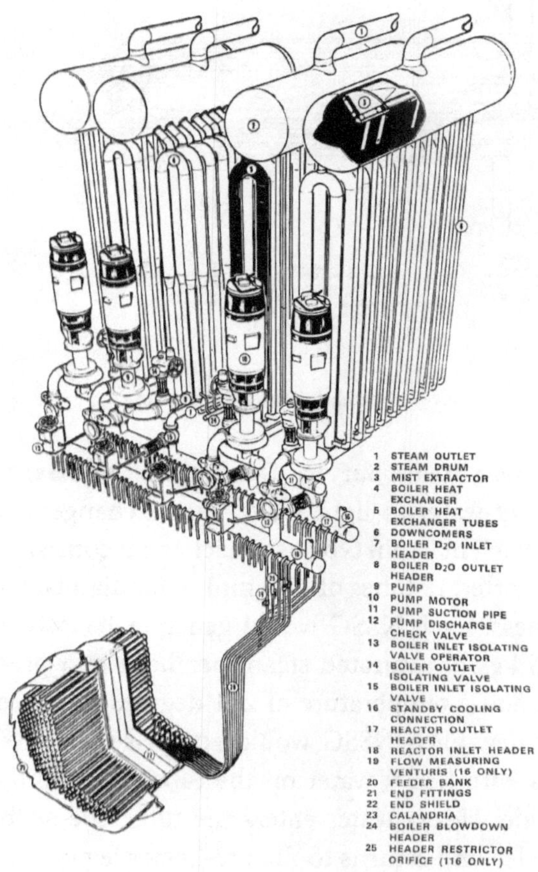

Main primary coolant circuit equipment.

7.9.6.1 *The cost of the NSSGs*

The cost of eight NSSGs for RAPP 1 was Rs. 1 crore and that for RAPP 2 was 1.2 crores.

7.9.7 The 220-KV Switchyard Equipment and Materials: the plant pump-motors and transformers

The 220-KV switchyard equipment and materials were entirely indigenous. All the generators, start-up and unit transformers, station service Transformers and 95% of the six hundred odd number of pump-motor units, required for each reactor unit, were indigenous. It could be stated that the indigenous value and content of the equipment and materials was 36% for RAPP 1 and 60% for RAPP 2.

7.10 Co-operation with the State Government of Rajasthan, the RPS Dam Project and the Rajasthan State Electricity Board (RSEB)

The co-operation of the RSEB was essential for the provision of power-supply at the site for construction purposes and for the housing colonies. This co-operation was amply forthcoming. This co-operation was mutual as illustrated below. We had discussed in para 7.1.3 that the Rana Pratap Sagar power station was in an advanced stage of construction at the commencement of work at the RAPP site.

Sri. K.S. Sivaprakasam, the additional Chief Engineer of the RPS Dam power project, during one of his several meetings with Sri. V. Surya Rao, the Chief Project Engineer of RAPP, had requested the services of the skilled welders available at the RAPP site to weld the embedded parts of the penstock of the RPS dam power station. These services were duly made available.

Note: The author recollects at this point of time that the then Madras Electricity Board—at which he had served for four-and-a-half years (May 1954-November 1958)—had availed the services of Japanese welders for carrying out specialised welding for the penstock pipes of the Kundah hydro-electric project. This was way back in the 1950s. In the intervening period of ten years, sufficient indigenous skill in welding had been built up to take care of the most demanding welding work in this field and elsewhere.

7.11 `Co-ordination of the Rajasthan State Administration for the movement of over-dimensioned consignments (ODCs)

7.11.1 General requirements

Over-dimensioned consignments (ODCs) ranging in weight from 40 metric tonnes to 150 metric tonnes were required to be moved by road from the ports of Kandla and Mumbai. Constant and concerted interaction between the RAPP authorities and the Public Work/Highways Depts. of the concerned State Governments, such as Maharashtra, Gujarat and Rajasthan, was rendered necessary. The roads and bridges in these States, through which the ODCs would pass, had to be strengthened to take care of the considerable axle loads imposed by these consignments. By-pass roads were also required to be constructed to ensure the safety of the ODCs in several places.

7.11.2 Mock-up Tests

The RAPP conducted the movement of mock-consignments of crucial ODCs, such as the Calandria, Generator Stator, Boilers, etc. Measurements of deflection under the positions of the axle-loads over the concerned bridges were also conducted with mock-consignments of the ODCs. Such tests were necessary to establish the design adequacy of bridges while passing such critical consignments.

7.11.3 Security arrangements for the RAPP Project: co-ordination of the State utilities

Apart from the technical requirements of establishing the suitability of roads and bridges for meeting the axle loading imposed by the ODCs, several logistic requirements had to be organised as well. Some of these included the arrangement of police security, co-ordination with the utilities such as the Telephone/Telecommunication Departments, State Electricity Boards and Municipal/Local authorities for the safe passage wherever power, telephone and tele-communication lines had to be crossed, among others. This involved writing to these authorities well in advance of the date of passage of these loads, arranging for the electrical line and telephone line clearances by way of shut-downs, the removal or raising of the utility lines, etc. Personal contact with these authorities at the actual time of passage expedited the safe passage of the ODCs as per the schedule drawn up for the movement of the same.

7.11.4   Site organization: the ODCs

The entire work related to the movement of ODCs was overseen by the ODC section at the RAPP site. The section had the responsibility of maintaining and operating an entire fleet of vehicles and equipment required for the ODC movement work. The section had the required help of fully articulated tractor-trailers, each with their power train for the entire range of weights of the ODCs mentioned in para 7.11.1. A usual ODC convoy would consist of an Engineer of SC/SD grade, a rigging superintendent, a rigging foreman, two assistant rigging foremen and about 10-15 riggers dictated by the nature of the ODCs. The vehicles required, apart from a tractor-trailer, would be a jeep; one or two pick-up vans; a lorry containing tools, tackles and materials required for the ODC movement and a lorry which would house the provisions, utensils, stoves, etc. which are needed for meeting the food requirements of the convoys during transit. A convoy would consist of 30 to 35 personnel and the mess arrangements would cater to this number.

Note: The arrangement of a mobile kitchen has been discontinued after the Narora Atomic Power Project.

7.12   Headquarters for the RAPP site

7.12.1   Details of the organisation's strength

When the RAPP unit was set up in 1963, a skeleton staff of ten people, consisting of a planning engineer and supporting administrative and accounts personnel, was all that was available at the Headquarters Office in Bombay for doing the spade work for the project. The staff was located in the Old Yacht Club building near the Gateway of India and later moved to Jehangir Villa, Wodehouse Road, Colaba, Bombay. Sri V. Surya Rao was appointed the Chief Project Engineer (CPE) for the RAPP site in 1964. Sri. S.S. Jaggia was appointed the Rajasthan Project Administrator (RPA) in the same year. The personal staff of the RPA and CPE and the administrative and accounts staff to run the Headquarters office were appointed and took up their positions at the Jehangir Villa. SD and SC Grade Engineers belonging to the Civil, Mechanical and Electrical disciplines also took up positions to deal with the various facets of the project work. These Engineers would initially provide support to the planning Engineer in Bombay and would later be moved to the site to look after the construction work. The strength of the headquarters was about 40 in 1964.

Note: The author of this book was one among the personnel assigned to the headquarters in 1964.

7.12.2 Responsibilities of the Headquarters

The major responsibilities of the headquarters are listed as follows:

1) Liaising with MECo and Ontario Hydro for the efficient and economic execution of RAPP
2) Reporting to the DAE and co-ordinating with the same in relevant matters
3) Liaising with the concerned State Governments and other agencies as required
4) Settling matters regarding customs clearance; arranging foreign-exchange as required for the project and obtaining DGTD clearances for imports and other related matters
5) Arranging the deputation of project personnel abroad and arranging the visits of foreign personnel to India in connection with project work
6) Arranging the lease of heavy water as required for the project

7.13 Site organisation

The corresponding staff positioned at the site, as of 1964, was a total of about 100 people, comprising about 20 Engineers and 80 supporting staff. This would gradually increase to a maximum of about 1000 employees, comprising about 250 Engineers and 750 supporting staff, technical and non-technical, by the year 1970.

Note: The author of this book joined the construction group in 1964.

7.14 Major plant enabling work

Job shacks, warehouses, concrete-testing laboratories, garages, temporary office buildings, construction power-supply and tele-communication systems were the very first enabling workshops to be constructed at the site. A diesel engine-driven power house, with two 415-volt 600-KW 'Skoda' make sets, was also established to provide emergency power to the plant construction work. All these works cost about Rs. 1 crore according to the prices in 1964.

7.15 Components workshop

This was an important facility that was required even at the early stages of the project construction. The shop was a 1000-sq. metre built-up area with facilities for handling large equipment like the Calandria, the dump tank, the end-shields, etc. It was provided with a 75-metric ton over-head electrically operated travelling crane. Most of the site work related to the nuclear portion of the plant such as the Calandria dump tank, pressure tubes, the Calandria tubes, etc. was carried out in this shop. A photo of the end-shields is enclosed in P-7

End-Shields for Unit II under fabrication in India

7.16 Movement of heavy equipment: access way to the reactor building (RB)

Another important enabling facility was the provision of a sloping access-way from the components shop to the RB through an opening on the East side of the Reactor building perimeter wall to the loading bay of the RB—to hoist large nuclear equipment and parts to the various floors of the RB. The access-way was a sloping two-rail steel track-way emanating from the component building complex and ending in the loading bay of the RB.

All the major nuclear equipment and parts were thus hoisted onto the upper floors with the use of the boiler room over-head travelling crane. The power for driving the track car, on which the equipment was loaded, was provided by an electrically operated winch with wire rope and drum arrangement. The necessary safety measures were undertaken for handling large heavy loads. All these operations were carried out under the supervision of the Rigging Superintendent and staff of the rigging section. Representatives of the section to which the Nuclear Equipment pertained and representatives of the Safety Engineering section were the other parties necessarily associated with the rigging operations. Peak ODC operations began in 1965 and lasted until 1969.

7.17 The Fuelling Machine Rehearsal Facility (FMRF)

This facility had to be established mid-way through the equipment installation schedule. This facility would utilise the spare fuelling machine head (one for each unit). Two fuelling machine heads are required for the operation of each reactor unit. This facility would also utilise the fuelling machine carriage with its X, Y and Z axes drive-mechanism. This drive-mechanism would enable the fuelling machine head to align itself with any of the 306 fuelling channels of the Calandria for on-load re-fuelling. Such a rehearsal facility was of the utmost essence for commissioning purposes. Fuelling machine operations involved 600 automatic sequential operations from the time the initial command is given from the Main Control Panel of the station to the time the spent-fuel is ejected through a chute to the spent-fuel storage bay. The rehearsal facility required the rigging up of the local control panel to simulate the actual operations during re-fuelling; the simulation was a crucial one to validate the procedures for the actual operations after the installation and commissioning of the fuelling machines.

7.17.1 Electrical Relay Testing Laboratory/Instrumentation Testing

This facility was established for testing and calibrating the various relays. This was very important as the relays constitute the very heart of the protection of the Electrical Equipment. This was the case for the Instrumentation equipment as well.

7.18 The Co-ordination of the Plant Work

We have already discussed how co-ordination was required to be carried out with the State Government and other agencies to expedite the project work.

Of greater and more immediate importance was the co-ordination that was required to be maintained between the different work sections of the project, and the agencies charged with the execution of the multi-faceted plant construction work at the site.

It is to be noted that at any point in time, during the major portion of the execution of work, an average number of eight or more different agencies would be deployed at the same work site. To ensure that the work of one agency did not clash with, damage or negate those of the others co-ordination was a must. This was carried out with the help of schedules, made out in advance, for the execution of different work in consultation with and concurrence of the concerned agencies. Even with the drawing up of closely co-ordinated pre-approved schedules, damage to work could never be eliminated. The best of efforts ensured that such damages were minimal.

7.19 Erection work on the Reactor dome: Fire accident

We have seen earlier that the erection of the pre-stressed concrete work for the dome of RAPP 1 was awarded to Gammons India Pvt. Ltd. Gammons have been carrying out specialised Civil Engineering work over the past several decades and their experience in the erection of critical civil work, particularly the pre-stressed engineering work, was one of the major reasons for awarding this work to them in 1966-1967. The scaffolding work had begun in the first quarter of 1967. In the middle of 1967, the form-work was ready for the first pouring of cement. Around 8 PM, when the welding of re-bars was being carried out, a spark from one of the welding electrodes ignited a fire on the plywood form portion of the work. The form-work was entirely made up of plywood, and the fire that was ignited spread quickly. The dome at its springing point was 40 metres in diameter and its height at the maximum point was 15 metres from the base of the dome. The entire inner surface of the form-work for the dome was already completed at the time, so the gigantic volume of inflammable plywood could be gauged. The entire form–work of the dome was thus reduced to ashes and the re-bar work in place was completely twisted and rendered out of shape.

The repair work took one year to be completed. The accident led to considerable litigation between the DAE and Gammons Pvt. Ltd. The settlement took a considerable amount of time and effort.

Before we move on, the herculean efforts made by the fire-fighting section of the RAPP site in containing the dome fire have to be recorded. It may be stated herein that the lowest point of the fire was 28 metres from the grade level and the highest point 43 metres from the grade level. The primary aim of the firefighting section was to prevent the burning embers of the dome from causing further fires in the nearby buildings. The site control store was one such building which was near the fire. The fire-fighting section removed all combustible and other valuable materials to safer places. It also removed such hazardous materials from all nearby areas. Once this was achieved, efforts were made to contain the dome fire. This was, of course, an impossible task since the entire form-work was on fire. The fire subsided only when the entire frame-work had burnt out.

It is well-known that high temperatures can weaken concrete structures. The entire boiler floor was subjected to high temperatures during the fire accident; hence measures had to be taken to strengthen the same. Ultra-sonic tests and other tests were carried out to assess the strength of the 1263-metre elevation boiler floor at various locations. The boiler floor was then strengthened by the addition of structural steel supports at locations considered necessary as a result of the tests.

7.20 The 220-KV Switchyard work

This work was taken up at an early stage of the project to ensure that the start-up power for the commissioning of the various systems was in place in time. The contract was awarded to Voltas Ltd. between late 1966 and early 1967. Almost all of the switchyard equipment was indigenous, made and supplied by HBB (now ABB), BHEL and others. The scope of the switchyard work awarded to Voltas comprised the following:

a) The supply and installation of all switchyard steel structures and bus-work including support insulators, clamps, connectors, etc.

b) The supply and installation of 220-KV disconnect switches, 229-KV potential transformers, 220-KV current transformers and earthing switches

c) The installation of carrier communication equipment, 220-KV air blast circuit breakers and 220/3.3 KV start-up transformers

d) The supply and installation of lighting and grounding of cable ducts

e) The supply and installation of compressed air piping, pipe supports, valves, pipe connections, etc., from the air receivers to the 220-KV ABC breaker control panels

Voltas completed all their work on time. The power from the RSEB/Northern grid was available with the commissioning of the switchyard in 1969. About 15 MWe of power from the grid would be required for commissioning purposes. The switchyard work covered both RAPP 1 and RAPP 2 units and the work was awarded to Voltas for Rs. 15 lakhs.

7.21 Piping Work

These works were awarded to Dodsals who had considerable experience in carrying out large-sized mechanical and piping work. Almost the entire piping work and most of the Mechanical Engineering work, including the erection of pump motor units, were covered under this contract. The piping work covered the entire contract. The piping work covered the entire range of diameters from 100 mm to 1200/1500 mm and included all types of pipes like carbon steel, cast steel, black pipes, stainless steel, etc. One important part of this work comprised the site preparation, cleaning up and installation; the work was made complete with the welding of the heavy-water feeder tubes, 612 in number, for each reactor unit which transports the heavy water to and from the 'cold' and 'hot' heavy-water headers and the pressure tubes of the PHT system. The site preparation of the feeder tubes was carried out by Dodsals in a special area of their site shop since this work called for 'clean conditions' to be maintained during the pre-erection processing of the tubes.

7.22 The critical nature of the piping work: incentives offered to contractors

Most of the engineering work carried out at the site, particularly the nuclear engineering work, was the first of its kind in the country to be done by Indian Engineers. The quality control requirement was quite demanding as well. Even with the pre-qualification of welders, it was seen that the same re-work, particularly on the welded joints, was required to be carried out to conform to the stipulated standards. The contractors had not provided for the costs necessary to conform to the stringent quality-control requirements. As a result, the schedule of progress could not be kept up due to financial constraints on the part of the contractors. The project schedule for the commissioning was affected as well. Thus, a proposal for financial incentives to the contractors was mooted at the site. The DAE's approval of the proposal was necessary since such payments were outside the scope of the contract. This was duly obtained with incentive payment for earlier completion. As a result, the work picked up tempo and began to conform to the schedule drawn up.

The cost of the erection work, carried out by Dodsals for RAPP 1, was about Rs. 150 lakhs. A normal period of three years, from the date of awarding of the piping contract to the date of the first criticality, is normally provided for the company but Dodsals completed their contract in four years. This time over-run did not significantly affect the date of first criticality for RAPP 1 since there were similar over-runs in other site work systems as well.

7.23   The electrical system works

The philosophy regarding the planning and carrying out of electrical system work was decided upon at the very initial stages, as part of the overall strategy for the construction and commissioning of the project. This philosophy is detailed in the upcoming sections.

7.23.1   The erection of transformers

All the large-capacity transformers, like the Generator Transformers (GTs) and Unit Transformers (UTs), were to be erected and commissioned departmentally. The erection of auxiliary station service transformers and rectifier transformers was to be included in the electrical system erection contract. The erection of start-up transformers (SUTs) was included in the 220-KV switchyard erection contract.

7.23.2   The erection of other electrical equipment

The erection of all other electricals was included in the electrical system erection contract. Some of the electrical equipment are as follows: 21-KV isolated phase bus duct; PT cubicles; Neutral Grounding Transformers (NGTs); lighting and receptacles system; conduit system; cable pans; HV and LV cabling and cable terminations; 3.3-KV, 415-volt and 258-volt DG switchgear; 415-volt Motor Control Centres and Load Centres; control cabling; control cable terminations at electrical equipment (ends only), etc.

The other items included in the erection contract were the erection of control panels for the 415-volt DG Sets and 415-V Class-III Motor Generator sets; fire alarm junction boxes; smoke and fire detectors; aluminium and copper conduits; pyrotenax cables; gas-proof cable penetrations through the Reactor Building for power and central cables (about 500 in number); the splicing of 33-KV PILC cables to PVC cables and a grounding system for cable pans only (other grounding systems were carried out departmentally).

7.23.3 The planning of work

The commissioning of all electrical Systems would be carried out departmentally and the electrical erection contractors would render all assistance to the department in the commissioning of the erection work carried out by them. All the systems in a Nuclear Plant are remote-controlled from the Main Control Panel in the Control Room. Several interlocks and safety features were built into the Control System. This was the reason for the department assuming all commissioning responsibilities.

7.24 Electrical system erection contract

The contract for RAPP 1 was awarded to Western India Erectors at a tender cost of Rs. 8 lakhs during 1967-1968. The provision for additional and extra items of work was built into the contract. The cost of the work, as finally carried out, was Rs. 10 lakhs. The contract was time over-run but this did not affect the commissioning of the project in any way.

7.25 Instrumentation work

These works were carried out entirely by the different departments. The testing out and commissioning of all field instrumentation panels, connections of control cables at the Control Distribution Frame, data logger and annunciation system panels and the Main Control Panel were some of the most important works carried out by the instrumentation group of the project.

The extent of indigenisation of instrumentation equipment was above 40% for RAPP units 1 and 2.

7.26 Nuclear equipment works

Cut-Away view of Reactor

7.26.1 The Reactor and Plant Group

These works were carried out entirely by the various departments of the Reactor and Plant group. They comprised the Calandria (reactor pressure vessel), reactor end-shields, reactivity control systems, spent-fuel discharge and transportation systems, pressure tubes, end-fittings, end closure systems, the shield tank system, reactor building air-lock systems, Biological and Thermal Shield systems, dump tank, moderator and ion-exchange systems, break-out panel system, slow and fast pump-up systems (of heavy water), etc. These constituted the most crucial and, at the same time, the most demanding of quality-control requirements of all the Plant systems.

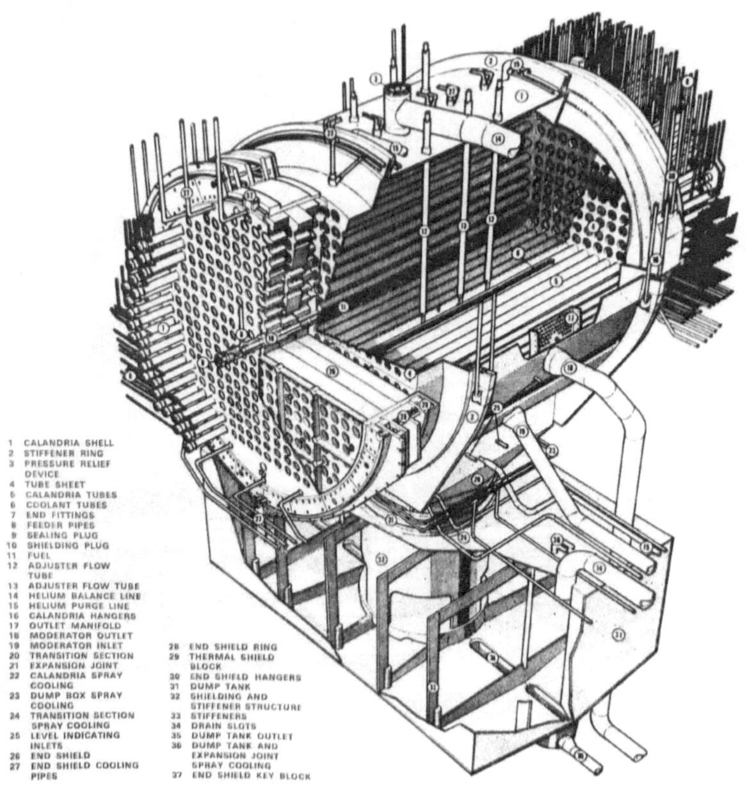

7.26.2 Fuelling Machine/Fuel Handling work

These works were also carried out by the different departments of the Fuelling Machine group. The work carried out, under the aegis of their group, comprised the erection and commissioning of the re-fuelling machine with the drive mechanism fuelling machine vault door system; fuelling machine hydraulic drive system, with its control of the fuel loading; spent-fuel ejection system, spent-fuel storage system, etc. Photo nos. P-13 and P-13(a) show the fuelling machine mechanism in detail.

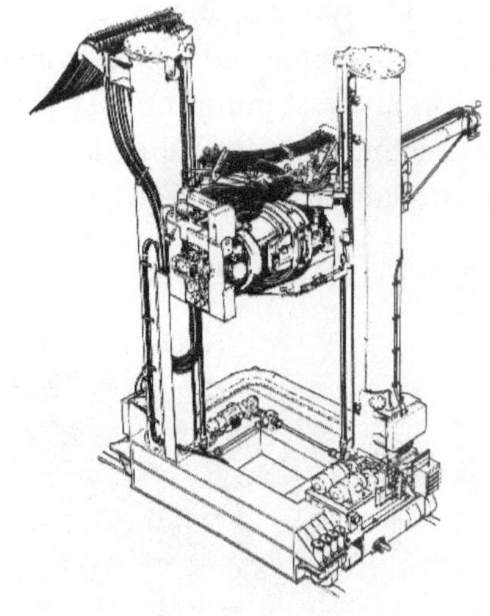

Fuelling Machine.

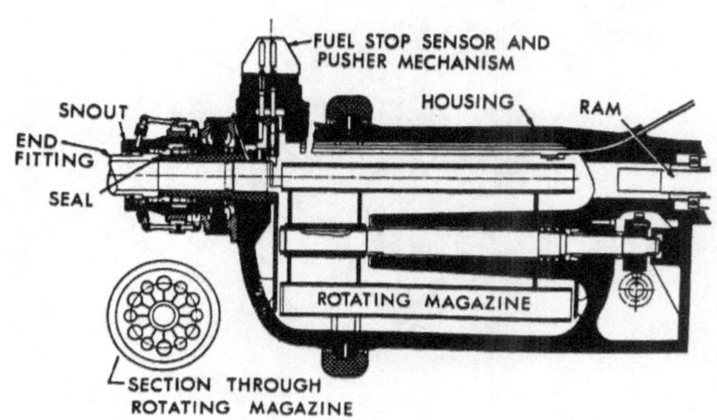

Fuelling Machine head, schematic view

7.27 Quality Control Work

7.27.1 General

These works comprise quality control and surveillance over every discipline of the work carried out at the site. The work on the concrete control laboratory was finished first, as it established the work requirements for all aspects of the Civil work and conducted tests over the specimens of the re-inforced cement concrete and other Civil Engineering work carried out at the sites. The scope of the quality-control work, regarding the various disciplines, will be explained in the upcoming sections.

7.27.2 Inspection of materials and equipment

The inspection of materials, equipment and other parts—concerning the specifications stipulated in the purchase order and relevant contract documents—is carried out as soon as the intimation is received from the stores.

7.27.3 Inspection Acceptance Report

The inspection and acceptance report is issued after the inspection is completed. The mention of re-work and other details thereof are mentioned in the report. The test certificates of the equipment are to be perused and comments, if any, are to be included in the report.

7.27.4 Quality-control association with the completion of work and the readiness for commissioning

Representatives of the quality-control section are to be associated closely with the erection of the work, falling under the scope of the various disciplines and their concurrence of the completion of the erection and release for the commissioning of the relevant work.

7.28 The erection of the TG sets and the TG set auxiliary systems

7.28.1 General: the contract

The contract for the erection and commissioning of the TG set system and its auxiliary system was awarded to John Inglis, contractors for the manufacture and supply of GEC, Canada. The TG set and its auxiliaries cost 3,54,130 Canadian dollars (approximately Rs. 25

lakhs as per the 1964 exchange rate). The living costs in India of John Inglis and the GEC Canada personnel were taken care of with a payment of Rs. 7,18,700 in Indian currency. The contract provided for the purchaser (Department of Atomic Energy) to supply all the skilled, semi-skilled and unskilled labour, free of charge, to John Inglis/GEC, Canada.

7.28.2   Terminal Points

The terminal points for the TG set system work are explained in detail in the upcoming sections.

7.28.2.1   *The Electrical System*

The electrical system consists of six terminals of the Turbo-generators, two terminals of the shaft-driven exciter, exciter amplidyne panels and the local TG supervisory panel.

7.28.2.2   *The Mechanical System*

This system comprises the exhaust glands of the turbine with the neck of the condenser, HP turbine glands, lube oil storage tank, centrifuge, lube oil purification panel, bled steam re-heater and the piping governing system.

7.28.3   Receipt of equipment for the site and the erection programme

The various components of the TG system started arriving at the site by the end of 1967, after due inspection and clearance by the quality-control representatives of the MECo in Canada. The components were stored on the 1182-metre elevation operating floor of the Turbine Building (TB). The components were released for erection after necessary inspection by the TG erection group of RAPP, the local representative of John Inglis and the quality-control group of RAPP.

One of the most critical erection works was the 'threading' (exact positioning) of the 30-tonne rotor of the TG set into the annular portion of the stator winding. The rotor was 6 metres long and the process was difficult, but the high-tech job of 'threading' was successfully carried out in the end.

7.28.4   Completion contract: mothballing

7.28.4.1   The Contract completion

The erection job was completed by early 1970. Steam from the auxiliary boiler house, erected specially for the purpose, was made available to roll the turbine. The turning gear system was tested for satisfactory operation.

7.28.4.2   Mothballing

7.28.4.2.1   General

Steam from the steam generators of the reactor was yet to be made available for running the TG on load. Mothballing (preservation for quite a period) of the TG set was hence required to be carried out. Details of the mothballing carried out are provided in the upcoming sections.

7.28.4.2.2   Turbine and blades

The shaft of the turbine and the blades of the turbines were given an application of protective coating. The generator was filled with Nitrogen at a pressure of 0.75 to 10 psig.

7.28.4.2.3   The Turbine parts

The air-tightness of oil had to be ensured in the oil system of the turbine as part of preservation. It took one month to achieve this process. The preservative coating had to be applied to the TG system equipment which was done as follows:

a)   Tecyl 506 for turbine diaphragms

b)   Tendyl 610 for re-heaters, separators, etc.,

The electrical turning gear was to be run twice a month (i.e. two days) with jacking oil pumped for 8 hours per day. The oil purifier was to be run for six hours every fifteen days.

The costs of mothballing according to the 1970 prices were as follows:

(1) RAPP 1

Canadian dollars: 40,000 (approx. Rs. 3 lakhs)

Indian Rupees: Rs. 2 lakhs

(2) RAPP 2

Canadian dollars: 28,000 (approx. Rs. 2 lakhs)

Indian Rupees: Rs. 55,000

7.29   The erection of the RAPP 1 condenser

The work involved in the supply and erection of the condenser for the Turbo Generator (TG) set of RAPP was carried out by BHEL at a cost of about Rs. 60 lakhs. The free-supply component to BHEL was 2,50,000 (Rs. 19 lakhs approx.) in Canadian dollars and Rs. 36 lakhs in Indian rupees. The spring-loaded support system for the condenser was also supplied and erected by BHEL. The condenser was a 20,446-sq. metre twin-pass single-shell twin-nest equipment.

7.30   The main feed-water system of RAPP 1

The system consisted of six re-heaters, including the de-aerated re-heater. The erection was carried out by various agencies like the manufacturers (the IAEC), the piping contractors (Dodsals) and others. The total cost of the erection of the system was Rs. 55 lakhs. The cost of the free-supply components was 4 lakhs (Rs. 32 lakhs) in Canadian dollars and Rs. 21 lakhs in Indian rupees.

7.31   The Nuclear fuel of RAPP 1

About 50% of the critical charge of the nuclear fuel for RAPP 1 was obtained from Canada. The remaining 50% of the fuel for RAPP 1 was fabricated at the nuclear fuel complex (NFC) in Hyderabad. The entire charge of nuclear fuel for RAPP 2 was fabricated at NFC, Hyderabad.

7.32   The progress of work of various agencies: the crucial role of piping contractors and the Reactor and Plant work

The workshops of the piping, electrical and instrumentation groups were spread throughout the plant—whether these be in the reactor building, turbine building, service building or elsewhere. These workshops needed to be closely co-ordinated on a shift-to-shift basis throughout the duration of these works, over 36 to 48 months.

The piping contractor Dodsals had sound financial resources. They were well-experienced in the project work; their exposure to nuclear project work was, however, scarce and hence the RAPP project was their first such work. They were well-organised and had enormous industrial clout. It was a general observation at the site, that it was difficult to get the better of them while over-seeing their work. Sri. V. Surya Rao, CPE of RAPP, once jocularly observed: 'The only defeat of Dodsals was on the RAPP cricket field when their team was defeated by the RAPP cricket team' (the author of this book was the Captain of the RAPP cricket team during the period 1965-1975 when he served at the RAPP site).

Dodsals had staff that could accustom themselves to the rigorous quality-control requirements of the work more or less quickly. This was very much reflected in the firm being able to adhere to the scheduled progress during the later stages of the completion of the piping work.

The Electrical and Instrumentation work were more conventional. The exposure to the quality-control requirements have been more wide-spread and the quantum of departmental work in these disciplines have been considerable; as a result, it was easier to control the progress of work and to deploy the required flexibility to adopt site resources by way of men and materials to suit the project schedule. This requirement was established at an early stage of the project, subject to the availability of terminal points for the work and the receipt of design information from the consultants.

The same could be said for the Reactor and Plant Group under the aegis of which the erection of the Reactor and allied Plant equipment was carried out entirely through the various departments. The restraint in the completion of these works was imposed by the delayed supply of quite a few crucial nuclear plant equipment and components, the delay in the receipt of design information from consultants, etc.

7.33   Pre-commissioning and progress towards commissioning

7.33.1   The 220-KV Grid Supply

The 220-KV power-supply from the RSEB/Northern Grid required for commissioning purposes was made available by the middle of 1969. The lighting system of the plant was the very first permanent plant system to be commissioned. This was soon followed by the other systems.

7.33.2   Commissioning of the sub-systems: RAPP 1

The pump house water system was the next system to be commissioned. This system would provide cooling water for the condenser circulating system and the low-pressure and high-pressure process water required for the cooling water systems of the nuclear equipment in the Reactor building. The processed air system and ventilation system for the plant buildings followed suit. The water treatment plant, to provide de-mineralised water for the boiler water system, was the next system to be commissioned.

7.33.3   TG set Rolling: RAPP 1

One of the milestones towards the commissioning of the TG system was achieved when the turbine was rolled by utilising steam from the auxiliary boiler house, which was established for the very purpose.

7.33.4   Pressurising the PHT System

Another milestone was the pressurising of the primary heat transport system by using light water in place of heavy water, and also of importance was the testing out of the pressure relief interceptor and throttle valves of the main turbine steam system by using the heat of the primary heat transport pumps. This was carried out before nuclear fuel loading and hence the availability of nuclear heat. The exercise was a big march towards the first-criticality.

7.33.5   Filling of heavy water: RAPP 1

An inventory of 230 tonnes of heavy water—200 tonnes of 90% isotopic purity for the PHT system and 30 tonnes of 99% isotopic purity for the moderator system—is required for each Reactor unit.

Quite some time was taken in building up this inventory of heavy water for RAPP 1 and filling the same in the two systems.

7.33.6 Fuel loading: RAPP 1

Once all the systems came online, fuel loading was carried out. The nuclear fuel was in the form of uranium dioxide pellets clad in zircaloy. The pellet assemblies were contained in a cylindrical element (bundle) that was 495 mm long and 86mm in diameter. Each element weighs 15 kg and twelve such fuel bundles were loaded in each of the 306 pressure tubes of the reactor. Out of the 3672 fuel bundles required for RAPP 1, 50% or 1836 bundles weighing about 23 metric tonnes were manufactured at the nuclear fuel complex in Hyderabad. The other 50% (1836 bundles) were imported from Canada. The entire nuclear fuel, required for all the subsequent heavy-water cooled and moderated reactors commencing from RAPP 2, was manufactured at the NFC.

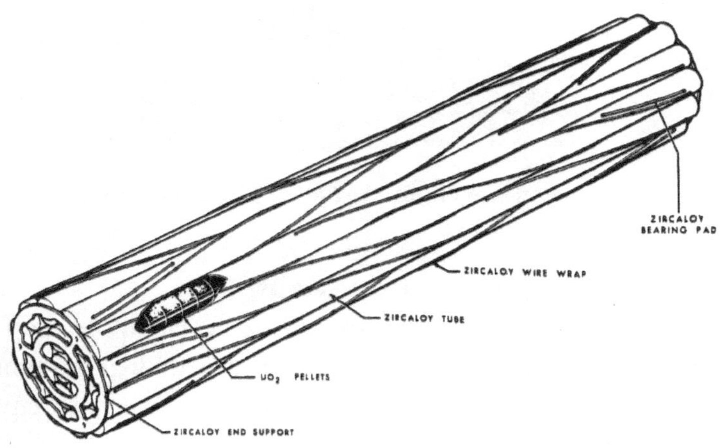

Fuel Bundle

Fuel bundle fabrication at the Bhabha Atomic Research Centre, Trombay.

7.34 The first approach to criticality

All eyes were now turned towards the achievement of the first-criticality. The Health Physics staff were ready with their monitoring of radiation counts, neutron power and other measurements necessary for keeping a pulse on the health and integrity of the Reactor safety systems. All the adjuster rods were fully out and Boron was added to the moderator system. Heavy water was gradually filled into the moderator.

7.35 Attainment of first-criticality

It was indeed a red-letter day for everyone involved in the RAPP project, when RAPP 1 became critical at 10 hours 58 minutes 10 seconds on 11.8.1982 at a moderator level of 215.1 cm, with the radiation counter indicating 19,200 counts moment. The involved personnel had been looking forward to this phenomenal event ever since the construction work had started at the RAPP site on 28.12.1964.

Many dedicated individuals have contributed their might to this significant development in the annals of nuclear power production in India. This was a small step in the history of global nuclear power production, but a giant step indeed for the Indian nuclear power production programme. There was cheering among the Engineers and staff gathered in the control room and the passing of mutual congratulations among them.

7.36 The synchronisation with the Northern Regional Grid: declaration of commercial operation

RAPP 1 was duly synchronised with the Northern Regional Grid after checking out the live testing of the various systems including the emergency transfer (of power-supply) system. The unit was declared commercially operable on 16.12.1973.

7.37 The erection work of RAPP units 1 and 2

7.37.1 The Civil Works

As discussed earlier, HCC was awarded the Civil Engineering work for RAPP 1. Weathered sandstone laterite was used as the sub-soil at the RAPP site, and controlled blasting was carried out wherever the same was required. Line drilling, using the wagon drilling method, was done wherever the depth of rock had to be excavated. Controlled blasting was carried out after wagon drilling to the required depth had been completed. This was quite an innovative piece of Civil Engineering work which contributed to quick and efficient excavation work.

Blasting work was required to be carried out extensively at the Reactor sites of both units. Since even-controlled blasting would affect the foundation, was all blasting work needed to be carried out before the foundation work for the Plant buildings had commenced. HCC was already working on RAPP 1, so it was found expedient to award the excavation work for RAPP 2 also to the same company. This would save valuable time and would also ensure that all excavation work involving blasting was completed before the commencement of the foundation work.

Extensive blasting work was also required for the extension of 220-KV switchyard bays to accommodate the RAPP 2 portion of the switchyard. This work was also carried out by HCC before the commencement of concrete footing work for the structures of the switchyard.

Negotiations were subsequently carried out with HCC, and the contract for the remaining Civil Engineering work for RAPP 2 was also given to them.

7.37.2   The piping, electrical and instrumentation work: RAPP 2

The piping work for RAPP 2 was awarded to Dodsals after the completion of negotiations with them about the rates, completion time and other conditions of the contract. The cost of piping work carried out by Dodsals for RAPP 2 worked out to be Rs. 180 lakhs.

The reactor plant and instrumentation work for RAPP 2 were carried out departmentally and were completed in time for the commissioning of RAPP 2.

The electrical work for the RAPP 2 unit was awarded to Bombay Suburban Electricity Supply Ltd. after the receipt and scrutiny of offers against the work put to tender at an estimated cost of Rs. 13.57 lakhs during the late 1972. The contractors for RAPP 1 had completed their job, demobilised their staff and set up the site. It was hence found advantageous to put the electrical work for RAPP 2 to open tender.

7.37.3   The philosophy of approach towards carrying out the erection work of RAPP 2

The basic philosophy of approach in awarding the erection work was that it was preferable to negotiate with the contract agencies, involved in the erection of RAPP 1, about those cases where the track record regarding the carrying out and completion of the work was good, in terms of the Civil work, piping work, etc. This approach was sound in the case of work of large magnitude such as Civil work and piping work; the mobilisation of resources such as manpower and equipment; the setting up of the site erection facilities, etc. which altogether take 6 to 9 months.

Quite some time is taken by the erection agencies to get the feel of the quality-control requirements governing the erection work. Thus, 12 to 15 months could be saved if the contract is awarded to the agencies already working at the site. This is of course only one aspect of the approach to the planning of the erection work. Other pertinent aspects also need to be considered and provided for before finalising the approach for the awarding of such erection work of the site.

The contractors for the erection of the RAPP 1 electricals had completed their work and demobilised their staff and erection facilities at the site. Open tenders for the electrical erection work of RAPP 2 were given out as already discussed in para 7.37.2.

7.37.4   The TG erection work of RAPP 2

These were carried out by GEC Canada. The erection work was completed and the TG set was kept in mothballed completion, as was done for RAPP 1 earlier, until the attainment of criticality of RAPP 2.

7.37.5   Other erection work of RAPP 2

All the other erection work for RAPP 2 was completed. All the systems were commissioned and kept in 'poised' condition waiting for the criticality of RAPP 2 to be attained.

7.37.6   The delay in building up the heavy-water inventory

In para 7.33.5, we have seen that an inventory of nearly 230 tonnes of heavy water had to be built up for the commissioning of each unit of a 235-MWe heavy-water moderated and cooled Nuclear Power Reactor. Very few countries in the world, including India, produce heavy water. Canada and Soviet Russia are the major producers of heavy water. Other countries such as Norway and Sweden produce very low quantities of heavy water (D_2O) and these quantities are of no significance for commercial supply to nuclear reactors.

The heavy-water production unit at Nangal produces about 15 metric tonnes of nuclear-grade heavy-water per year—the production depends on the availability of Ammonia from the

Nangal fertilizer factory. Ammonia is an input feedstock for the Nangal heavy-water unit. The Nangal unit has been in production since the mid/late 1950s. The 60-tonne capacity Baroda heavy-water production unit is also utilising Ammonia as feedstock and has been in production since the mid-1960s. But this plant has had teething problems. An explosion in an Ammonia tank of the plant in the early 1970s severely affected the production for some time. Thus, the indigenous production of heavy water was insufficient for filling up the PHT and moderator circuits. Hence, an agreement was signed with Soviet Russia for the import, on lease use and later return, of 200 tonnes of D2O. All the nuclear inspection requirements of the IAEA were accepted by India for this import.

7.37.7 Readiness for Criticality

The stage was now set up for RAPP 1 to take its first step towards criticality.

7.37.8 First criticality: synchronisation with the NREB: commercial operation

7.37.8.1 First criticality

Around 6.43 ppm of Boron was added to the moderator system before taking the first step for criticality. There were 656 depleted nuclear fuel elements used against the natural uranium fuel used for RAPP 1. This was to flatten the positive reactivity curve of the fuel core and hence ensure uniform nuclear heat generation. RAPP 2 went critical at 14:14 hours on 08.10.1980 at a moderator system level of 305.5 cm.

7.37.8.2 Synchronisation: commercial operation

The unit was synchronised with the Natural Resources and Environment Board (NREB) on 1.11.1980 and declared for commercial operation on 01.04.1981.

7.37.8.3 The evacuation of power from RAPP units 1 and 2

The 400 MWe of power generated by RAPP 1 and that which is to be generated by RAPP 2 will be evacuated through the following 220-KV EHV lines:

a) 220-KV double-circuit Kota-1 and Kota-2 lines to the Industrial area 220-KV grid sub-station. This area is steadily growing in industrial and manufacturing activity. The DC lines will cater to this.

b) 220-K SC-line Kota-3 to the grid sub-station in the city to cater to the increasing needs of the expanding city.

c) 220-K S-line to Udaipur. This line will meet the city's growing power needs.

7.37.9 Time of completion of RAPP 2: reasons for delay

The civil work for RAPP 2 commenced in 1967. The main reason for commencing the work at this point in time was to ensure that all the blasting work, required for the excavation of hard rock, was completed before concreting and commencing the reactor and other buildings of RAPP 1. Around 60% of all the Civil work of RAPP 2 was completed by 1970. The erection work for various systems was awarded and the work commenced in 1972-1973. All the erection work was completed in 1978. Peak construction work for RAPP 2 lasted for about six-and-a-half years, a period that was more or less the same for RAPP 1. The intervening period from 1978 to the first-criticality date of 8.10.1980 was due to the arrangement of necessary inventory for heavy water which had to be imported from Russia.

7.37.10 Preciousness of heavy water: sources of leakage

Heavy water is a very precious ingredient in a HWPT reactor. According to the 1980 prices, it cost about Rs. 10,000 to produce 1 kg of heavy water. It is hence essential to control and conserve the inventory of heavy water in a HWPT reactor plant. Heavy water may escape or leak in the following ways:

1) The PHT system

2) The Moderator system

3) During re-fuelling operations

4) Mechanical seals of pumps, gland packing of valves and screwed tube fittings

5) Gaskets between the flanges

6) Cracks in pipes and heat-exchangers

7) Worn-out or cracked components like elastomers, rubber diaphragms and 'o' rings

7.37.11 Heavy-water recovery/upgradation

7.37.11.1 General

All methods are adopted to recover heavy water in vapour form through desiccant driers. Accidental spillages are recovered by the vacuum mopping system.

7.37.11.2 Rate of recovery of D2O

The escape rate of heavy water was seen to be 200 kgs per day per reactor. It was found possible to recover about 180 kgs of heavy water per day per reactor. The loss rate was thus 20 kgs of D2O per day per reactor. Efforts were continuously made to reduce this loss in spite of many constraints regarding further drastic improvement.

7.37.12 The D2O-upgrading plants at RAPPs

Four distillation-type heavy-water upgrading plants have been erected at the RAPP site to upgrade the recovered heavy water of varying purity to PHT grade (90% isotopic purity) and moderator grade (above 99% isotopic purity) of heavy water. These plants are capable of upgrading about 2500 kgs per day of degraded heavy water. These upgrading plants go a long way in ensuring the economic inventory of heavy water which is very much essential for the commercial operation of HWPTRs.

7.37.13 Other facilities at the RAPP site

7.37.13.1 The Meteorological Tower (Met Tower)

A Met tower, 137 metres high, has been erected at the RAPP site. The facility associated with the towers has the means of measuring wind velocities, directions of air current, air drifts, air pressure measurement and connected physical measurements of ambient temperature at RAPPs. This facility supplies vital inputs to the station authorities.

7.37.13.2 Environmental Survey Laboratory

An environmental survey laboratory has been established at Rawatbhata. This laboratory is responsible for establishing and recording all radiation dose-measuring values of all the facilities at RAPPs and the surrounding environmental areas. Radiation doses in livestock, pasture, and water sources—such as the RAPPs lake, potable water-supply streams, water-run off, among others—are routinely monitored, measured and recorded at this lab. Routine radiation dose measurements of the public, ambient air and other natural sources, as dictated by public safety requirements from the radiation dose safety point-of-view, are taken and recorded.

7.37.13.3 *Cobalt rods facility*

A facility for the complete monitoring and maintenance of cobalt rods, which are used as control rods for the control and regulation of reactor power, has been established at the RAPP site.

7.37.13.4 *The proposed resin fixation facility*

In 1970, a spent-resin fixation facility was being engineered to fix the radioactive resin in the solid form. This was to ensure that no radioactivity would leak out from the spent-resin. These resins result from the heat exchangers of the moderator system in a nuclear power reactor. This facility aimed to ensure that no leakage of radioactivity occurred in the nuclear power stations. The facility was later constructed and has been in operation since.

7.37.13.5 *Cost escalation of the RAPP units*

7.37.13.5.1 RAPP 1

The cost of RAPP 1, as approved financially in 1963, was Rs. 37 crores. The as-built cost, according to the 1980 prices, was Rs. 73.27 crores with a foreign exchange component of Rs. 33.36 crores. The final as-built cost included the cost of required modifications carried out after the commissioning of the plant. The retro-fitting was dictated by further design requirements which turned out to be quite extensive. This was the principal reason for the cost over-run of almost 100%.

7.37.13.5.2 RAPP 2

In 1967, the cost of RAPP 2 was Rs. 55 crores. The as-built cost, in 1982, was Rs. 102.54 crores with a foreign exchange component of 25.43 crores. The cost escalation was 86% and the reasons were the same as that of the RAPP 1 unit.

7.37.16 Revenue Generation

RAPP units 1 and 2 have together generated over 34 billion (34,000 mega units of electricity) till the end of 2003. They generated 39.56 billion as of June 2008. The NPCIL had charged the RSEB (part of the Northern regional electricity grid) a rate of 51 paise per unit when RAPP 1 began commercial operation. Both units have generated Rs. 1,730 crores of revenue as of 2003. This is nearly 10 times the combined cost of RAPP units 1 and 2.

7.37.17 Capacity Factor (CF)

CF has shown distinct improvement over the years. RAPP 2 has had CF between 78.91% and 91.36% during the years 2001-2004.

7.37.18 Problems encountered during the commissioning and operation: adopted solutions

7.37.18.1 Heat-exchanger failures

Several heat-exchangers in RAPP 1 and RAPP 2 had failed during commissioning and operation. These heat-exchangers have heavy water on the tube side and light water on the shell side. The failures were caused by flow-induced vibrations; corrosion/pitting caused by selective oxidation due to stagnant process water on the shell side, during and before commissioning; erosion and fretting of tubes near baffles and defects in manufacturing, design and other material.

Several tubes in the shut-down cooling system heat-exchangers had to be plugged. RAPP 2 was derated to 100-120 MWe for a year till the plugged heat-exchangers were replaced with new ones. Various design and manufacturing improvements were carried out in the new heat-exchangers to ensure satisfactory operation.

7.37.18.2 TG set bearings: RAPP 1

The main shaft bearings of RAPP 1 registered high temperatures during the first six months of operation. The unit had to be shut down to open up the bearing seats for inspection of the same. Extensive scoring of Babbitt material was observed. Loose lumps of foundry sand were found sticking to the inside body of the governor valve castings. The oil system had to be flushed thoroughly for several days as well. The load-bearing areas of the bearings were increased by 20%. These modifications were made in the TG set of RAPP 2 during ASP (annual shut-down period). No bearing problem has since been encountered either in RAPP 1 or RAPP 2.

7.37.18.3 *End-shield failure and repair work*

When RAPP 1 was shut down in September 1981, there was a leakage of heavy water from the F-10 lattice tube position into the adjacent areas of the fuelling machine vault and the Calandria vault. The reactor cooling was stopped, and the heavy water in the PHT system was drained up to the headers and the F-10 channel was also drained. The channel was isolated and made free from the rest of the devices. Remotely operated cutting tools were also developed.

The leakage was confirmed to be from under the bearing sleeve of the F-10 lattice tube. The bearing sleeve was cut in four segments using the spark erosion technique and remotely operated tools. Further inspection revealed that two longitudinal cracks were running parallel to the centre line of the F-10 lattice tube. This necessitated the removal of the lattice tube. It was later seen that cracks had appeared in a few other areas of the end-shield as well. The identification of the leakage spots, the development of removal and repair techniques and the completion of repairs took nearly 3 to 4 years. Thus, the RAPP 1 unit had to be taken out of service for almost 4 years.

7.37.18.4 *The TG Rotor blade failures*

Instances of blade failures in both the LP and HP turbines of RAPP 1 occurred on five occasions between 1972 and 1983. The failures were traced to high vibrations and eccentricity. The reblading was done at BHEL work whenever the failures occurred. Cannibalisation from RAPP 2 was also resorted to keep RAPP 1 going. Modifications to the design of rotor blades were also carried out. A new HP turbine rotor was also ordered from GEC UK and then erected/commissioned at the site. With the improvements carried out, the operation improved considerably.

7.37.18.5   *Problems with the MG sets of RAPP 2*

The originally supplied motor speed controllers did not have enough load-limit control and, hence, had to be replaced by a new controller incorporating better operating characteristics.

An AC voltage controller of the generator showed problems in the flashing circuit of the field and the voltage regulator. Changes were made in the operating voltage of the flasher circuit and the regulator control circuit. These changes improved the performance of the MG sets.

7.37.18.6   *The first full-scale de-contamination of the primary coolant system*

The first full-scale de-contamination of the primary coolant system of a Nuclear Power unit in India was carried out at the RAPP 2 unit on 12.08.1992 and 13.08.1992, respectively.

7.37.18.7   *Examination and replacement of the pressure tubes in the RAPP 1 and RAPP 2 units*

Inter-granular stress-corrosion cracking (IGSC) has been noticed in the pressure tubes of reactors the world over, especially in cases where the reactors had been in operation for about 10 years or more. Given this fact, and also the experience gathered during the operation of RAPP 1 and RAPP 2, the NPCIL decided to examine the pressure tubes of RAPP units 1 and 2. This work was taken up in August 1994. As of 1994, RAPP 2 had completed 8.5 full-power years of operation. A thorough out-of-service examination revealed the need for replacing the zircaloy-2-based pressure tubes. Special remotely handled tools were developed for cutting the tubes which were later replaced.

7.37.18.8   *The replacement of the pressure tubes in RAPP 1 and RAPP 2*

Four tight-fitting garter springs were used in place of the two loose-fit springs between the coolant tube and the Calandric tubes, as was the case earlier. The reduced thickness of the new pressure tubes would result in fuel economy.

In early 1998, repairs to the coolant channels of RAPP 2 were completed and the unit was re-commissioned. Colossal saving in total man-rem for the work was achieved, the count having been 16.8 against an estimated 120. The replacement of the coolant tubes of the RAPP units was carried out during the long upgrade shutdown between 01.05.2002 and early February 2004. The unit has been operating at 100 MWe re-defined capacity rating since then.

The cost of the removal and replacement work in both the RAPP 1 and RAPP 2 units, complete with all materials, tools and table, was estimated to be Rs. 250 crores.

7.37.18.9   RAPP 1

7.37.18.9.1   The elimination of leakage from the over-pressure relief system and the coolant channel

The RAPP units faced the problem of leakage from the over-pressure system in 1994. The seal of the system was damaged due to corrosive radioactive products in the Calandria; the damage went on to cause minor leakage of helium and heavy water. The repair work was taken up in February-March, 1997. A cast in-situ metal seal made out of 99.8% pure Indium, a rare metal with low melting temperature, was developed by the National Fuel Complex (NFC) at Hyderabad to plug the leakage from the over-pressure relief device. The seal was developed over 6 months, with the RAPP repair team working day and night on the job. The cost of the entire work, including the development of remote working and maintaining tools, was Rs. 1 lakh.

The leakage from eight coolant tubes was also noticed. Six of the damaged tubes were replaced; special gaskets, designed and developed by the NFC, were inserted into the above eight channels with the aid of robots. The leakages were plugged into all the channels. The repair work cost about Rs. 20 crores. This unit was restored to service with the capacity limited to 100 MWe.

The output was later raised to 150 MWe. The operation at this level will be stabilized for 6 months, after which the capacity will be raised to 180 MWe. It is interesting to observe, at this stage, that foreign nuclear scientists had suggested the de-commissioning of the Rajasthan reactors on the grounds of the irreparability of the coolant channels. A Canadian team had estimated the cost of repairs at nearly $1 billion and had also stipulated politically unacceptable conditions for undertaking the repairs. Some of these conditions were:

(1) Full-scope safeguards on the part of India
(2) Search and pursue clause for the nuclear materials to be supplied by Canada

Despite such problems, Indian Engineers completed the repairs by fully utilizing indigenous 'know-how' and technology at a cost that is only 10% of the cost quoted by the Canadian scientists.

7.37.19 Further performance improvement of RAPP 2

7.37.19.1 The en-masse feeder tube replacement

The RAPP 2 unit was shut down to replace the en-masse PHT feeder tube and improve the performance grades with indigenously developed technology, tools and tackles. All safety aspects were taken care of during this work.

The work was completed well within the estimated cost. The unit was re-synchronised with the Northern Regional Electricity Board (NREB) grid at 10:22 hours on 10.09.2009. The unit has since been operating at the full-power level of 200 MWe.

7.37.19.2 Highlight of the operation of RAPP 2

The highlight of the operation of RAPP 2 was the generation of more than 5 mega units of electricity on 31.1.1987. This happened for the first time at the site, without taking into consideration the steam supply to HWPK (Heavy-water plant Kota).

7.37.19.3 Overview of the execution of the RAPP 1 and RAPP 2 units

For those of us who were privileged to have been associated with the erection and commissioning of RAPP 1 and RAPP 2, this association was one of a lifetime. RAPP 1 was the first Nuclear Power Project, the erection of which was the complete responsibility of Indian Engineers. The Engineers felt privileged to be involved in the project and shared an added sense of responsibility to the legacy of Pandit Jawaharlal Nehru and Dr. H.J. Bhabha, who had the sagacity and foresight to put India on the path of this frontier technology. The Indian Engineers had thus taken full advantage of this exposure.

As we have seen in the previous chapters, several steps have been taken to ensure the quality of work and the adherence to safety standards. Indian Engineers had taken over the entire responsibility of commissioning the RAPP 2 units from the Engineers at MEIL and Ontario Hydro.

The RAPP units 1 and 2 had successfully been commissioned despite the sanctions imposed by the Governments of the United States and Canada, following the peaceful nuclear explosion (PNE) carried out by India on 26.5.1974.

7.38   The scope of indigenous supply

It would be of interest to mention here that the scope of supply of indigenous equipment and materials was 68% for RAPP 1 and 80% for RAPP 2. It is to be noted that the scope of indigenous supply for RAPP 2 was much higher than that of RAPP 1, as there was a time delay for such supply, compared to the imported equipment and materials made available for RAPP 1. This problem was solved for later projects by the process of series ordering for long-gestation critical equipment. This ensured the timely receipt of equipment at the site, and hence the timely erection and commissioning of the reactors.

7.39   Staff recruited locally: the quality of work and required training

Most of the staff, from the level of Tradesmen to Asst. Foremen/ Foremen were recruited locally. Tradesmen were put to a trade test before recruitment. It was seen that the tradesmen picked up skills in their particular trades fairly quickly. Some of the work—like earthing joints, using pre-formed moulds, pyrotenax cable joints, among others—was deemed as a specialization that the tradesmen excelled at. Training was necessary for the tradesmen/foremen before taking up these jobs.

With the training imparted, such specialized jobs have since become routine.

7.40 Awards and Certificates for RAPPs

RAPP was the recipient of many awards. Some of these awards are listed below:

(1) ISO-4001 certificate for environmental management in March 2002 for RAPP units 1 and 2

(2) ISO-9001 certificate for the quality management system of its nuclear training centre (NTC)

(3) 'Golden Peacock' National award in the area of environment in 1999; for training in 2001, 2002 and 2003 and quality management in 2003

(4) 'Greentech' award for environment and safety for the years 2001-2002 and 2002-2003

(5) The NPCIL's 'Rajbhasha' shield for excellence in the use of 'Hindi' for official work in 2002-2003

(6) The RAPP site also received the AERB industrial safety award for 2011-2012

7.41 CSR Work in nearby areas of the RAPP site

7.41.1 CSR work

As part of its CSR activities, RAPP has involved itself in the work of education, health and well-being of the people living in nearby villages like Tamlao. The electrification of Tamlao was carried out and link roads were constructed. About 150 solar street lights were installed in 20 Panchayats. Seven-metre-wide cement concrete roads were constructed over an 11-kilometre stretch. Concrete approach roads were constructed from RAPP/HWPK residential area in Rawatbhata to the Charbuja Puliya area.

7.41.2 Health measures

RAPP provides consultation and free medicines to about 2,500 patients from nearby villages every year through mobile medical services within a 16-km radius of Rawatbhata. Emergency medical services are also provided at the RAPP Medical Hospital in Rawatbhata. RAPP also provides financial and infrastructural support to blood donation camps, eye treatment camps and ENT (Ear, Nose, Throat) camps that were organized for the villagers. A free Primary Health Centre (PHC) for the public was established at Rawatbhata which was opened on 29.12.2003 by the Chief Minister of Rajasthan and the Chairman of the AEC. The PHC was part of the referral hospital building constructed by RAPP. A Tata-410 ambulance and a Maruti van were also donated to the hospital.

7.41.3 Provision for drinking water

A drinking water-supply system has been provided for the nearby villages.

7.41.4 Education and Vocational Training

Education is provided to talented children of the villagers at the RAPP CBSE School in Rawatbhata. Vocational training is provided to all categories of students in the RAPP plant. The construction of School buildings and the provision of furniture at these Schools were also made possible by RAPP.

7.41.5 Horticulture and tree planting: bio-diversity

Quality seeds are provided for sowing healthy and high-yield crops. Clean and healthy environmental conditions are maintained by selective tree planting. It will be of interest to nature lovers to learn that 'Gypsy' vultures, which were officially declared to be endangered in India, have been seen at Rawatbhata in recent times.

7.41.6 Further support given to schools

The RAPP CSR unit has been providing material and financial support to schools in the Chittorgarh District. The Minister of Education of the State of Rajasthan has recognized these contributions to primary education by awarding the CSR unit of RAPP with the 'Bhamasha' award, in the form of a trophy, to Sri. J.P. Gupta, Director of the RAPP site on 28.06.2013.

7.41.7 Living amenities for the RAPP employees

7.41.7.1 Living accommodation

We will now look at the amenities that RAPP has been providing for its employees. The station has provided accommodation for its staff and officers at the permanent townships 'Vikram Nagar' and phase-II colony at Rawatbhata. The permanent township provides accommodation for about 300 families while the phase-II colony provides accommodation for about 800 families. About 150 to 200 families have made their accommodation at Rawatbhata and the nearby villages.

7.41.7.2 Transport facility

Bus transport is provided by the station authorities on a subsidized basis. The shuttle bus service to the bazaar at Rawatbhata is provided on Sundays and holidays. Daily bus service to Kota City is also made available to employees and their families on a first-come first-served basis, under nominal charges.

7.41.7.3 Education: common services

A CBSE high school has been established at Rawatbhata. This school caters to the requirements of the HWPK as well.

The costs of common services such as educational facilities, health facilities, mutual accommodation facilities, etc., are shared between RAPP and HWPK.

7.41.7.4 Medical facilities

A 30-bed hospital at Rawatbhata and a peripheral dispensary at the permanent township cater to the medical requirements of the employees and their families.

7.41.7.5 Recreational and Sports Activities

These facilities were provided at the very beginning of construction at the site. The RAPP recreational club was set up in 1965. The club had its temporary buildings at the phase-I and phase-II residential colonies. Tennis courts were provided initially at phase-I and phase-II colonies and later at the permanent township as well.

The sports club building at the permanent township had facilities for indoor games like table tennis and shuttle badminton. A swimming pool was also set up, as part of the club, at the permanent township known as Vikram Sarabhai Nagar.

RAPP organized the all-India Atomic Energy tournament for Kabaddi in November 1992, and table tennis in 1993 at the RAPP site.

Note: The author of this book served as Secretary and Treasurer of the above-mentioned recreational club for several years during his tenure at the RAPP site from 1965-1975. A RAPP cricket team was later formed. The author served as Captain of the Cricket team during his tenure at the RAPP site.

This team became a member of the Kota District cricket association, and later participated in the Kota District championship and scored a few notable victories as well.

7.42   Development activity in Rawatbhata and Kota City

Much visible improvement has been seen in the 'service sector' of Rawatbhata village after the commencement of construction activities at the RAPP site. This sector has grown over the years to cater to the day-to-day living requirements of about 2000 families living in the RAPP and HWPK colonies.

7.42.1   More development activity in Rawatbhata and Kota City

An Arts College and Industrial Training Institute (ITI) have come up at Rawatbhata to cater to the educational requirements of the growing population.

The scarcity of water and power and the lack of other essential infrastructural requirements have come in the way of industrial development at Rawatbhata.

The picture at Kota City is quite different. There is an industrial estate in the city where several industries are located. The availability of enhanced quality nuclear power from RAPP units 1 and 2, thermal power from Kota station and the ready availability of water have contributed to making Kota City the principal industrial city in the State of Rajasthan. The city has also proved to be the training capital of India for IIT students. About 40,000 students stay in the city and study for one or two years and simultaneously take training for IIT entrance examinations. The training industry, by itself, is worth Rs. 400 crores per year.

7.43   Seven cases of radiation-related illnesses in villages near the RAPP site

There have been reports of cases of cancer-related illnesses, deformed births, etc. in some of the villages within a 10-km radius of the RAPP site like Tamlao, Deeppura, Malpura, Bhakshpura and Jharjhani. Tamlao, the village nearest to the RAPP site (about 3 km distance from the station) has a population of about 1000 in 3 hamlets. Some of the non-governmental organizations functioning at Rawatbhata, about 13 km from the RAPP site, are studying these cases. Their conclusions are awaited with interest. The DAE has stated that these cases were studied by their medical teams and that the reported cases of illnesses, in the above-mentioned villages, are not directly attributable to any effect of radiation from the station.

7.44   Further health promotion measures undertaken as part of the CSR during 2016-2017

So far, we have seen the measures taken by RAPP to promote the health of people living within a 15-km radius of the site. These measures received an immense boost with further long-term plans for promoting health in about 20 villages, coming within 15-20 km of the site.

This unique medical support reach is expected to cost Rs. 1.6 crores. The medical set-up will be done through select Non-profit organisations (NGOs) with experience in the field and with active co-ordination with the State Government. The chosen NGOs would be responsible for constructing the proposed primary health centres (PHCs), employing qualified medical personnel and operating the PHCs. These PHCs will operate at pre-fixed hours in the various villages. The RAPP CSR group will oversee their operations. Funds would be made available only after the NGOs start the construction of the PHC buildings and desirable progress is made in the construction. The RAPP CSR group has made a list of medicines, medical equipment and other medical necessities which would have to be made available at all times by the NGOs. It is to be expected that this seeding effort by the RAPP CSR group would, in due course, lead to the upgradation of these medical facilities in these villages. Once the work is complete, it will hopefully be taken over and run by the Medical Dept. of the Govt. of Rajasthan in due course.

7.45 ISBTI (Indian Society for Blood Transfusion and Immunology) award to RAPP

ISBTI, during its 34th National Conference on 20.11.2009 at New Delhi, awarded RAPP the 'ISBTI Appreciation award-2009' in recognition of its vital contribution to the field of blood transfusion, donor motivation and the organization of blood donation camps. The award was received by Dr. Anil Mathur, a pathologist and the doctor in-charge of the blood bank at the RAPP hospital.

7.46 RAPP CSR group: 'Project Nuclear 90'

The RAPP CSR unit along with a social service group called 'Alter Kota' started a unique educational project called 'Nuclear 90' in 2017 with the following objectives:

(1) Identify talented 8th to 10th grade students residing within 16 km from the RAPP site.

(2) Provide a competitive and conducive environment for increasing their knowledge through well-organized educational programmes.

(3) Make them capable of taking various State/National-Level Exams such as NTSE, NSTSE, IRO and NSO.

(4) Skill-development Committee of RAPP (SDC) will implement the above-mentioned initiatives through a project implementation agency (PIA).

(5) Ninety students will be selected for the training programme through an open test conducted by the PIA. Reservation policy as per the State Govt. (Rajasthan) protocol, once directed by the NPCIL, will be adopted for the selection. A waiting list will also be maintained.

(6) The duration of the course is 120 days; classrooms will be made available for 20 days for the preparation for examinations.

(7) The training hours will be from 16:00 hours to 19:30 hours with suitable breaks.

(8) There will be three classes, and a minimum of six faculties will be made available to cover subjects like Physics, Chemistry, Biology, Social Sciences, English language and mental abilities. The PIA will charge Rs. 10000 for students who seek tutelage by the faculty.

(9) The assessment of each student's performance should be sent by the PIA to the NPCIL PR site. The PIA should encourage students to participate in various District, State, National and International competitions.

(10) The examination centre should be made available by the PIA at a central location in Rawatbhata at their own cost.

(11) The venue will be provided by the NPCIL by requesting some of the local schools to make some of their classrooms available for this program.

(12) Three classrooms and one staff room shall be made available to students during the evening hours. Utility and other charges may be of the order of Rs. 500/600 per day.

7.47 Independent assessment and suggestions for the direction of future CSR work

It is estimated that the RAPP site spent about Rs. 50 crores on various developmental and social work in the 136 villages situated within a radius of 16 km from the RAPP site. These fall under the following categories:

(a) Social/Medical/Cultural: Holding medical camps that deal with the ophthalmologic, hearing and general medical conditions of patients; providing spectacles for the needy, minor ear surgeries and wheelchairs for the physically handicapped.

(b) Educational: Providing free scholarships to the meritorious students of higher secondary education schools; blackboards, amplifier equipment and loud-speaker kits for school events; books and educational kits for students; skill-development training for students/general public; building of school structures, separate toilets for girl students, rooms for dining/ preparation of mid-day meals, fans in school/headmaster rooms, etc.

(c) Economic: Providing 'haat' facilities for small traders like vegetable vendors/tailors/day-to-day providers of public needs.

(d) Infrastructural: Providing water-supply, electricity supply, roads, bus shelters, solar and other non-conventional energy sources.

All of the above facilities would have benefited about 2 lakh residents of the 136 villages within a radius of 16 km from the RAPP site.

7.48 The need for the assessment of facilities provided: suggestions for future CSR work

So far, we have explored the considerable CSR work that the RAPP CSR committee has carried out in the nearby villages. These works cover a variety of spheres and have made the villagers more conscious about the needs for improvement in their day-to-day lives. Notwithstanding the above, the RAPP CSR committee has invited tenders from NGOs, professional bodies and even individuals to interact intensively with all sections of the beneficiaries, including those belonging to the SC and ST castes. They wish to come up with more focused measures to improve the quality of life in these villages.

There was no tender fee and no earnest money. The intending contractors would have to set up work sites near the RAPP site at their own cost and further employ their officials to cater to the tendering requirements. There would be no payment; payment would be phased out depending on the progress of work. The estimated cost of work was Rs. 12 lakhs and the expected time of completion was 6 months. This novel method of ascertaining the beneficiaries' views on their development and improvement in the quality of their life speaks of the total commitment on the part of the RAPP CSR committee. They have thus played a very meaningful role in improving the quality of life in the nearby villages.

The Madras Atomic Power Project (MAPP) Units 1 and 2

8.1 The site in general: seismic zone

This project was the second HWPTR that was taken up by the DAE. The site chosen was Kalpakkam, which is at 12.4° North Latitude and 80° East Longitude, located about 105 km from Chennai and almost lying on the shore line of the Bay of Bengal. The nearest town, Chengalpattu is about 30 km from the site. The site lying on the upper Eastern coast of the State of Tamil Nadu is one of great scenic beauty. One could go on gazing for hours at the serene waters of the Bay of Bengal, which at this location is almost unlike its normal state of fretting and fuming.

Tamil Nadu had an installed capacity of about 1.5 GWe in the late 1960s. The State was very badly in need of base-load thermal (or nuclear) power stations to shore up the large preponderance of its Hydro-Electric power stations. The then mix of Thermal and Hydro power stations was in the ratio 1:5, while the ideal mix would have been 3:2. The lead source of coal was 1000 km away. Coal linkages for additional thermal power stations were becoming more and more difficult to tie up. These were the circumstances that went into the selection of Kalpakkam as the site for the second HWPT Nuclear power station in India. The site falls under seismic zone-III (IS 1893-2002). The required safety-related buildings, structures and systems have been designed for zero-period ground acceleration of 0.1 g.

8.2 Site-specific details

8.2.1 Cooling water requirements

The site is very near to the coastline of the Bay of Bengal. Sea water is hence readily available for plant cooling water requirements. The Marine intake and out-fall structure had to be constructed to avail sea water for plant cooling requirements. The pumps and ancillary equipment had to be suitably designed to enable sea water to be used as a cooling water agent.

8.2.2 Drifting of sea sand

The drifting of sand—which at certain depths in the reactor site was of the order of 1 million tonnes per hour—called for the required strengthening of the outer perimeter wall of the reactor. The requirement accordingly called for the strengthening of the surrounding perimeter earth of the reactor containment wall. This was done by way of re-inforced concrete anchors drilled deep and long into the surrounding perimeter earth soil at various locations of the circular perimeter and concrete grouted right through the anchors.

8.3 Design changes from RAPP:

8.3.1 The RB containment building and pressure-relief system

There were a few significant changes in MAPP as compared to RAPP. RAPP had a single perimeter containment wall of RCC with 0.9-metre thickness. The quality of the RAPP RB raft was M21/M20. The MAPP design called for two RB perimeter containment walls, the inner one of RCC with 0.6-metre thickness and the outer of rubble masonry with 0.6-metre thickness. The space between the two walls was 2 metres in width and kept at a slightly negative pressure as compared to the outside atmosphere to ensure that there was no leakage of ambient air inside the above space to the outside atmosphere. If there was any leakage at all, then it would only be from outside air to inside of the reactor thus building ambient atmosphere and not vice-versa. The annular space ambient atmosphere would be vented through the stack.

8.3.2 The suppression pool

Another important design change was the scrapping of the dousing system, consisting of the dousing tank and the dousing distributor piping, adopted at the RAPP site. This was replaced by a pressure suppression chamber located in the basement of the Reactor building. Any explosive radioactive gas formed in the boiler floor would be led through suitably sized piping to the suppression pool which is full of water. The pressure of the explosive gases would be dissipated into the pool water and rented out. The suppression pool would be provided with its own circulation and purification system. The scrapping of the dousing system has resulted in a flatter dome for the reactor building of MAPP. The dome is almost 9 metres lower in height as compared to RAPP. Further details of the dome are as follows:

(a) Design pressure: 1.6 kgm/cm sq.
(b) Test pressure: 1.44 kgm/cm sq.
(c) Seismic zone-III
(d) Seismic coefficient/zero-period seismic coefficient: 0.19
(e) Grade of concrete: M-21
(f) Pre-stressing wires: 12 wires of 8 mm diameter each
(g) Pre-stressing wire capacity: 92 tonnes (UTS)

8.3.3 The 220-KV switchyard

The switchyard would be an indoor one. The change from RAPP was dictated by the humid and salty atmosphere prevalent at the MAPP site. This atmosphere constituted an adverse environmental factor for the switchyard equipment necessitating an indoor housing for the same.

8.4 Construction work

8.4.1 Preliminary work

The enabling works like warehouses, job shacks, concrete control laboratories, etc. commenced in 1967.

8.4.2 Main plant civil work

These were awarded to Engineering Construction Co. (ECC), a subsidiary of Larson and Toubro (L&T).

8.4.3   Other plant erection work

8.4.3.1   *Splitting up of work*

A different strategy was adopted at MAPP regarding the carrying out of some of the major erection work like Electricals. These works were farmed out to different partners concerning separate work like lighting, cable pans, cabling, etc. as and when these floors became available for taking up of the work. While the risk of unsatisfactory performance or other exigencies was spread out by this method, it resulted in greater departmental involvement due to a larger number of contracts being overseen. All in all, the method adopted appears to have been worth it.

8.4.3.2   *General philosophy of approach to erection work*

The philosophy was the same as adopted in RAPP. The scope of the departmental work was also more or less the same as at RAPP.

8.5   Organisational changes between RAPP and MAPP: further progress towards indigenisation and strong heaving of the headquarters technical group

8.5.1   Organisational changes.

We had earlier seen that when the preliminary work on RAPP commenced at the site during 1963-1964, there was only a selective staff of 10 in Bombay to offer organisational support to the RAPP site work. The strength of the headquarters has grown steadily over the years and this strength was about 750 (300 Engineers and 450 supporting staff) during the years 1970-1975. The headquarters was now known by the name Power Projects Engineering Division (PPED). The PPED was formed in 1967 with Sri. H.N. Sethna as its first director. The mandate for the PPED was the provision of design, procurement, construction planning and commissioning back-up for all the Nuclear Power Projects to be set up in the country. The increase in strength of the PPED was matched by improvements in the infrastructural facilities available therein. MAPP could be said to have scored over RAPP as regards the back-up it was able to get from the PPED for project construction and commissioning.

8.5.2 Further progress towards indigenisation

MAPP marked a further stage in the progress towards the indigenisation of technology for the design, construction and commissioning of Nuclear Power Projects. MAPP was the first Nuclear Power Project to employ 100% indigenous technology for the design, construction and commissioning of the project. Indian consultants were also employed for the first time: Tata Consulting Engineers (TCE) for the Nuclear, Electrical and Instrumentation Systems of the Project; and Development Consultants Private Limited (DCPL) for the Civil and Mechanical systems of the project.

8.6 Problems regarding the procurement of critical equipment

Problems were faced regarding the procurement of a few critical equipment from Canada. The Canadian Government imposed a complete ban on the export of nuclear equipment to India, following the Pokhran PNE carried out by India in May 1974. The single biggest lot of equipment thus affected was the PHT pump motor units for MAPP 1 and MAPP 2. The equipment finally had to be ordered from Pump Gerard, France. This switch-over caused by Canada's unilateral action resulted in a severe set-back to the scheduled progress of MAPP units 1 and 2. A delay of over two years was caused in obtaining equipment elsewhere. These pump units were one of the very few equipment that had to be imported for MAPP 1 and MAPP 2. A few other ancillaries and materials, which were not indigenously available, also had to be imported. The extent of indigenous equipment and materials used in MAPP 1 and MAPP 2 was of the order of 80%. This was a very big advancement compared to the corresponding percentages for RAPP 1 and RAPP 2 which were 35% and 60%, respectively.

8.7 Details of supply of the other critical equipment

8.7.1 The 235-MWe TG sets

BHEL built the 235-MWe TG sets and supplied them to MAPP 1 and MAPP 2 for the first time in the history of Indian Nuclear Power Projects.

8.7.2   Coolant tubes and Calandria tubes

The NFC at Hyderabad supplied the zircaloy Calandria tubes and zirconium alloy coolant tubes for MAPP units 1 and 2.

8.8   Staff housing at the MAPP site

MAPP had made a head start over RAPP regarding its plan-of-action for housing the project personnel. It had planned permanent housing at the very beginning of the project. This was an important change; it also avoided the heart-burn that was caused among the early-bird personnel at RAPP by the provision of inadequate and off-norm housing accommodation in the early stages of the project construction. Apart from meeting the essential requirements of reasonably adequate housing accommodation, quite some cost saving was affected by the choice of going in for permanent accommodation at the very incipient stages of the project. A saving of at least one crore rupees would have been effected at MAPP by the avoidance of temporary accommodation for the project personnel.

8.9   Test facility for fuelling machines

A test facility, common for both units of MAPP, was set up. This would be a useful adjustment for the operation and maintenance of the station.

8.10   Mothballing: the inventory of heavy water

Critical equipment like the TG sets had to be mothballed as carried out at RAPP since there was to be a long wait for building up the required inventory of heavy water. This waiting period was almost three years.

8.11   The evacuation of power from MAPP

The following 220-KV lines were provided for evacuating power from MAPP:

1)   220-KV MAPP Chengalpattu double-circuit line
2)   220-KV MAPP Villupuram single-circuit line
3)   220-KV MAPP Arni single-circuit line
4)   A future 220-KV single-circuit line has also been provided

8.12 The completion of the erection work of MAPP 1 and MAPP 2: readiness for criticality

The erection work of MAPP 1 was completed in all respects by 1981. All systems were a 'go' and kept in readiness for the first-criticality.

Similarly, all the erection work for MAPP 2 was completed in all respects by 1982 and all systems were poised for first-criticality.

As seen already in para 8.10, a further period of three years from the state of readiness for criticality was taken for building the required inventory of heavy water and hence for criticality, concerning both MAPP 1 and MAPP 2.

8.13 Pre-commissioning: innovations

8.13.1 The 235-MWe TG Sets

We have seen the need for mothballing in the earlier sections. The generators of the TG sets of MAPP 1 and MAPP 2 were filled with nitrogen under pressure for their preservation. It was found expedient to improve the insulation and polarization index values of the stator windings, as we have seen in para 5.9.1.1. The stator windings are formed by hollow copper tubes through which de-mineralized water is circulated for the cooling of the stator windings. The hollow copper tube conductors were taken advantage of when hot compressed air was blown through these tubes. This procedure enhanced the insulation and PI values of the stator winding.

8.13.2 Positioning of the garter springs

The garter springs are positioned between the coolant tube and the pressure tube. These springs maintain uniform spacing between the above tubes. Precise positioning of the springs is essential for maintaining this uniformity of the spacing during reactor operation. Remote cooling for the locality and, if required, changing the position of the springs was feasible. This proved to be a distinctly useful facility for commissioning, operation and maintenance.

8.14 The D2O upgrading Plant at MAPP

We have seen earlier, that MAPP was provided with four upgrading plants for upgrading the isotropic purity of heavy water mopped up and collected as a result of plant operations. These were of the distillation type. These upgrading plants are a must for being in operation, before criticality, as the leakage of heavy water and its mopping up starts from day 1 (date of first-criticality).

MAPP was provided with an electrolytic-type upgrading plant. Ten electrolytic banks have been installed with 14 cells in each bank. Degraded heavy water mopped up from the plant is mixed with an electrolyte of 5% potassium hydroxide (KOH) and continuously electrolysed. Heavy Hydrogen (H3), Oxygen (O2) and Nitrogen (N2) are the by-products. This gas mixture is burnt in a re-combination burner to yield low-isotropic purity heavy water (D2O) which is recycled till the desired level of isotropic purity of heavy water is reached. KOH is removed by steam evaporators and the purification is done through heat-exchangers.

The desired feed for the plant is 526 metric tonnes (mt) per year of 30% isotopic purity and the end-product is a quantity of 150 mt per year of reactor-grade heavy water (90% IP and above).

During the operation of these electrolytic banks, over-heating of the anode collars of these plants was noticed by the operating personnel. This issue was even brought to the notice of the author during one of his visits to the MAPP site. The problem was that the collars were of heavy-current carrying capacity. This was duly taken up with the suppliers, and the problem was rectified.

8.15   First-criticality of MAPP 1 and MAPP 2 and synchronisation with the grid: commercial operation, energy generated and revenue earned

MAPP 1 went critical on 2.7.1983 and was first synchronized with the Southern Regional Electricity Grid (SREG) on 23.07.1983. It was declared commercially operable on 27.1.1984.

MAPP 2 went critical at 13:25 hours on 12.8.1985 with a Calandria level of 319.75 cm and with 7.29 ppm of Boron in the moderator. It was first synchronized with the SREG on 20.09.1985 and declared commercially operable on 21.3.1986.

Another milestone was reached on the road of the history of Indian Nuclear power projects with the commercial operation of MAPP 1 and MAPP 2.

MAPP 1 generated 22.20 billion units from January 1984 till June 2008 and earned revenue of Rs. 13.30 billion at a sale rate of Rs. 0.60 paise per unit.

Similarly, MAPP 2 generated 22.72 billion units from the period of March 1986 to June 2008, and earned revenue of Rs. 13.62 billion.

8.16   MAPP project costs

The as-built costs of MAPP 1 and MAPP 2 are as shown below:

MAPP 1 : Rs. 119.30 crores; CFE component : Rs. 10.29 crores

MAPP 2 : Rs. 134.84 crores; CFE component : Rs. 12.50 crores

8.17   The period of construction of MAPP 1 and MAPP 2

The preliminary work for MAPP 1 began in 1967. The permanent plant work commenced during 1969-1970. The first concreting could be said to have commenced between 1970-1971. As we have seen earlier, the cancellation of the order for the PHT pump motor units by Canada and the transfer of the same to France resulted in a delay of about 2 years each concerning MAPP 1 and MAPP 2. The actual time for the commencing of the erection work from the first concreting was seven years each concerning MAPP 1 and MAPP 2—the same period that was taken for each of the RAPP units 1 and 2. The ban imposed by Canada, following the PNE of May 1974, made the authorities look for greater indigenisation even at a late stage in the erection of MAPP 1 and MAPP 2. Self-sufficiency in the Nuclear power field was taken a stage further via MAPP 1 and MAPP 2.

8.18   Energy and Revenue Generation

MAPP 1 and MAPP 2 generated 17 billion units of electricity till June 1993. The revenue generated at a tariff of 60 paise per unit was Rs. 1002 crores till June 1993. The revenue generated by both units within 9 years from commissioning was nearly four times the combined capital cost of MAPP 1 and MAPP 2.

8.19   Operational Problems and Solutions

8.19.1   Moderator inlet manifold

The manifold distributes the flow of the moderator inside the Calandria to ensure the optimum temperature of the rolled joints of the Calandria tubes and the Calandria tube sheet. The manifolds of MAPP 1 and MAPP 2 failed during 1988-1989. Internal inspection of the manifold was carried out with the active involvement of BARC and firms from the UK and France. Following the inspection, the training of the crashed position of the manifold was carried out. Necessary repair at the Calandria end cavity was also carried out. The units were started with a 50% power level and raised in stages to 75%, after evaluation and approval of the AERB. Technological development was initiated for the development of a sprayer system to substitute for the inlet manifold. Remote cooling was developed for the removal of the Calandria tube to make room for the sparger tube, the boring and grooving operations on the Calandria tube sheet at both ends, rolling off the sparger onto the tube sheet and the extended insert of the SS304 material onto the sparger tube. The inserts extend into the fuelling machine vaults (FM) and accommodate the shielding mechanism and whirler for ensuring an even-flow through the sparger. Three full-length spargers cold-worked and made of zircolay-2 material, with a wall thickness of 4 mm, are provided with the flow of the heavy-water moderator at both ends. A central blocking device is provided in the sparger tube. The sparger tubes are perforated at both ends; the holes are 2820 in number and of 8-mm, 10-mm and 12-mm diameter. Supply tube connections are made to the extended inserts in both the FM vaults. These operations were engineered and completed satisfactorily for the first time in the nuclear field of the world.

8.19.2   Turbine SP Rotor

Failure of the SP turbine blades was one of the other challenges faced in the operation. The help of the manufacturers, BHEL, was taken. Modifications to the LP stage-5 and other such solutions were suggested and carried out by BHEL.

8.19.3 Boiler Tube leaks: replacement of legs

Boiler tube leaks were affecting the performance of MAPP 2. All 88 legs, each leg weighing about 3.5 tonnes, were replaced en masse.

8.19.4 En-masse coolant channel replacement (ENCCR)

En-masse coolant channel replacement (ENCCR) was executed at MAPP 2; several records concerning bolt removal and replacement of channel tubes were established. The entire process was completed in 191 days, including zero clearance rolled joints during 2002-2003.

8.20 Upgradation of MAPP 2/MAPP 1

The long outage for carrying out ENCCR in MAPP 2 was utilised for life-extension jobs, safety upgradation, ECCS (Emergency Core Cooling System), cable segregation, fire prevention/detection, fire system upgradation, etc. Around 113 jobs in all were carried out to bring the unit within the current standards.

All the above jobs at 19 and 20 were planned for Rs. 413 crores and scheduled for completion in 24 months. The jobs were, however, carried out in 18 months for Rs. 248 crores (60% of the estimated cost). This upgrade is expected to enhance the life of the plant by 30 years. MAPP 2 was made critical after ENCCR on 13.7.2003, synchronised to the grid on 23.7.2003 and declared re-commercial on 1.8.2003. The gaps between criticality, synchronisation and commercial operation were the shortest in the history of the NPCIL. Similar en-masse replacement of coolant channels in MAPP 1 was subsequently carried out and the upgradation of the unit was also carried out. The unit is now expected to function till 2035. The unit re-commenced operations on 18.01.2006.

8.21 Highlights of the operation

The units at MAPP were the first to establish continuous operation which exceeded 6 months at a time on five occasions.

8.22 Tsunami: Safety of MAPP

People are all now generally aware of the 'Tsunami' that swept Chennai and other coastal areas of Tamil Nadu on 26.12.2004. The 'Tsunami' was brought about by an earthquake of 8.5 on the Richter scale centred at the North of Sumatra Island in the Indonesian Archipelago. This caused the Sea in the Bay of Bengal to rise by a height of 10 metres. MAPS 2 was then in operation and was safely shut down due to flooding of the pump house. MAPP 1 was already in the shut-down condition for major renovation and refurnishing that was being conducted at the time. MAPP units thus demonstrated their inherent safety against natural hazards.

8.23 Proposed facility in MAPP and the facility in IGCAR

8.23.1 The Immobilisation Plant

This plant has been proposed for the immobilization and storage of the solid form of radioactive waste of the intermediate level from the fuel re-processing plant which was commissioned recently.

8.23.2 The Fuel re-processing plant in IGCAR (Indira Gandhi Centre for Atomic Research)

This plant is a part of the IGCAR complex situated next to MAPPs in Kalpakkam. The plant has been touched upon due to its importance in the development of nuclear fuel for the on-going and future nuclear power reactors.

This plant is the third of its kind, after those at Trombay and Tarapur, and the most sophisticated of the three. The plant has been designed to re-process the spent-fuel of all the power reactors in India. The Zirconium–Thorium separation facility, for the separation of uranium-233 from irradiated thorium fuel on a plant scale, has also become operational. The plant will help build the necessary plutonium base for meeting the initial fuel requirement of the future fast-breeder reactors. It will help meet the mixed oxide fuel requirements of future advanced thermal reactors as well as the enriched fuel requirements of the Tarapur boiling water reactors. A feature of the plant is the concept of hybrid maintenance by which maintenance of radiation-affected hardware will be carried out in shielded cells using remote handling equipment. This plant was commissioned and dedicated to the Nation by the Prime Minister of India on 15.9.1998.

8.24   Nuclear Training Centre (NTC)

The NTC was established at MAPP in 1926 with a view of imparting hands-on training to the operators and maintainers of MAPPs. Higher-level training for the supervisory and shift engineer staff of the station at the lines of NTC at RAPPs was also envisaged. The centre moved to its permanent building on 21.9.1992. The centre has in-house facilities for an annual 150 man-years of training for Engineers/Diploma holders and IIT/H.S Certificate holders. The building houses the training staff office, lecture halls, the examination hall, library, computer facilities etc. The centre not only caters to MAPP requirements but also those of Kakrapar APP, Kaiga APP etc.

8.25   Contribution to Society: CSR

The MAPP CSR units have contributed to a significant extent in improving the infrastructure of the surrounding villages, especially in the streams of education, healthcare and water-supply. MAPP has built 8 schools and upgraded 3 schools to higher secondary level and 2 schools to middle/high school level. These were carried out for Rs. 8 crores. MAPP has conducted 13 medical camps for the benefit of about 15,000 persons and 6000 children for Hepatitis-B.

The station spent Rs. 1.6 million during 1998-1999 for the renovation of the primary school at Sadurangapattinam and the construction of a new building for the higher secondary section of a Government school at the same place.

The NPCIL donated Rs. 5 crores to the Prime Minister's relief fund for the relief of victims of the 'tsunami' that hit the TN coast on 26.12.2004. Many residents of Kalpakkam were victims of this natural calamity. Over 23000 NPCIL employees contributed their 1 day's salary for the relief and rehabilitation work taken up at Kalpakkam. The NPCIL also arranged for the supply of food and clothes, organised medical camps and provided temporary shelter to the victims.

8.26 Activities of MAPP to mark the diamond jubilee of the establishment of the DAE

As part of its outreach to the general public, and to further mark the diamond jubilee of the DAE, the MAPP made special efforts towards community outreach and awareness programmes. They invited students, journalists, community leaders and opinion makers to acquaint them with the various facets of the Nuclear Power Projects programme, and most of all acquaint them with the various safety measures and in-built features of the plant to ensure fool-proof safety. The programme was also conducted to acquaint them with the effects of natural and man-made radiation and to bring home the very low degrees of plant operational radiation as compared to natural radiation. Around 50,000 persons attended the out-reach programme during 2014-2015.

A nuclear carnival was held by the MAPP site in conjunction with NESCO between 6.9.2014 and 14.9.2014. Around 20,000 people visited this carnival.

8.27 MAPP 2: Continuous run

The station has clocked the continuous run of MAPP 2 for 392 days as of 23.7.2018. It was still running.

The Narora Atomic Power Project (NAPP) Units 1 and 2

CHAPTER
09

9.1 Site and its general features: estimated cost

The Government of India, which has been continuously vetting the recommendations of the site selection committee, decided during 1973-1974 to locate the third heavy-water moderated uranium-fuelled Reactor Power Plant of 2×235 MWe capacity at Narora (27° North Latitude, 77.8° East Longitude) in the Bulandshahr District, about 20 km from Dibai for Rs. 209.83 crores according to the 1974 prices (FE cost Rs. 27.73 crores). These two units would be the 7th and 8th reactor power units in the country.

The site is about 110 km from New Delhi and is situated in the upper-middle North Western part of Uttar Pradesh (UP). This was the first nuclear project to be located in the Northern part of the country, the others having been located in the Western, Central and Southern parts of the country. The site is situated near the barrage on the right side of the sacred river of Ganga at Narora. The cooling water and process water requirements of the project are drawn from the river Ganga and are tapped from the barrage nearby.

9.2 Special features of the site

9.2.1 Seismic factor: suitable design

The site lies in zone-4 of the seismological classification and is within a distance of about 30 km from the Moradabad fault line which is seismically active. The engineering design features took into account the seismic forces likely to be encountered in this zone. This zone is seismically equivalent to the Po Valley reactor site in

Italy. Required safety-related buildings, structures and systems have been designed for a zero-period ground acceleration of 0.3 g.

9.2.2 Hydraulic cooling towers

This project was the first in which hydraulic natural-draft cooling towers have been provided for meeting the circulating coolant water requirements of the condenser of the plant. This was also the first Nuclear Power Project which has incorporated a forced cooling draft system with heat-exchangers for meeting the process water (PW) requirements. The PW pumps consist of separate high-pressure and low-pressure pumps, each with their loops. The water-circulation requirement for the PW system is 20% of the circulating cooling water requirement of the condenser. The adoption of these cooling water systems has limited the total water requirements to about 10% of what would have been required without the draft cooling systems (150 cusecs against 1500 cusecs of water).

9.3   Civil works: details

9.3.1   General

Civil work for both units of NAPP 1 and NAPP 2 commenced simultaneously.

9.3.2   Foundation details

9.3.2.1   *Reactor Building*

The soil was made up of alluvial sand and the design for the foundation is based on this characteristic. The mud-mat (foundation raft) for the Reactor building was 4.6 metres thick and 17 metres below the grade level, which was 100,000 mm. The raft foundation was 50 metres in diameter and the quality of concrete for the raft was MRO.

9.3.2.2   *Turbine Building and Service Building*

The RCC piles for the turbine building and service building were driven to 20-23 metres depth. Groups of 500-mm diameter piles of 4, 6 and 11 each were driven at the designated locations—the number in each group being determined by the static dynamic-motive-seismic load each group was to carry. The pile caps of each group of piles were dovetailed into a common footing foundation. The columns rested on this footing foundation. About 1048 piles were driven for TB-1 while 600 were driven for TB-2. Each pile was designed for a test load of 240 metric tonnes (twice the working load of 120 metric tonnes).

9.3.3 Hydraulic Cooling Towers

Two cooling towers were provided, one for each reactor unit. The cooling capacity of each tower unit was 363×10 to the power of 6 litres/hour. The other design parameters are as stated below:

1) Cooling water circulation: 42,600 M^3/hour.
2) Hot water temperature: 42.8 degrees C
3) Re-cooled water temperature : 32 degrees C
4) Design atmospheric wet bulb temperature : 26.5 degrees C
5) Design ambient relative humidity : 60%
6) Design wind speed for thermal performance : calm speed
7) Electrical power requirement for pumping head from the basin water level : 14 MWe maximum
8) Design wind speed : 178 km/hour at 30 metres above grade ground level
9) Seismic design values [designed for OBE (Operating Basis Earthquake)] : Nuclear safety Class-I with peak horizontal at 0.15 g and peak vertical at 0.10 g.
10) Dimensions of the towers :
    Height : 139 metres (above ground grade level of 135 metres)
    Base : 114.5 metres diameter
    Top : 75 metres diameter
    Thrust : 67.5 metres diameter

9.3.4 Reactor building containment domes

The reactor building containment domes comprise double containment: ICD and OCD. The inner containment dome (ICD) is designed to withstand 1.25 kg/cm^2. The test pressure is 1.44 kg/cm^2. The annular space between the ICD and OCD is 2 metres wide. The ventilation of this space is of the same design feature as that of MAPP units 1 and 2. The inner diameter of the inner wall is 44 metres. The outer wall is 5.8 metres high and topped by a segmental dome with a rise of 7.9 metres. The outer wall is made up of RCC and the inner one of pre-stressed concrete. The quality of the concrete of the inner wall is of 1727.5 grade while the outer is of grade M-25.

The important difference between NAPP and the earlier projects is that an outer containment dome (OCD) has been provided for the first time. This feature has resulted in an order of magnitude of safety which is higher than that of earlier projects. The inner containment wall (ICW) and inner dome structure are made up of pre-stressed concrete, while the outer containment wall and outer dome structure are made up of ordinary concrete and are designed for an internal pressure of 0.07 kg/cm^2. Some of the other details of the ICD are as follows:

a)   Seismic co-efficient (zero-period ground acceleration) : 0.3 g
b)   Grade of concrete : M-20/M-25/M-35
c)   Pre-stressing wires system : 12 T 13 (twelve wires with 12.7 mm diameter)
d)   Pre-stressing wire capacity : 200 tonnes.

A pressure suppression pool has been provided in the basement of the RB as in MAPP.

9.4   Important design/construction/equipment manufacture changes as compared to RAPP and MAPP

9.4.1   Reactor shut-down and regulation

Two independent fast-acting shut-down systems have been provided as against one system in RAPP and MAPP. Fourteen mechanical shut-off rods for the primary shut-down system and twelve liquid poison tubes for the secondary shut-down system have been provided.

Four slim rods are provided for coarse power control.

The Reactor is of the integral moderator-type which operates on the fixed level of moderator. Hence, there is no dump tank for emergency shut-down as seen in RAPP and MAPP.

9.4.2 Garter springs

Four garter springs have been provided between the pressure tube and the Calandria tube to provide uniform concentric spacing between the two tubes. Only two springs were provided in earlier projects.

9.4.3 Calandria/end-shield system: Calandria vault cooling system

An integral Calandria/end-shield has been provided as against the separate Calandria and end-shields that are present in RAPP and MAPP. The integral Calandria/end-shield rests on the Calandria walls; stainless steel balls replace the stainless-steel slabs used earlier. The lighter end-shields have resulted in a great advantage in terms of transportation, installation etc. Water-filled Calandria vault housing has been adopted, which has resulted in the elimination of shield tank, shield-tank cooling and thermal and biological cooling systems. The emission of Argon-41 has also been reduced as a result of the provision of a water-filled Calandria vault.

The Calandria was filled with water and submerged in water. The Calandria vault has to be isolated from the fuelling machine vaults. This is done by inserting a bellows assembly at each end of the pressure tube and then rolling it onto the end-fitting body; the bellows are connected by 6-mm thin-walled tubes at the stub. CO_2 gas is maintained in the annular space between the pressure tube and the Calandria tube, thus effectively isolating the Calandria vault from the fuelling machine vault.

9.4.4 The light-water auxiliary system

This system has been shifted out of the RB; this has reduced the frequency of entry into the RB and hence will reduce radiation exposure to the operating personnel. Separation of light-water and heavy-water system areas has helped to facilitate larger heavy-water recovery.

9.4.5 The heavy-water drier system

D_2O recovery driers have been shifted out of the RB. This will result in easier maintenance and increased availability of the driers.

9.4.6   In-site welding of the Calandria and end-shield

On-site welding of the two end-shields and Calandria was undertaken for the first time with the construction of the Indian Nuclear Power Projects. The required tri-junction welding process was developed by Larsen & Toubro and duly qualified by the quality-control group at NAPP before the commencement of work.

9.4.7   Heavy water-to-light water heat-exchangers (boilers)/ Primary circulating pumps: valves in the PHT system and some equipment costs

The number of heat-exchangers per Reactor was reduced to four as against eight in each of the RAPP and MAPP units. The number of valves in the PHT main system was reduced to 16 as against 24 in each of the RAPP and MAPP units.

Each of the boilers at NAPP was of 2.7 metres diameter and 20 metres in height with the tubing made up of Inconel 600. The weight of each boiler was 120 tonnes. These boilers were transported by rail on the DAE's 16-axle rail wagons. The journey time for the 2080-km distance between BHEL (Tiruchirappalli) where the boilers were manufactured and the NAPP site was 3 to 4 weeks. The eight boilers for NAPP 1 and NAPP 2, with Inconel 600 tubing, cost Rs. 1522 lakhs according to the 1989 prices. In comparison, the 16 boilers of RAPP 1 and RAPP 2 cost Rs. 225 lakhs according to the prices of 1964-1968. The cost of the boilers of MAPP units 1 and 2 included Rs. 75 lakhs towards capital investment, the amount is 50% of the capital investment envisaged by BHEL for undertaking the manufacture of the boilers. The manufacture of the PHT pumps was entrusted to KSB pumps, the first lot for NAPP 1 being imported from Germany and the second lot for NAPP 1 being indigenously manufactured by KSB for the pump portion and NGEF for the motor portion.

9.4.8   Site Construction Workshop

Both RAPP and MAPP were provided with departmental construction workshops for fabricating the various embedded parts and other mechanical components required for the projects. In the case of NAPP, Triveni structurals—located at Naini near Allahabad in UP—were entrusted with the responsibility of setting up a site workshop for fabricating the various mechanical components required for the project on a cost-plus basis. This proved to be a good alternative to the project setting up its shop with the accompanying over-heads, running costs, etc. Since Triveni Structurals was a Government of India undertaking, the possibility of containing the fabrication cost was quite distinct.

9.4.9 Separation of Class-II power and control supply

RAPP and MAPP were provided with two 415-V 300-KVA Motor Generator (MG) sets for each reactor unit to provide an uninterruptible Class-II power-supply. Each MG set was designed to supply both power and control power systems through the respective Class-II basis.

In the case of NAPP units 1 and 2, it was decided as a design improvement. To provide better safety and integrity to the Class-II systems, power and control systems should be segregated and fed by separate MG set units. It was decided that two 415-V 325-KVA power MG sets and four 415-V 55-KVA control MG sets should be provided for each reactor unit.

9.4.10 Auxiliary loads

All auxiliary loads including NAPP township loads, part pump-house loads, waste-management plant and clarifloculator facility loads are fed from the 66/11 KV UP SEB grid S/S. In the case of RAPP and MAPP, only the township loads are fed from the utility grid sub-station.

9.4.11 Emergency Control Room

Another design improvement which has considerable impact on the Safety Engineering aspect of the Indian Nuclear power stations was the 'Emergency Control Room' concept introduced in NAPP. This emergency control room has been designed from the point-of-view of ensuring continuous capability of operating all the safety and safety-related systems of the reactor plant in the event of the entire main control room of the plant becoming incapacitated for any reason. This is a giant step, by far, in the field of Safety Engineering of Indian Nuclear power projects.

This emergency control room is physically separated from the main control room. All the cabling and other engineering systems of the emergency control room are also in the redundant route (physically and system-wise separated from the engineering systems of the main plant and hence the main control room) so that the continued operability of the emergency control room is ensured under all conditions.

9.4.12 Special cooling

If the containment in the reactor building gets pressurized, special cooling fan units run by 415-V Class-III power will help in cooling the RB ambient atmosphere.

9.4.13 The DG set system improvements: measures for industrial safety

Ensuring industrial safety required particular importance after the tragic Bhopal gas tragedy of 1984. Safety aspects of the 415-V Class-III DG set system were thoroughly reviewed from the point-of-view of fire and explosion hazards. As a result of the discussions with the Chief Inspector of Explosives of Nagpur and Consultants from Tata consulting Engineers, certain improvements were implemented. We will discuss the same in the upcoming sections.

9.4.13.1 The DG steel chimney

Self-supporting steel chimneys were provided for the exhaust gases of the DG sets. These chimneys would have a height of 1 metre more than the highest part of the TB, where the DG sets are located. In the case of NAPP, it was 11.2 +1 metre or 12.2 metres approximately.

9.4.13.2 Day tanks: physical isolation

The areas accommodating individual day tanks would be physically isolated from each other using a sufficiently high RCC wall between the adjacent day tanks.

The above safeguards are being adopted for all current projects and will be adopted for all future projects as well.

9.4.13.3 Underground diesel storage tanks

These are provided with anti-corrosion protection. The protection extends to supports, nozzles, manholes and other appurtenances of the tanks.

The minimum temperature at the NAPP site goes below 3 degrees C and may be expected to reach near sub-zero temperatures under extreme weather conditions. These conditions may result in the waxing of the stored diesel oil. The storage tanks have hence been provided with surface heating and insulation pads to ensure non-waxing conditions of the diesel oil under all operating conditions.

9.4.14   The Gravity Addition Boron System (GRABS)

An additional system called 'GRABS' is present in the plant. This system was first introduced in NAPP. GRABS is a process-independent, manually-operated system which is designed to ensure the safe shut-down of the reactor in case of total electrical black-out.

9.4.15   Evacuation of power from NAPP

Five 220 KV extra-high-voltage transmission lines have been proposed for the purpose.

9.5   Progress of the construction work: equipment erection costs

9.5.1   Preliminary work

We have seen earlier that financial and administrative approval for the construction of 2×220 MWe units was accorded in 1973-1974. The preliminary site work commenced during 1974-1975. The UP State Electricity Board (UPSEB) established a 66/11 KV sub-station at the site with an installed capacity of 6 MVA (6000 KVA) for feeding the site construction loads as well as for its customers nearby. Permanent plant work commenced during 1975-1976 while the first concreting commenced in 1976. The electrical and piping erection work was put up to tender during 1979-1980 and was awarded thereafter. Some of the costs of various work are furnished below for comparison with earlier projects.

(1) Switchyard installation cost : Rs. 65 lakhs
(2) Electrical Equipment erection cost : Rs. 105 lakhs
(3) Piping and mechanical equipment erection : Rs. 375 lakhs (NAPP 1 cost)

9.5.2 Construction status as of 1984

The construction of main plant buildings for NAPP 1 was completed; construction was nearing completion for NAPP 2. Many of the major equipment like Calandria, end-shields, turbo-generator components, heavy-water headers, feed-water heaters, generator transformers etc. required for NAPP 1 and NAPP 2 were received at the site and under installation. The 220-KV switchyard work was completed. Main piping work had progressed substantially; the overall progress on NAPP 1 and NAPP 2 was 60% as of early 1984.

9.5.3 Nuclear Plant/TG Plant work: status as of 1987

9.5.3.1 *Nuclear Plant Work as of 1987*

Eight heavy-water headers supplied by BHEL for NAPP 1 and NAPP 2 were received at the site. The four headers required for NAPP 1 were erected.

The erection of the fuel handling system of NAPP 1 was completed; the drive system was commissioned and the system was prepared to receive the fuelling machine head as of early 1987.

The end-shields of NAPP 2 were grouted after the completion of field welding and alignment of the end-shield Calandria assembly in early 1987.

Three steam-generating units out of the eight required for NAPP 1 and NAPP 2 were ordered from BHEL and received at the site in February 1987. Two units of the three units were erected while the other one was under erection as of early 1987.

Zircaloy-Calandria tubes and Zirconium-alloy pressure tubes were manufactured and supplied by NFC, Hyderabad.

9.5.3.2 *The 235-MWe TG plant work*

Both the 2 × 235-MWe TG sets required for NAPP 1 and NAPP 2 were manufactured, supplied and erected by BHEL. The turbines were manufactured at the Bhopal workshop and the generators at the Haridwar workshop of BHEL. BHEL in turn bought out 560 metric tonnes of the TG plant stores from 'Prommesh Exports', Moscow.

9.5.3.3 *NAPP 1: hot-conditioning and readiness*

NAPP 1 was well on the way towards the attainment of hot-conditioning as of the end of 1987.

9.6 NAPP 1: Approach to Criticality

9.6.1 Phase A

This was achieved by the removal of Boron with the Calandria full. Starting with 32 ppm of Boron in the moderator, the concentration was reduced gradually in steps by the ion exchangers, and at each stage, the neutron population in the reactor was monitored. The correlation of this parameter with the Boron concentration was a painstaking process and called for a rapid and accurate analysis of a large number of samples. As criticality was being approached, as indicated by increasing neutron count rates, the reactor period was large and gave assurance of a safe and smooth start-up.

9.6.2 Phase B

9.6.2.1 *First-criticality*

The first criticality was attained on 12.03.1989. This unit was the 430th power reactor in the world to be commissioned.

9.6.2.2 *Physics observations*

Some important Physics observations were made which are listed as follows:
1) Calibration of reactivity control rods
2) Calibration of primary and secondary shut-down systems
3) Dynamic testing of all shut-down systems
4) Testing of automatic liquid poison addition system

The objective of the above experiments was to validate the correction of positive reactivity worth introduced into the reactor with the status of the shut-down devices as predicted by calculations.

9.6.2.3 *Flux profile*

The axial neutron flux profile in the reactor core was measured using irradiation of a copper wire in the core after the ice plugging of feeders and the removal of irradiated fuel bundles.

9.6.2.4 *The thermo-syphoning test*

It was demonstrated for the first time in the history of Indian Nuclear power units that in the event of loss of Class-II electrical power, reactor cooling can be maintained by a temperature difference called thermo-syphoning (natural gravity flow cooling). This was a very important development in demonstrating the inherent safety feature of Indian Nuclear power units.

9.6.3 Phase C

9.6.3.1 *The Power run-up test*

The performance of the process and safety systems was closely monitored at every stage. Of interest are some of the operational tests carried out to study various modes of operation and transients of the system. These are listed below:

1) Pressure control valve failure
2) Relief valve (CIESV) operation
3) Grid power-supply failure
4) Testing of protection and emergency core cooling system

9.7 NAPP 1: further milestone dates

(1) First synchronisation with the Northern Regional Grid : 28.07.1989
    (25% full power)
(2) 50% full-power operation : 26.09.1989
(3) Declaration for commercial operation : 01.01.1991

9.8 NAPP 2: progress of erection work as of the end of 1990

The erection work of NAPP 2 was completed by the end of 1990. All the systems were checked out and the pre-commissioning test was carried out; fuel loading was completed. The general approach to the first-criticality of NAPP 2 was about the same as that for NAPP 1.

9.9 NAPP 2 Test Criticality: Revenue earned

NAPP 2 became first critical on 24.10.1991. The unit was declared commercial on 1.7.1992. The reactor had been operating at full power since November 1993. NAPP 2 has generated a total of 20.14

billion units up to June 2008 and has generated revenue of Rs. 46.5 billion during this period.

9.10 NAPP units 1 and 2: costs

9.10.1 Cost of construction

The sanctioned cost of NAPP 1 and NAPP 2, as we have seen earlier, was Rs. 209.83 crores according to the 1974 prices. The cost, as per the 1990 prices, was Rs. 745 crores. Many improvements were carried out in the construction of NAPP 1 and NAPP 2; most of them were done during construction. The reasons for the increase of 25% over the sanctioned cost are stated below:

1) Design changes during construction.
2) Additions, improvements and increase in the scope of work due to design changes during construction, among others. Many of the changes were done to bring the systems/work in conformity with state-of-the-art safety requirements, the AERB requirements, etc.

9.10.2 The final cost of construction

The erection work of NAPP 1 commenced at the end of 1980 and was completed by the end of 1987. The time taken for the erection is about the same as that in RAPP and MAPP. This was quite an achievement since many improvements were effected in NAPP as compared to RAPP and MAPP (these have been detailed in para 9.4); the time taken for the completion of NAPP 2 work was a year longer than that of NAPP 1. The continuing efforts toward indigenisation fully justify the slightly longer time taken for the completion of the erection work.

9.11 Scope of supply of indigenous equipment/materials

The scope of supply of indigenous equipment/materials achieved in NAPP is 95%. This is the highest yet attained in the history of construction of Nuclear Power Projects in India.

9.12 Proposed facilities in NAPP

An environmental survey and micro-meteorological laboratory are proposed to be set up at the NAPP site. This would help carry out environmental studies to establish all the parameters on which the disposal of radioactive material depends. It will also carry out the monitoring of radioactivity in the ambient atmosphere in the operational stages of the plant. Such studies would help establish guidelines for enforcing discharge limits and other design and operational parameters for nuclear power stations.

9.13 Significant Nuclear incidents in NAPP: the IAEA classification and the AERB recording

9.13.1 The IAEA classification of nuclear incidents

The IAEA has classified nuclear incidents on an ascending scale of 0 to 7. Zero being the least serious and seven the most serious.

9.13.2 The AERB's recording of Indian nuclear accidents

The AERB has taken steps to implement the recording of nuclear events in Indian Nuclear power stations, as per the 'International Nuclear Event Scale', since 1990. A total of 105 events had occurred in all the Indian Atomic Power Plants; all these events had neither safety significance nor off-site or on-site impact. No event with potential safety consequences (level 2 and above) occurred in India since 1990 with the sole exception of the accident at NAPP 1 on 31.03.1993. All the observed events belonged to level-0 to level-1.

9.13.3 Nuclear incident in NAPP 1 on 31.03.1993

9.13.3.1 Description of the accident

We now come to a very significant nuclear event that occurred on 31.3.1993 at 3:31 AM in the NAPP 1 unit. A disastrous fire in the 220-MWe TG set of the NAPP 1 unit occurred at the above-mentioned time. The fire has been attributed to the leakage of hydrogen gas from the unit and the occurrence of an electrical fault within the set. The resultant explosion generated significant volumes of smoke which engulfed the TG bay and entered the control room. The alarm appears to have been set off first in the control room. The second-in-command of the control room initiated the shutting down of the reactor and also the crash cooling down of the reactor once he heard the explosion, and the issuing of smoke from the turbine hall was noticed. A local emergency was declared in the plant. It took nearly four hours for the fire-fighting forces from the plant and the nearby areas to put out the fire. All these operations were carried out from the emergency control room. The damage to the TG plant and indirect damages have been tentatively estimated at Rs. 20 crores for capital assets like equipment/building, and several tens of crores of rupees towards indirect damages like the loss of revenue due to the incident. There was no loss of life or injury to operation and maintenance personnel, nor release of activity to the plant ambience and or to the environment.

The scale of the nuclear event at the NAPP 1 unit on 31.03.1993 has been put at level-3. This event classifies the incident as a 'serious incident' in which a further failure of safety systems could have led to accidental conditions. This incident could be described to be akin to the 1939 nuclear incident at the Vandles Power Plant in Spain, which did not lead to any escape of radioactivity but significantly degraded the plant safety system.

9.13.3.2 *NAPP 1 nuclear incident: further details of the cooling system of the TG set*

Hydrogen gas at a pressure of 38 psig (3.76 kg/cm^2) is used in the NAPP TG sets for cooling the generator. This system has the advantage of ensuring lower friction and windage losses (hydrogen is one of the lightest of noble gases and has a very high heat-transfer coefficient; the disadvantage with the use of this gas is that it has a very high explosive characteristic as well). The gas cools the rotor using two 100% hydrogen to de-mineralised water heat-exchangers. De-mineralised water is on the tube side and hydrogen is on the shell side.

The stator has its independent cooling system. The hydrogen content has to be contained between the ranges of 10% to 70% in the volume of ambient air to eliminate any possibility of an explosive mixture being formed. There are monitoring systems to indicate this content, and an alarm system to warn about deviation from these limits also exists. It appears that the leakage of hydrogen gas from the turbine generator resulted in an explosive mixture being formed in the ambient atmosphere. The mixture ignited causing a gas fire and extensive damage to the TG set, TG ancillaries and the turbine building.

9.13.3.3 *Investigation into the incident*

The AERB set up an expert committee to investigate the cause of the incident. The NPCIL set up its panel to investigate the causes of the incident. Specialists from the AERB, National Thermal Power Corporation (NTPC), Bharat Heavy Electricals Ltd. and Defence Research and Development Organisation, besides nuclear experts from the NPCIL and BARC, were members of these committees.

The AERB's conclusions about the cause of the fire were based on the reports of these committees and were released during the second week of July 1993. The conclusions are summarised below:

1) The failure of two turbine blades in the fifth and the last stage of the low-pressure turbine was the initiating cause of the incident.

2) The failure of the blades resulted in a severe imbalance in the dynamics of the 3000 rpm (high speed) turbine and resulted in extensive damage to the turbine bearings as well as the various accessories and components of the turbine and generator.

3) The extensive damage to the turbine accessories resulted in the loss of leak-tightness of the turbo-generator's hydrogen gas seals.

4) The loss of leak tightness of the hydrogen gas seals led to the leakage into the ambience of the inflammable gas and triggered gas fire in the ambient atmosphere.

5) The fire spread rapidly, aided by oil spills from the ruptured oil lines of the turbine lubricating oil system.

6) The fire caused extensive damage to the power and control cables system of the TG and resulted in the total disruption of all classes of power; this caused a total black-out of the nuclear power station lasting over 17 hours.

7)   GRABS (see 7.4.14) worked perfectly.

8)   The damage to the plant was estimated at Rs. 25 crores.

9.13.3.4   *Conformance to the accepted design basis practice*

It is an accepted design-based practice of the NPCIL that segregation of power and control cables, provision of fire barriers and independent routing of safety systems' cabling are adopted for nuclear power stations. The AERB had also re-iterated conformance to these requirements. However, according to the AERB, these requirements had not been adhered to, especially concerning the cabling system for the TG set in question (NAPP 1). If these requirements had been met, a total black-out of the station, which occurred on 31.03.1993, could have been avoided.

The segregation was subsequently carried out both in NAPP 1 and NAPP 2. Fire barriers in the cable routes were provided and the application of fire-retardant paint to cables was carried out both in NAPP 1 and NAPP 2.

9.13.3.5   *Lessons to be learnt from the failure*

The failure of the TG set at NAPP 1 points out the need for the AERB, the apex regulating body for all nuclear activities in the country, to evolve stringent procedures within the AERB itself to ensure that all its recommendations are implemented by the operating nuclear stations within a reasonable time frame.

9.13.3.6   *Follow-up action by the AERB and the NPCIL*

9.13.3.6.1   The AERB's follow-up action

It is a matter of gratification that the AERB ordered an intensive check-up of all the turbo-generators in service in the nuclear power stations of India. It is worthy of commendation that this was done despite the crying need to keep as many generating units in India as possible, to meet the increasing power demand and reduce the perennial peaking power shortage of the country. The inspection required planned outages of the respective stations. This rigorous inspection has paid off and has averted a potential incident in the turbine of MAPP 1. Four cracks were noticed in this turbine at the same places as the turbine of NAPP 1.

9.13.3.6.2   The NPCIL's follow-up action

The LP turbine of NAPP 1 was replaced by a new one manufactured by BHEL with improved blades. The defective LP turbine of NAPP 1 was also replaced; similar root-blades in the last stage of the LP turbines of all operating TG units in the HWPTRs were suitably modified.

9.13.3.6.3   Re-criticality of NAPP 1

NAPP 1 went critical again at 15:05 hours on 22.12.1994, after 21 months taken for due repairs to the TG units which called for renovation. A sum of Rs. 8.15 crores was spent on the rehabilitation of NAPP 1 during 1993-1994. Out of this sum, Rs. 6.17 crores was capitalised and Rs. 1.98 crores was taken as revenue expenditure.

9.13.4   Operational history of NAPP 1 and NAPP 2

9.13.4.1   Capacity factor: record generation of NAPP 1

NAPP 1 has been in commercial operation since 01.01.1992. The performance of the unit has been consistently improving over the years since then. The monthly capacity factors during January 1998 of NAPP 1 and NAPP 2 were 105.36% and 103.03%, respectively. Both the units together generated 341 million units during January 1998, breaking their previous record of 335 million units during January 1997. The capacity factor during December 2002 was 106.30% and during March 2003 was 103.40%. The annual capacity factor of the two stations improved from 72.5% in 1998-1999 to 93% in 2002-2003 as would be seen from the figures in the graphs attached in this section. NAPP 2 generated 154.8 million units during December 1992, a record in a single month of all PHWRs in India as of then.

The NAPP 1 monthly capacity factors for 2002:
(in percentages)

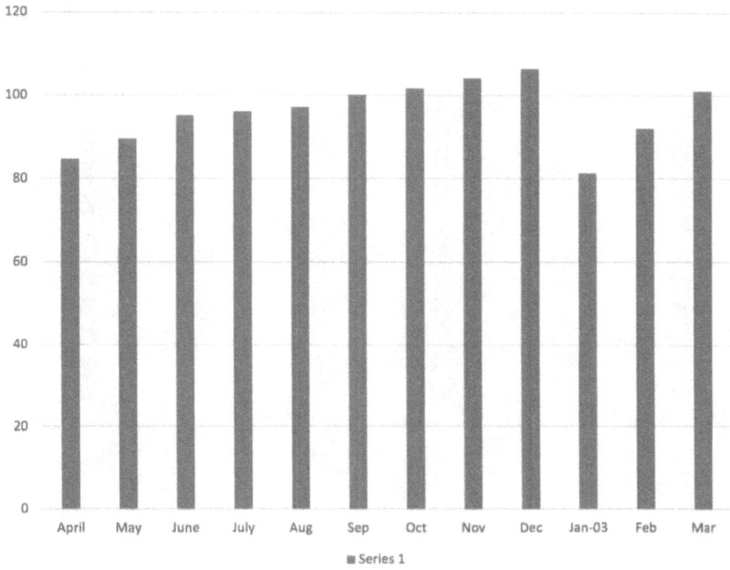

The NAPP 2 monthly capacity factors for 2003:
(in percentages)

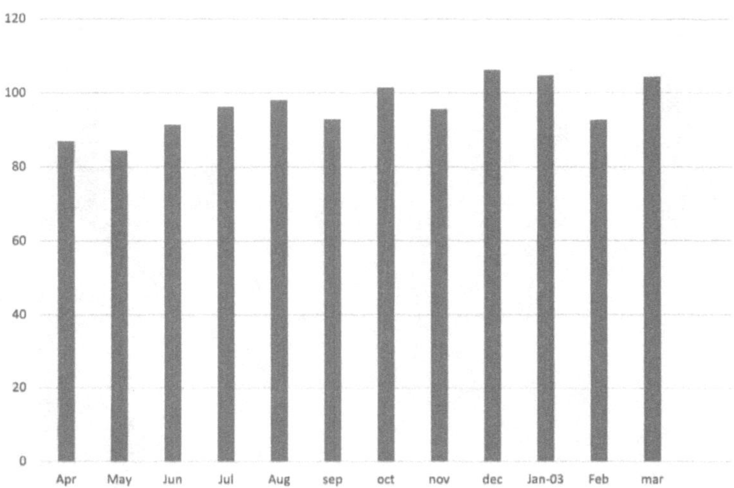

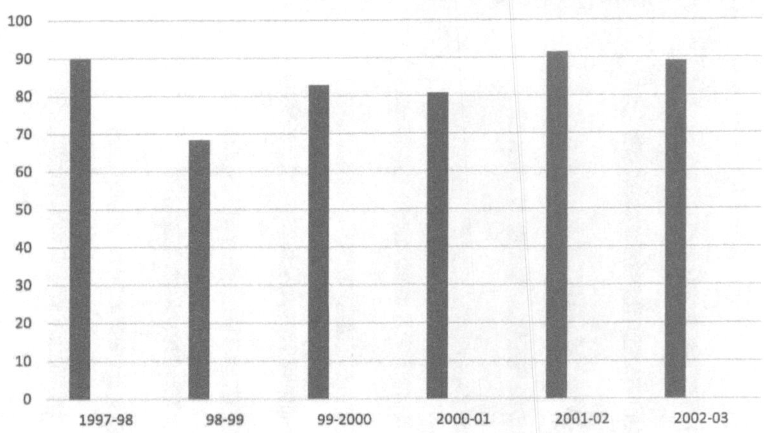
NAPS-1
Capacity factor
1997-98 to 2002-03 percentages
100
90
80
70
60
50
40
30
20
10
0
1997-98
98-99
99-2000
2000-01
2001-02
2002-03

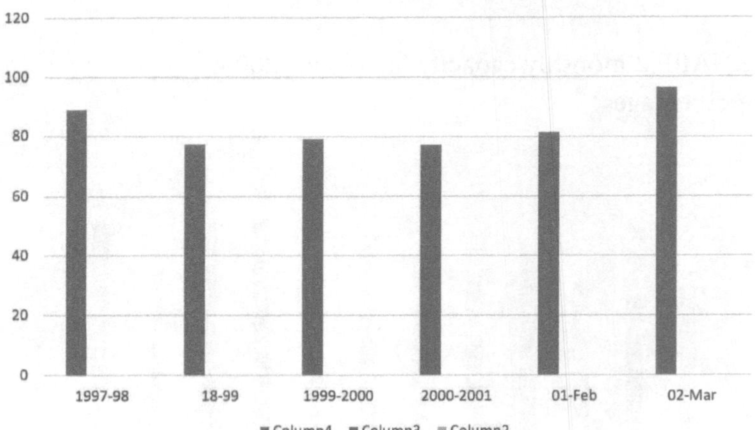
NAPS-2
Capacity factor
1997-98 to 2002-03 percentages
120
100
80
60
40
20
0
1997-98
18-99
1999-2000
2000-2001
01-Feb
02-Mar
Column4
Column3
Column2

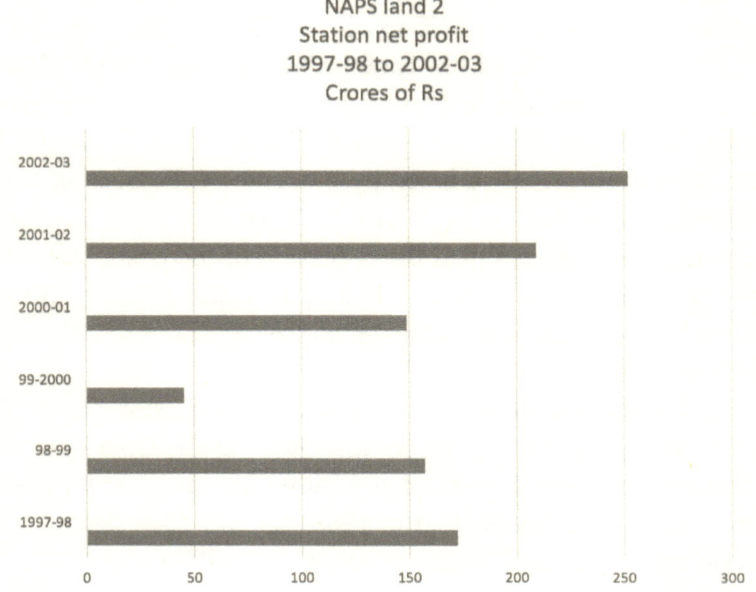

(1999-2000 decrease due to adjustment of increase in loan interest from 8.5 to 15% of Rs. 195.37 crores)

9.13.4.2   *Improvements carried out to units as of mid-2005*

9.13.4.2.1   NAPP 1: modifications to HP feed-water heaters no. 5 and 6

Insufficient drainage of steam condensate was noticed in the above heaters. This was causing undesirably high levels of condensate in the heaters. Design modifications to improve the draining and venting arrangements were carried out. This improved the performance of the heaters and increased the outgoing feed-water temperature by 6 degrees C, a considerable improvement over the previous temperature. The gain in generated power was 2.5 MWe and the gain in annual revenue was Rs. 5.05 crores, at 100% capacity factor and rate for energy at Rs. 2.31 per unit.

9.13.4.2.2   NAPP 1 and NAPP 2: cleaning work

Cleaning of the basins of the Natural Draft Cooling Tower (NDCT) of NAPP 1 and NAPP 2 was carried out. Around 1500 tonnes of muck were removed from the basin. Double screens with a modified design were installed to prevent debris from getting into the CCW pump intake. Cleaning of the hot water duct of the NDCT was carried out; nozzles with a modified design were installed to improve the spray of the hot water for better cooling. The CCW intake temperature was reduced by 2 degrees C because of these modifications carried out in NAPP 1. A reduction of 1 degree C in this temperature results in the reduction of the condenser back pressure by 5 mm of Mercury. The gain in power-generation is about 3 MWe; the corresponding gain in revenue is Rs. 6.06 crores at a CF of 100%.

Similar modifications are proposed to be carried out in the NDCT of unit 2 during the annual shut-down. Chemical and mechanical cleaning of the condenser tubes was taken up for NAPP units 1 and 2 in 2003. A total of 1948 kgs and 1008 kgs of scale material were removed from the tubes of NAPP 1 and NAPP 2, respectively. Both the units could now be operated at 220 MWe even during summer months.

9.13.4.2.3 Reduction in the annual shut-down (ASD) period

The period of ASD has reduced progressively with the latest ASD of NAPP 1 witnessing a drastic reduction to 26 days during July 2002 and 19 days during 2003 against a planned outage of 25 days.

9.13.4.2.4 Efficient fuel management

A continuous increase in the fuel burn-up rate has been noticed due to efficient management. The increased burn-up rate was observed in 543 fuel bundles. A new re-fuelling scheme which was adopted for low/excess reactivity in the fuel core and for catering to the required power load initiated during April 2003 resulted in a saving of 1,607 fuel bundles, and Rs. 37 crores in fuel consumption over 12 months. The fuel failure rate was maintained at 0.03% of the number of spent-fuel bundles discharged into the spent-fuel cooling and storage bay.

9.13.4.2.5   Longest continuous run: accident-free record

NAPP 1 had set a record for longest continuous run of 174 days surpassing the earlier record of 125.5 days. NAPP 2 established 192 days of longest continuous run as of 31.03.2004.

NAPP units 1 and 2 further established a record for simultaneous continuous operation for 238 days during 2004-2005. NAPP 2 further increased this record by operating uninterruptedly for 272 days. The excellent record achieved by NAPP 2 as well as that of NAPP 1 stations has been touched upon in para 9.13.4.1 of this Chapter.

9.13.4.2.6   Life-extension work

Life-extension and upgradation of crucial systems of NAPP 1 were taken up and completed on 02.12.2007. NAPP 1 was re-commissioned on this date. This work would enable NAPP 1 to operate till 2035-2040. The en-masse replacement of coolant tubes (EMCCR) was carried out as part of the life-extensive work. EMCCR becomes necessary after the completion of 10 effective full-power years of operation. The replacement was carried out after life-assessment studies and in-service inspections were carried out from time to time. A technology developed by the Raja Ramanna Institute of Advanced Technology at Indore (MP) was adopted for the laser cutting of critical nuclear components as part of the EMCCR work. The whole work was carried out well within the radiation dose limits for the workers imposed by the AERB.

9.13.4.2.7   Milestone dates

Some of the milestone activities and dates for the above NAPP 1 upgradation are given below:

a)   Unit shut-down date : 31.10.2005

b)   Re-fuelling of the entire core transfer of 344 fuel bundles in NAPP 2 : 08.04.2006

c)   Removal of old components : 15.05.2006 to 02.09.2006

d)   The PHT system pneumatic and Helium leak test : 18.05.2007

9.13.4.2.8   Other activities

Some of the other works are highlighted below:

a) Around 55 end-fittings were removed and disposed of in the Waste-Management Plant (WMP) on a single day.

b) Nearly 32 pressure tubes were removed and disposed of in the WMP on a single day.

c) Eighteen coolant tubes were installed and rolled on a single day.

d) 50% of coolant channels were installed in 19 days.

e) The installation of the first half of the PHT circuit was carried out in 50 days.

9.13.4.3 *CSR work of the NAPP site and awards received*

We will now look into the social welfare causes undertaken by the NAPP team under CSR. We will also learn about the awards received by the team in the upcoming sections.

9.13.4.3.1 Safety and other awards

NAPP won the AERB industrial safety awards for two consecutive years, 2001 and 2002.

9.13.4.4 *Social Welfare Schemes: CSR*

Some of these are listed below:

1) The construction of a community centre in Nai Basti village.

2) The renovation and construction of two rooms in a primary school in Silan village was in progress during mid-2004. Utility items for study were distributed to the students.

3) One tractor-trolley was donated to the Nagar Panchayat for carrying out cleaning operations in the Panchayat area.

4) A medical camp was set up at Silan village on 01.01.2004. Around 679 patients were given free treatment as a result of the same.

5) The 18th free medical eye camp was organised at the NAPP Hospital between 11.03.2004 and 20.03.2004. Around 121 eye-surgery operations were carried out. A total of 1,649 cataract operations have been performed in such camps to date.

6) A NAPP co-operative farming society was formed at the station site. Apart from running a vegetable garden, the society organises tree plantations at the sports stadium in the NAPP township. The agricultural society has grown a new variety

of paddy (SAKET-4) for the first time on 10 acres of land with excellent results. The yield was better than that of maize normally grown on such types of soil. The society has also grown a hybrid variety of groundnuts supplied by BARC, again with excellent results.

7) The station proposes to plant one lakh trees within the 1.6-km exclusion zone of the plant. It has established a green-belt area over this zone. The Society has helped in training the local volunteers to preserve and promote wildlife in association with the Bombay Natural History Society (BNHS). NAPP has signed a memorandum of understanding (MOU) with BNHS for the above-mentioned purpose. The MOU envisages an environmental stewardship programme for the volunteers. Many rare species detailed in the following points were traced and are being protected.

8) A bird marathon was organised on 31.01.2011 in conjunction with the BNHS. Eight teams were constituted. Each team traversed a distance of 5.3 km, and thus a total distance of 42.4 km (a typical marathon run) was covered. Around 172 species of birds including 10 endangered species were located. Fifty-five new species, as compared with an earlier bird count of 117 species, were identified. The marathon was a significant contribution to the enumeration of bio-diversity in the environment in and around NAPP units 1 and 2.

9) The 5th bird marathon was held on 01.02.2015. A similar distance of 42.4 km was covered by the marathon. The run was flagged off by the Station Director, Sri. D.S. Choudhary at 7:30 AM. Sri. A.K. Dutta, Chief Station Superintendent, Sri. Chandra Sekhar DGM (HR) and Sri. Subhasis Patra, Maintenance Superintendent NPS were also present.

The eight teams of the marathon were dropped at their respective ranges. Each team was allotted 8 hours for the marathon. A team could do with less time if the range allotted to it was covered earlier than the time allotted. Each team was allotted a bird book, binoculars, a camera and a data sheet. The ranges allotted covered the distance from Bharia to Dinapur Bridge. It is of interest to mention here that of the 40 threatened bird species in the UP State, 16 were present at the Narora plant site.

The findings of the marathon are as follows:
1) Total number of birds counted : 7,349
2) Total bird species detected : 163
3) Number of bird species added to the checklist : 10 (from 286 earlier to 296 now)

The new bird species that were added to the checklist are listed below:
1) Sand martin
2) Common starling
3) Isabelline shrike
4) Long-tailed minivet
5) Red-breasted flycatcher
6) Yellow-throated sparrow
7) Chestnut-bellied Nuthatch
8) Brown hawk-owl
9) Desert wheatear
10) Grey-chinned minivet

9.13.4.5 *NAPP 2: En-masse coolant tube replacement (EMCTR) and Life-extensive work*

NAPP 2 work to record shut-down for EMCTR and life-extensive work resumed from 18.12.2000 in due course.

9.13.4.6 *NAPP's unique effort to further Climate change consciousness*

'Earth Day' observance was inaugurated by Australia in 2007. The observance is to drive home the immense harm to global climate caused by the emission of greenhouse gases. Nuclear plants emit very little greenhouse gases. Social responsibility impels them to join the movement of 'Earth Day' observance. The Narora colony houses (1000 in number), other colony buildings, office complexes and shopping centres voluntarily switched off electrical power from

8:30 PM to 9:30 PM. In this way, the Narora station site observed the 'Earth Day' on 27.03.2010.

9.13.4.7 *CSR: further work on the environment and bio-diversity*

We have seen in para 9.13.4.4 some of the environmental and bio-diversity promotional causes taken up as part of the station's CSR work. The cause was taken to a new height when an environmental set-up of the station established a turtle-rearing facility at the site. Three-striped hard-shell turtles, with the shell growing up to 25 cm in length, abound in the Ganges River basin of the Narora station site. A rare species, this type of turtle goes by the name of Batagur dhongoka (in the Hindi language). The Narora facility bred 190 fresh-water turtles, out of which 130 were released into the Ganges water basin in 2015. Over 60 eggs were being artificially hatched and would have been released into the above-mentioned water basin in due course.

9.13.4.8 *Seismic event of 26.10.2015: adequate safety features of the Narora station*

The NAPP event report of 26.10.2015 mentioned a seismic occurrence at 14:39 hours with a magnitude of 7.7 on the Richter scale. NAPP continued to operate satisfactorily under this high-magnitude earthquake.

9.13.4.9 *Further record of continuous operation of NAPP 2: record accident-free days*

NAPS 2 completed a record continuous run of 731 days as of 25.08.2020 and was still running on that day. It became the fourth Indian Nuclear power unit to set a record run after the Kaiga unit (962 days, which is a world record) and RAPP 3 (777 days). This NAPP 2 run is the 34th time that an Indian nuclear Power unit has recorded a continuous run of more than 1 year.

NAPP 1 and NAPP 2 have witnessed an accident-free period of 927 days (2.54 years) from 16.09.2000 till the latest available record period of 31.03.2003. No more than five accidents have taken place for 1,547 days (4.25 years) since 04.01.1999, and the latest available record date of 31.03.2003.

9.13.4.9 Further CSR work

Carrying on its CSR work, NAPP presented a new high-roof ambulance to the Chief Medical officer of Bulandshahr on 01.14.2016 at the site. This vehicle will augment health services to the residents of the Narora area.

Kaiga Units 1 and 2

10.1   Site details: Land acquired, trees felled and planted

The Govt. of India chose Kaiga, a village about 55 kilometres to the East of Karwar and located at 14.5°North latitude and 71.1° East longitude in the Uttar Kannad Dist. of the State of Karnataka, for locating the 9th and 10th Heavy-water Moderated Pressurised Tube (HWPT) Reactor Project units and the 6th Nuclear Power Project in India. This is the first HWPT reactor project to be located in Karnataka; the first to be located off the lower West Coast of India above the Malabar Coast, and the third to be located near the West Coast of India. The site is located on the left bank of the West-flowing Kali River. The reservoir formed by the Kadra dam on this river meets the cooling and the service water requirements of the project which is located up-stream of the dam. The project is the 2nd to be located in the Southern part of the country.

The project authorities say that 50 hectares of the forest land have been acquired for siting the first two reactor units which are under construction in the first phase. Twenty-five hectares of land will be utilised for the plant facilities. The remaining 25 hectares will be utilised for colony housing and other related facilities. Another 70 hectares will be acquired for housing the four additional units (Kaiga 3 to 6) planned at the site.

The project authorities have stated that 10,000 trees have been felled so far. Around 1,50,000 saplings have been planted elsewhere to ensure sufficient tree cover in the entire area.

10.1.1   Further greenery

An area of 9.1 hectares has been converted to lawns, rendering vast greenery to the area.

10.2 Safety aspects of the project

10.2.1 Flood protection

Safety aspects have been taken into consideration in the unlikely event of the failure of the nearby Kadra dam and the resultant flooding of the site. The grade level of the site has been chosen, considering the postulated occurrence of the above-mentioned event; the accessibility, maintainability and operability of the plant have been ensured by the design parameters.

10.2.2 Earthquake protection

The Kaiga site is situated in the seismic zone-III as per Indian standards. The nearest epi-centre is situated 20 kilometres from the site. The design of the plant has been based on peak ground acceleration (PGA) of 0.2 g for safe shut-down earthquake (SSE) and 0.1 g for operating base earthquake (OBE). (Chapter 13 may be referred to in this connection)

10.3 Evolvement of bench-mark: design parameters and criteria

10.3.1 General

We have seen in the foregoing chapters (commencing from Chapter 5), the evolution of the design and construction criteria for Indian Nuclear Power Projects. RAPP was the first Nuclear Power Project to be fully constructed by India. The design for RAPP was supplied by MECo/MEIL (Montreal Engineering Co./Montreal Engineering International Ltd.). MAPP was the first Indian Nuclear Power Project to be designed, constructed and commissioned by India. The design criteria have evolved over the last 22 to 23 years since the commencement of the construction of RAPP and have carried over to those of Kaiga 1 and 2. The design and construction of Kaiga 1 and 2 may be said to constitute a bench-mark for the on-going and future Nuclear Power Projects. Some of these criteria are described in the upcoming sections.

10.3.2 Unitised concept

The engineered layout is based on the unitised concept for all safety-related areas and facilities. This means that all these areas and facilities are independent and self-contained for each reactor unit; sharing of common facilities (for twin units as a module in general) for non-safety-related areas has been allowed on a module basis. In areas such as the control building where the safety and safety-support systems of both reactors of a twin model are located, the principle of physical separation of the two systems is ensured.

All safety-related systems and components have been grouped and located in structures of the relevant safety classes namely safety Class-I, safety Class-II and non-safety.

The layout of Kaiga 1 and 2 plants is shown in an attached sheet.

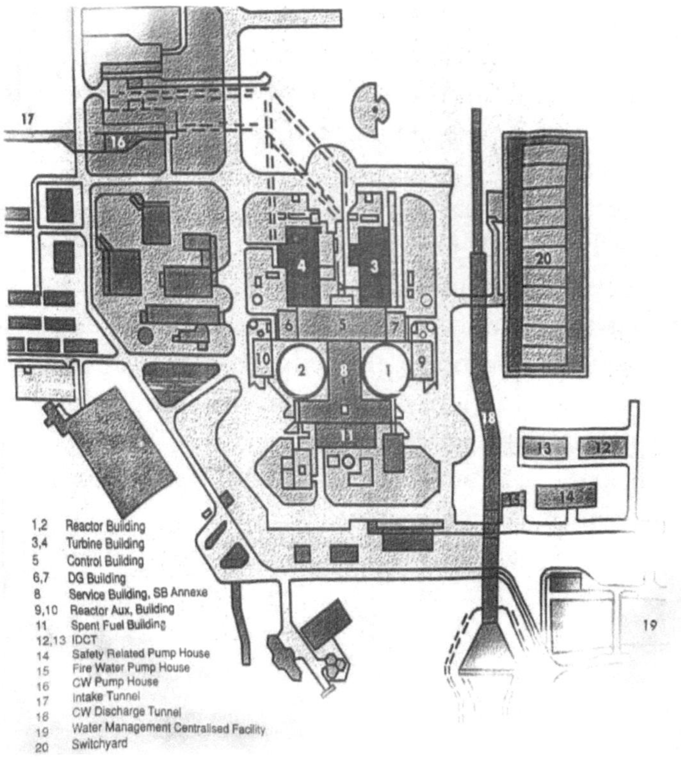

10.3.3 Safety against low-trajectory missiles

This is ensured by the positioning of the safety-related structures. The turbine building is located radially to the reactor building. Safety-related structures such as the reactor building, reactor auxiliary building, control building, diesel generator building, etc. are located to provide safety against low-trajectory missiles emanating from the turbine building. One such instance may be called one of the numerous blades of the 3000 rpm turbine flying loose; the positioning of the related buildings is such that a missile would fly away from the reactor and other safety-related buildings.

10.3.4 Safety-related buildings (SRBs): design and engineering concept

A separate building called the control building designed for SSE criteria has been built to house the Main Control Panel and other safety-related systems.

Two buildings called Diesel Generator (DG) buildings have been built, one each on the North and South sides of the control building. Each of these buildings houses three DG sets to cater to each reactor unit.

The reactor auxiliary buildings, one each for the two reactor units, have been built adjacent to each reactor building. These auxiliary buildings house the heavy-water systems and safety-related systems including the emergency core cooling systems, among others. The location of the buildings has been decided based on the economy of inventory of heavy water in the pipelines and the length of piping carrying active fluids.

A schematic diagram showing the important structure rooms, major equipment, etc. is shown below.

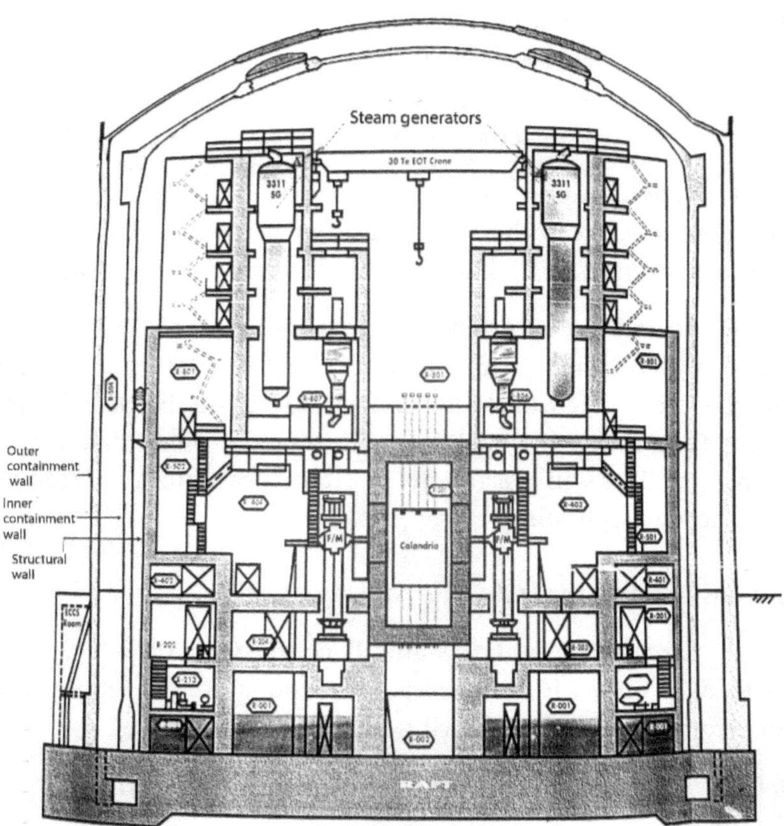

Reactor Building Schematic

10.3.5 Guard pipes between the ICW and OCW

Guard pipes have been provided around the main-steam lines and other high-enthalpy pipe penetrations between the ICW and OCW to prevent the over-pressurisation of the secondary containment in the event of a break in these lines.

10.3.6 Location of main-steam and Inter-isolation valves (MSIVS)

The isolation of nuclear and non-nuclear islands has been achieved using the location of mainstream isolation valves, including seismic anchors between the reactor buildings and turbine buildings; this has enabled the classification of turbine buildings as non-nuclear structures.

10.3.7 Survival Ventilation System

The survival ventilation system (SVS) has been incorporated to cater to the requirements of the control room and the computer room in case of high-radiation activation in the outside atmospheric air.

10.3.8 Vapour suppression

The annulus between the ICW and the OCW has been used for vapour suppression. This has eliminated the need for vent shafts.

10.3.9 Emergency heat removal from the reactor

During a station black-out situation, the SSE part of the water storage system is fed to the secondary side of the Nuclear Steam Service Generators (NSSGs) by making use of diesel-driven fire pump units. This will assist in decay heat removal from the reactor using thermo-syphoning.

10.4 Safety system specific to Kaiga APP

10.4.1 Circulating cooling water discharge (CCWD)

The CCWD tunnel is designed for SSE and is used as emergency water storage under the following conditions.

10.4.1.1 Make-up water to the induced draft cooling tower (IDCT)

Under the conditions of non-availability of Kadri Lake water, as a result of an earthquake of SSE level, the CCWD storage water is used as make-up water to IDCT. This water is likely to last for about a week.

10.5 Estimated cost

10.5.1 Design changes: conventional equipment and systems

The project has been envisaged as a 6-unit one. Initial financial sanction was issued for 2x235 MWe units as a first phase of the project for Rs. 730.72 crores in 1987. The project was included in the 7th five-year plan which commenced in 1984. The first phase was scheduled to be completed by 1995-1996. The anticipated cost of the first phase at 1994 prices was Rs. 2,275 crores. The cost over-run has been ascribed to the following reasons:

1. Escalation due to increase in raw material and equipment cost: Rs. 430.70 crores
2. Incorporation of additional safety requirements, increase in taxes and duties, foreign exchange price increase due to decline in value of Rupee and related reasons : Rs. 283.67 crores
3. Increase in scope of work due to design improvements : Rs. 144.91 crores
4. Interest during construction : Rs. 685 crores

The cost as of the end of 1999 was Rs. 2896 crores. The final cost at the end of construction was Rs. 3,100 crores.

Several design changes were adopted in Kaiga 1 and 2. Some of these are described in the upcoming sections.

10.5   1.1 The PHT Main System

Valves were eliminated. This has resulted in saving the heavy-water inventory by eliminating leakages in glands, valve seats, etc. It may be noted that there were 24 valves in each reactor unit in RAPP and MAPP and 16 in each unit in NAPP and KAPP.

10.5.1.2   *Excitation system for the turbine generator units*

A brushless excitation system was adopted for the generator. Excitation transformers and excitation cubicles were hence done away with.

10.5.1.3   *The 220-KV switchyard system*

The double-main double-bus system was adopted, which would provide added operation flexibility; the shut-down on one complete main bus could hence be taken without affecting the normal operation of the switchyard which in turn would also help operational activities under emergency conditions.

10.5.1.4   *Feed from start-up transformers*

It was decided to feed the output of the 220 KV/6.6 KV-6.6 KV start-up transformers to the 6.6-KV SwGr through the 6.6-KV bus ducts. This would eliminate cumbersome 6.6-KV cable box arrangements, cable terminations, etc. on the 6.6-EV side of the transformers. The best duct arrangement would also result in more reliable and flexible operation and maintenance conditions.

10.6 Chronological progress of work

10.6.1 Preliminary work as of 1986

This work commenced in 1986 after a token provision was made in the Union 1985-1968 budget. The work comprised job shacks, approach roads, warehouses, residential quarters, construction power-supply, water-supply, housing colonies, plant site work, etc. as was the case in earlier Nuclear Power Plants.

10.7 Permanent plant work: status as of 1987-1989

10.7.1.1 Circulating cooling water and other plant water requirements

These requirements were to be met with water from the Kadri Dam reservoir. The Kadri dam was part of the 132-MWe 6500-crore World Bank-aided Kali River project of the Govt. of Karnataka. The circulating water system is a once-through system. This work was taken up during 1987-1989.

10.7.1.2 Kaiga 1 Civil work: progress as of mid-1991

10.7.1.2.1 Excavation work: RB-1 and TB-1

The Civil work of Kaiga 1 commenced in 1988. The excavation for RB-1 was completed in late 1988/early 1989. A giant geodesic dome was constructed over RB-1 for the first time in India to carry out concreting without any interruption due to monsoon rains, which lasted for 4 months in this area.

10.7.1.2.2 Excavation work: concreting of the main plant

Mud-mat and raft foundation for Kaiga 1 RB were completed in 1990. The quality of concreting was of M-25 grade. Piling work for TB-1 and TB-2 was taken up during the end of May 1990. About 15% of the work was completed by March 1991.

10.7.1.2.3 Kaiga 1 and 2 Civil works: progress as of mid-1991

About 25% of the overall work was completed by mid-1991.

10.7.1.3 Permanent plant equipment: the ordering and receipt at the site and status as of 1991-1993

NSSGs for Kaiga 1 and 2 were ordered from BHEL and were manufactured at the company's Tiruchirappalli (TN State) work. The first NSSG was shipped to the site in December 1991. The weight of each NSSG was 105 tonnes: NSSG for Kaiga 1 was, for the first time in the history of the construction of Indian Nuclear Power Projects, transported by road covering a distance of 1,237 kilometres. The cost of each NSSG was Rs. 5 crores.

The first 220 KV/6.9 KV-6.9 KV start-up transformer for Kaiga 1 was received at the site in 1993. The second start-up transformer (SUT) intended for use at the KAPP 2 unit was subsequently diverted to KAPP 1 given the problems faced in this station unit.

10.7.1.4 Kaiga 1 Civil work: status as of early 1994

The ICW and OCW work were taken up during 1992-1993. The quality of concrete for OCW was M20 and that of ICW was M35. The concreting of ICW was completed on 31.01.1994. This signalled the completion of all major Civil work for Kaiga 1. Ground break for the induced draft cooling tower (IDCT) was done. Raft foundation work for the ventilation stack was completed.

10.7.1.5 Failure of ICD of Kaiga 1

The dome failed on 13.05.1994 during concreting with implications for other on-going containment dome work. The mishap was a major one and is dealt with in the following sections.

10.8 Kaiga 1: full details of the ICD failure during concreting

10.8.1 Physical details of the ICD

Before detailing the failure, let us first look into the physical details of the ICD. The dome is 42.6 metres in internal diameter. Its thickness varies from 0.34 metres at the base and up to the ring beam, and to 0.15 metres thereafter and up to the crown. The maximum height of the dome from the base of the dome is 7 metres.

10.8.2 Details of the failure

It was about noon on 13.05.1994. Workers of the piping system and the Civil contractors were working at the floor level of the inner pre-stressed concrete dome. Pre-stressing of the high-tensile strength of the dome had started from the beginning of May 1994. About 183 metres of steel cables housed in 80-mm diameter flexible steel sheaths were to be stressed to the design of 280 tonnes. The guaranteed ultimate strength of the rods was 156 kgs per mm sq.; 65 cables had earlier been pre-stressed one after the other in a planned sequence without any problem. The tensioning of the 66th cable was completed at 4:00 AM on 13.05.1994.

10.8.3 Details of the damage

Large pieces of concrete started falling from the concreted surface of the dome at a height of nearly 40 metres from the reactor grade level at 11:45 AM and again at 12:40 PM. The falling pieces caused injury to 13 persons belonging to the civil and piping contractor agencies, who were then working on the floor of the dome.

It was later estimated that a total of about 130 metric tonnes had fallen from an area of 40% of the dome's inner surface. It would further be seen later that a much major portion, namely 104 tonnes (8% of the total content of the 1300 tonnes of concrete poured into the dome and nearly 80% of the total weight of concrete that had fallen), had come out of 15% of the dome inner surface area.

10.8.4 Comparison between the design parameters of RAPP/ MAPP and the Kaiga domes

It would be useful at this stage to compare the design values of internal pressure for the domes built so far and currently being built. The Kaiga dome has been designed for an internal pressure of 1.73 kgs/cm sq. as against those of 1.16 kgs/cm sq. for MAPP and 0.42kg/cm sq. for RAPP units 1 and 2. It is thus evident that the design internal pressure for Kaiga APP was of a much higher order than that of RAPP and MAPP. The Kaiga dome was of course yet to be tested for withstanding this design pressure. This would be done after the dome is repaired or re-built.

10.8.5   Details of study of the dome failure

A very preliminary study of the failure of the Kaiga dome indicated that the lamination of the layer of concrete had taken place. The following steps were taken for further study of the failure:

(1) Samples of concrete from locations where it fell would be identified and taken for analysis and tests.

(2) Design calculations of Stupp India Ltd., who are design consultants for the Kaiga project and served for earlier nuclear containment domes as well, were to be checked.

(3) Computer studies of the behaviour of the domes till the occurrence of the mishap were to be made.

A 6-member expert committee headed by Sri. Ch. Surendar, then Director Engineer of the NPCIL, was constituted to go into the causes of failure and suggest remedial measures to guard against such occurrences in the future. The findings of the committee were submitted to the NPCIL in May 1995.

10.8.6   The AERB fiat regarding dome work on Kaiga 1 and other on-going projects

The AERB imposed a hold on all work on the safety structures of Kaiga 1, Kaiga 2, RAPP 3 and RAPP 4 and the pending completion of studies and analysis of the Kaiga 1 dome mishap. The AERB constituted its enquiry committee to go into the cause of the failure of the Kaiga 1 dome. This committee also submitted its findings to the AERB in May 1995.

10.8.7   Findings of the Enquiry Committees

Investigations of the two enquiry committees did not reveal any use of defective materials in the construction of the dome. Recommendations were made by the committees for the dismantling of the failed dome and re-building of a new dome with the following changes in the design of the dome:

(1) The thickness of the dome at the base is to be increased from 0.34 metres to 0.47 metres

(2) Radial re-inforcement to be provided to longitudinal and axial re-inforcement

(3) Upgrading of mix of concrete
   The failed dome was dismantled by the end of 1995.

10.9   Kaiga 1 Civil work: status as of June 1997

Almost all Civil work except the re-building of the dome of Kaiga 1 was completed.

10.10   Civil work: status as of 1998

Clearance from the AERB for the commencement of work on the re-engineered IC dome was received in October 1998. The work commenced by the end of 1998 and was completed successfully by early 2000.

10.11   Kaiga 1 conventional work: progress during 1993-1994

10.11.1   Installation work

The TB hall crane was installed. The TG accessories such as HP/LP heaters and main-steam condenser were installed. The LP turbine rotor was installed in position. The boiler feed pumps and the 415-volt SwGr and Mccs were installed.

10.11.2   Expenditure during 1993-1996

A capital expenditure of Rs. 273.94 crores was incurred on Kaiga 1 and Kaiga 2 during 1993-1994.

10.11.3   Kaiga 1 status as of June 1997/early 1998

10.11.3.1   Power and control supplied water system

AC and DC power and control cable supplies for equipment such as 6.6 KV/415 V SwGr, 250 V/48 V DC control system and 6.6 KV/415 V Class-III power-supply system were commissioned. A part of the process water-supply system and the de-mineralised (DM) water-supply system were commissioned.

10.11.4   The 235-MWe TG system

The erection of the system was completed. Several modifications on the turbine rotor cylinders, diaphragms, etc., as per the suggestions that were involved in the consultation with the engineering decision of the NPCIL, were carried out on the respective components in-situ. The set was put on barring gear on 6.1.1998. The TG now awaited nuclear steam for its commissioning.

10.11.5   Kaiga 1 Reactor Plant Work: chronological progress

10.11.5.1   *As of the end of 1993 or early 1994*

The integral Calandria end-shield assembly was aligned, welded and grouted in position. Calandria tube rolling was completed. Prefabrication of the PHT feeder piping and 40% of the nuclear piping were completed. The PHT feed pumps, bleed condenser, etc. was installed. Installation of the first NSSG manufactured by BHEL was taken up; filling of the SS balls in the end-shields was completed. Main-steam inlet and safety discharge valves (MSISDV), designed and manufactured by BHEL for the first time in India, were installed.

10.11.5.2   *As of June 1997*

All the major reactor plant equipment was installed. Nuclear piping was nearing completion. The PHT system was a 'go' and hot-conditioning was expected to commence. The primary leak test of R-3 was completed.

Kaiga-1, the second unit in the twin-project (Kaiga 1&2) at an advanced stage of completion.

10.11.6   The 220-KV switchyard: progress as of June 1997

The erection of all equipment was completed. The SUT-1 was prepared for commissioning. The commissioning of other equipment was in progress.

10.12 Kaiga 1: overall progress as of 1997-1999

10.12.1 As of June 1997
About 83% of the overall work was completed

10.12 2 As of early 1999
About 92% of the overall work was completed.

10.12.13 Power evacuation from the station
The already constructed 400-KV double circuit line would be charged at 220 KV initially for the evacuation of power. The above-mentioned DC line was ready by June 1999.

10.12.4 First-criticality of Kaiga 1
First-criticality was attained on 26.09.2000.

10.12.5 First synchronisation
The unit was synchronised with the Southern Regional Electricity Board (SREB) on 12.10. 2000.

10.12.6 Commercial operation
The unit was declared for commercial operation on 16.11.2000. The station generated 10.63 billion units as of June 2008.

10.12.7 Capacity factor (CF)
The capacity factor during 2001 was 88%. It jumped to 92% during 2002. It has attained 100% several times since then.

10.13 Kaiga 2: chronological progress

10.13.1 Civil work

10.13.1.1 As of the end of 1993
Over 60% of the Civil work in the TB and 90% of the work in the control building were completed. About 45% of the work in the service building was completed.

10.13.1.2   As of June 1997

10.13.1.2.1   Internal containment dome (ICD) work

The work of fabrication and erection of the supporting structure for the form-work of the ICD was taken up in November 1996. The structure consisted of 32 numbers of plate girders in the radial direction; these girders were interconnected by 22 numbers of pin-jointed trusses in the circumferential direction. The supporting structure was first assembled on the ground and the erection was carried out in place with the use of a 650-tonne capacity 'Liebherr' crane. The complete structure was erected in eight segments. The entire structure weighed 580 metric tonnes. The load of the supporting structure was transformed to 32 brackets fixed on the IC perimeter wall and also to 10 intermediate supports resting on the internal walls of the RB. The entire fabrication and erection work was completed in 90 days.

A mock-up of the ICD and ring beam was completed as a prelude to the concreting work.

The pumping method was adopted for the concreting of the ring beams. Two pumps connected to one concrete placer were operated simultaneously during the concreting process. Two additional pumps complete with piping were kept as stand-by.

Concreting of the ring beams was completed in three pours, the rate of concreting being 10 m3 per hour. The grade of concrete used was M-60/20/175/23 C with the content of micro silica being 71.2% of the weight of the cement. The cement content was 475 kgs/met3. The water binder ratio was 0.32; micro silica (in the form of silica fumes) used was 7.5%. The super plasticizer used was 2% by weight of cement.

10.13.1.2.2   Other Civil Work

Civil work of the induced draft cooling towers (IDCT), ventilation stack, domestic water tank and discharge tunnel were completed.

10.13   1.2.3 As of the end of 1998/early 1999

Both the ICD and OCD were completed after receipt of clearance from the AERB for proceeding further with the safety structure

Civil work. The secondary leak test on the OCD was completed. The integrated leak test was taken up.

10.14   Kaiga 2 conventional equipment work: chronological progress

10.14.1   As of the end of 1993
The work in the turbine building was taken up.

10.14.2   As of June 1997
The 6.6 KV/415 V Class-III power-supply system, complete with 415-volt diesel generators, was installed and tested. About 75% of the erection of mechanical, electrical and instrumentation equipment was completed. The erection of the 235-MWe TG set was in an advanced stage of completion.

10.14.3   As of the end of 1998/early 1999
The TG erection work and power and instrumentation cabling were completed. The process instrumentation system work was in an advanced stage of completion. The main communication system, the control power-supply system and the 220-KV switchyard were commissioned.

10.14.4   The Kaiga 2 reactor plant and associated equipment system: chronological progress

10.14.4.1   As of the end of 1993
Calandria end-shield alignment, welding and grouting work were completed. The installation of the PHT and moderated system equipment was taken up.

10.14.4.2   As of the end of 1997
More than 80% of the pre-fabrication of nuclear piping and more than 60% of the erection of nuclear piping were completed. About 75% of the erection of the nuclear plant and related equipment was completed.

10.14.4.3 As of the end of 1998/early 1999

Hot-conditioning of the PHT system was completed. The preliminary leak test of the RB was completed. Fuel loading was completed.

10.15 Kaiga 2: overall progress as of the third quarter of 1999

The overall physical progress as of the 3rd quarter of 1999 was 95%. With the commissioning of all the systems required for the first approach to criticality, steps were initiated to obtain approval for this approach. The project design safety committee visited the plant on 23.8.1999. This committee gave its findings to the Advisory Committee for Project Safety Review (ACPSR). The ACPSR gave its final suggestions and recommendations to the AERB.

10.16 Review by the AERB: approval for the first approach to criticality and dedication to the Nation

The AERB reviewed the first approach to criticality. This review comprised the following steps:

1. Cold and hot performance test
2. Fuel loading
3. Addition of heavy water to the moderator and the PHT system.

The AERB carried out low-power and further power tests on the reactor.

The AERB accorded its approval for initiating the first approach to criticality. The approach was initiated in the fourth week of September 1999. The Boron poison rods were fully drawn out to ensure the sustenance of the fission process. Criticality was achieved at 2:51 PM on 24.09.1999. To review the first criticality, the low-pressure test and further power tests were carried out by the AERB. The unit was dedicated to the nation by the then PM, Sri. A.B. Vajpayee, on 5.3.2000. Kaiga 2 has generated 11.67 billion units of electricity since it was declared commercial up until June 2008.

10.17 Design basis for the achievement of criticality: further details about the first approach to criticality

The reactors at Kaiga have been designed to operate at near full level of the Calandria. The Calandria was thus filled up to 94% of the full level during the run-up to first-criticality. Adequate neutron poison in the form of boron oxide (B2O3) in the Boron poison rods was provided in the moderator system to prevent inadvertent criticality during the filling up of heavy water. Spontaneous fission from U238 yields 11 neutrons per second per kilogram of uranium oxide. This neutron flux is too small to be detected by the ionisation chambers used for monitoring the neutron flux during power operation. Special Boron-coated proportional counters were installed in the central thimble as well as in the Calandria vault to continuously monitor the law values of flux during the processes involving reactivity additions.

These counters were independently connected to protective channels. The reactor period (time taken for neutron flux to multiply 2.717 times) was calculated and found to be 858 seconds. This is quite a safe value. The critical Boron concentration is a measure of core excess reactivity in the absence of fission products and other negative reactivity loads. The measured critical Boron concentration was 12.0 ppm which was in agreement with the predicted value and within the limits of inaccuracies in measurement.

10.18 Kaiga 2: quantities of work executed

We have in the foregoing sections dealt with the various design and construction facets of the Kaiga APP. Let us now look into the quantities of the work of some major items in the civil, mechanical, electrical inspection and instrumentation systems.

Item No.	Description	Quantity
10.18.1	Earth Work	3 Million met3
10.18.2	Concreting	3.5 lakhs met3
10.18.3	Welding	7.0 lakhs inch diameter
10.18.4	Radiography	5.0 lakhs in diameter
10.18.5	Electrical cables laid	1000 km

Item No.	Description	Quantity
10.18.6	Duct Erection (a/c and ventilation)	35000 sq. metres
10.18.7	Instrumentation Cabling	450 k metres
10.18.8	Tubing	125 k metres
10.18.9	Number of control wire terminations	10 lakhs

10.19 Time of completion of Kaiga 2: the effect of the failure of the dome of Kaiga 1

The first concreting of the Kaiga 2 reactor building began in January 1990. First-critically was attained on 24.9.1999. It would hence appear that the total time for completion from the first concreting was almost 9 years 8 months. It is, however, to be noted that a hold was put on all work on the safety structures including the containment dome work following the failure of Kaiga 1 in May 1994. The work on the containment structures commenced in November 1996 after the receipt of clearance from the AERB and was completed by the end of 1998. Thus, 2-and-a-half years was lost due to the hold imposed on the containment work. If this period is excluded, the actual time taken for the completion of Kaiga 2 is a little over seven years from the time of the first concreting, this compares favourably with the time of completion of the Kakrapar project whose completion time was 6-and-a-half years.

10.20 Capacity factor (CF)

The capacity factor of Kaiga 2 was 90.03% during the period 2000-2001. It was 86% during the year 2001-2002. It has since achieved a capacity factor of 100% several times.

10.21 Re-building of the failed Inner containment dome (ICD) of Kaiga 1

10.21.1 General

We have seen in para 10.8.2, the details of failure of the ICD of Kaiga 1, and in para 10.8.3 the details of the damages to the dome. The findings of the enquiry committees, which looked into the damage, and their recommendations for improvement in the design and construction of a new IC dome for Kaiga 1 and the IC domes of all future projects have been detailed in the previous sections. We will now go into the details of the construction of the new dome of Kaiga 1.

10.21.2 Mock-up for overcoming re-inforcement congestion: placement of re-inforcement steel

The problem faced by congestion of re-inforcement rods, particularly in a complicated safety structure like the ICD, was quite severe and hence required careful study before the taking up of concreting. The dimension of the problem could be visualised by the fact that the thickness of the concrete near the 4.2-metre openings (four) in the ICW Dome was 1.22 metres; the thickness was 0.47 metres in the general surface areas of the dome. Various mock-ups were conducted to address the congestion issues arising from the placement of re-inforcing and pre-stressing cables and arrive at the laying procedure. The mock-ups also helped in ensuring the normal flow of concrete in such congested locations.

10.21.3 Kaiga 1 ICD Mock-up for Overcoming re-inforcement Congestion

Cold twisted (TOR steel) bars are used as re-inforcement bars in re-inforced concrete. These bars are in a standard length of 10 metres. Four such rods were joined together by butt welding and mechanical splicing to reduce the congestion of rods during placement.

The total quantity of re-inforcing rods used in the ICD was 220 metric tonnes. About 1360 welded joints and 1350 numbers of mechanical splices were made. The average length of the rods as laid was 40 metres with four welded or spliced joints. Non-destructive and destructive tests such as dye penetration, radiography and mechanical tests were conducted on the joints before laying; only pre-qualified welders were employed for welding purposes. The welder output was 17 joints per day and that of splicing was 30 joints per day. The complete job of welding and splicing was carried out in 80 days. The work of lifting and placement of rods was carried out with the aid of a 35-metre-long space-frame truss.

21.3   Kaiga 1 ICD

12.21.3.1   Pre-stressed steel, pre-work and concrete placement

About 19 strands of pre-stressed low relaxation steel cable (K-13 grade) housed in an 80-mm internal diameter lead-coated dross batch of flexible mild steel sheathing were used in the inner containment dome. The capacity of the cables was 355 tonnes. Some of these cables called 'D' cables were anchored in the ring beam at both ends. These cables were threaded into the sheathing on the ground and lifted with the aid of the same space-frame truss used for the placement of re-bars. Pre-fabricated templates were used to lay and support the pre-stressed cables in their assigned locations. The cables which used to be anchored with one end in the ring beam and the other end in the stressing gallery housed in the base raft called 'J' cables were lifted without the sheathing with the aid of a de-coiler. The sheathings of 3-metre unit length each were later threaded onto the pre-stressed cables and the joints scaled with couplers and 'M-Seal'. On average, 10 to 12 'J' cables and 12 to 15 'D' cables were placed in one day. The complete operation of cutting, placement and alignment of cables took 60 days. The total number of cables in the ICD was 170.

10.21.4   Kaiga 1: ICD construction joint preparation

The complex nature of the ICD form-work does not provide proper conditions like space for chipping of already cast concrete to ensure proper bonding with subsequent pours of concrete. The following method was adopted to ensure proper and effective bonding.

A surface retarder which was 0.1% by weight of cement was used to retard the setting of concrete to a depth of 5 mm from the surface. The exposed construction joint was then green-cut by using an air–water jet at a pressure of 7 kgs/cm sq. to expose the concrete aggregate and to provide a proper rough bonding surface between successive concrete pours.

10.21.5   Kaiga 1: ICD development of high-performance concrete and regulation of temperature of pour

The design requirement called for a characteristic cube compressive strength of 60 MPa and a split tensile strength of 3.87 MPa. Various parameters with adhesives to concrete were studied and the concrete grade of M-60 was successfully developed with the use of silica fumes in concrete. A total quantity of 1950 cu. metres of concreting was involved in the IC dome work. The grade of concrete used was M-20 for the outer dome and M-60 for the inner dome. The temperature of the pour of concrete was controlled to less than 23 degrees C; this was ensured by replacing water with ice flakes to prevent cracking of freshly laid concrete by temperature grading arising from the release of heat of hydration by cement.

10.21.6   Kaiga 1: long-term structural monitoring of ICD

An innovation in the ICD concreting work of Kaiga 2 was the measure taken for long-term structural monitoring of the dome and for studying the loss of pre-stress in steel with the passage of time; vibrating wire strain gauges (VWSGs) were embedded in two diametrically opposite locations in the ring beam and 13 locations in the dome. The wires connecting all the VWSGs to the data logger were routed along the pre-stressing cable sheathing and brought out of the dome at the springing (base raft) level. These wires were further routed in a 100-mm NBGI pipe to the control room situated at the ground level of the RB for data logging; strain and temperature measurements were taken every 30 minutes after the placement of concrete in a particular pour.

10.21.7   Kaiga 2: time for OC dome concreting and total quantity of concreting

The time taken for OC dome concreting, only excluding connected work like the placement of re-bars, splicing welding, form-work, support steelwork, etc., was 70 days. It may be noted that the inter-pour gap was seven days. This period was included in the total time of 70 days. As a design requirement, concreting of pour-1 in the ring beam was started after the complete placement of re-bars including the pre-stressing cables.

A total of 1950 cu. metres of concreting was involved in the dome work.

The total time taken for concreting of the dome was 16 months; this period included 4 months of monsoon during which no work could be carried out.

10.21.8   Progressive financial expenditure strength and site organisation

10.21.8.1   Financial expenditure

A total amount of Rs. 182 crores was expended on Kaiga 1 and 2 from 1986-1987 to 1989-1990—of this, the site expenditure was Rs. 22 crores. An expenditure of Rs. 76.44 crores was incurred from 1989 to 1990 against a budget provision of Rs. 110 crores—the utilisation during the year working out to 70%. An amount of Rs. 126.1 crores was expended during 1990-1991, bringing the progressive expenditure to Rs. 308.1 crores. An amount of Rs. 976.13 crores was spent till 30.6.1994. The budget provision for 1992-1993 was Rs. 220 crores. Provisions continued to be made annually to meet the increasing tempo of work. Sums of Rs. 2500 crores, from 1999 to 2000, and Rs. 2900 crores, from 2000 to 2001, were expended.

10.21.8.2   Kaiga 1 and 2: Strength of personnel at the site

A total complement of 784 personnel was deployed at the site as of early 1992—of this, 620 personnel were appointed as a result of direct recruitment. A large proportion of this complement comprised local labour. The project authorities had stated that one member of each of the 85 families displaced by the project had been employed permanently at the plant.

10.22   Concluding remarks of Kaiga 1 and 2 construction

The design and construction of Kaiga 1 and 2 have been unique in many ways. We have seen in para 10.3.2 how the design and construction of the twin units have constituted a bench-mark for the on-going and future 235 MWe Nuclear Power Projects. Another important occurrence in the construction of Kaiga 1 and 2 was the mishap to the dome of Kaiga 1. The mishap was an unforeseen and unexpected event which had, on first impression, seemed to constitute a road-block to speedy construction of the other Nuclear Power Projects. The committees constituted for looking into the failure of the dome did not find any flaw in the design, nor use any defective materials in the construction; such failures may not be entirely unexpected in the progress of projects involved in cutting-edge and frontier technologies such as nuclear projects.

It is to the credit of the Engineers involved in the Nuclear Power Project construction that they have come up with improvements in the design and construction of the dome in a fairly reasonable period. These have already been dealt with in detail in earlier paras of this chapter; that the design of containment domes has been sound is borne out by the fact that the RAPP 1 dome has been in satisfactory service for more than 32 years, while other domes have been in service for upwards of two decades. These facts would serve to help us view the failure of the Kaiga 1 dome in the proper perspective.

The design and construction of Kaiga 1 and 2 have proved to be invaluable from the point-of-view of the challenges faced in this endeavour and have infused a sense of achievement and hope for the future in the minds of those involved in these projects. More importantly, the endeavour has served to reassure the general public in India, and abroad, that our nuclear power programme is headed in a sound and assured direction. Kaiga 1 and 2 have generated over 13.35 giga units of electricity up to December 2004.

Kaiga 2 set up a record in continuous operation by an uninterrupted run of 529 days during the period 19.08.06 to 31.07.08. It has also generated 11.67 billion units since going commercial as of June 2008.

10.23   Certification as per the ISO-14001 award inspection requirement

Kaiga 1 and 2 received the certification for ISO-14001 from 2002 to 2003. The station received the 'Industrial Safety Award' from the AERB for 2005 and the 'Yogyata Praman Patra' from the National Safety Council for 2005. Optimised in-service inspection requirements were issued.

10.24   Corporate Social Responsibility (CSR)

10.24.1   Contribution to Education

Utility educational items were distributed to needy students in the nearby localities.

10.24.2   Further CSR work

The Kaiga Station authority's 'Kaiga welfare committee' constructed a laboratory building for the 'deaf and dumb' school at Siddar village, which is 18 km from the site, and handed it over to the school authorities at a function held on 19.09.2009.

10.24.3   Water-supply work and CSR scholarship to students

The Kaiga Welfare Committee (KWC) undertook the drinking water-supply for Kunnipet village at the request of the village Panchayat. The work carried out in synergy with the Panchayat envisaged the KWC's scope of construction work of 75 kilo-litre capacity over-head rec-storage tank, pump-house, submersible water pump and motor power-supply and discharge pump connections of the tank. The Panchayat's scope of work comprised the distribution of pipe connections to the village consumers. The welfare committee also disbursed scholarship amounts to meritorious students of the adjoining Talukas pursuing Industrial Training Institute (ITI) and diploma courses under its 'Gyana Gangotri' scheme on 11.02.2010.

10.25   Preserving the Environment and Fostering bio-diversity

The Kaiga station has been nurturing the local environment and contributing to the fostering of bio-diversity in the vicinity of the plant. The nearness of Anshi National Park has furthered the above causes.

Avid bird watchers of the station aided by similar enthusiasts from the North Karnataka network and students from the forest college at Sirsi surveyed an area of 43 kilometres around the plant for bird species during 2011-2012. Around 152 species of birds were sighted. Some important species were:

1. Grey-haired lapwing (indigent to North Eastern China and Japan)

2. Laughing dove
3. Indian silverbill
4. Jungle Owlet
   (It may be of interest here to mention that the State of Maharashtra is making efforts to declare the Owlet as a State bird).
5. White-bellied Sea eagle
6. Malabar pied hornbill
7. Red-vented bulbul

The following winter migratory birds were also spotted by bird watchers in Kaiga and other States like Goa and Tamil Nadu.
(1) American toucan
(2) Ruddy-breasted crake
(3) Indian Thick-knee
(4) Common teal
(5) Eurasian coot
(6) Eurasian wigeon
(7) Gadwall
(8) Garganey
(9) Indian spot-billed duck
(10) Whiskered tern
(11) Green warbler
(12) Mottled wood owl
(13) Indian nightjar
(14) Striated heron
(15) Pallid Harrier

These species of birds were photographed at close quarters in Bengaluru, Hubli, Gadwal, Goa and Tamil Nadu.

10.26   The 'WANO' award for the Station Director

We now go into awards received by the station and site directors. The World Association of Nuclear Operators (WANO) conferred the 'nuclear excellence' award on Sri. Guntur Nageswar Rao, Station Director of NAPP, for the year 2007. Sri. Nageswar Rao was later appointed as the Executive Director of Operations at the NPCIL headquarters, AS Nagar, Mumbai.

10.27 National Award

RAPP units 1 and 2 and Kaiga 1 and 2 have been conferred with the 'Powerlines Award 2012' for having been the best-performing Indian nuclear power stations during 2011-2012. Sri. J.P. Gupta, the RAPP site director, and Sri. R.R. Dave, Chief Superintendent of Kaiga 1 and 2, received the awards at the hands of Sri. Sushil Kumar Shinde, Union Power Minister on 15.05.12 at New Delhi.

10.28 KAGs 1 and 2 generation details, availability factor and dedication

Kaiga (KAGs) 1 and 2 have generated 27.21 billion units of electricity since their commissioning. The revenue generated was Rs. 68 billion (6800 crores), approximately. The availability factor of KAGs 1 and 2 during 2011-2012 was 91.24%. KAGs 2 was dedicated to the nation by the then PM, Sri. A.B. Vajpayee, at the site on 05.03.2000. The former Chairman, the then Chairman of the AEC and senior officials of the NPCIL were present during the function.

(L to R) Mr. J.K.Singh, Mr. Anant Kumar, Mr. R.C. Rawal, Mr. N Rajsabai, Dr.R.Chidambaram, Mr.V.K.Sharma, Prime Minister Mr. Atal Bihari Vajpayee, Mr.Brajesh Mishra, Dr.Raja Ramanna, Dr. Y S R Prasad at the dedication ceremony.

Kaiga unit-2 after dedication to the nation.

10.29   Severe grid disturbance on 03.04.2012

There was a very severe disturbance on 03.04.2012 that resulted in a total loss of evacuation of power from KAGs 1 to 4. All 4 units continued to operate on the in-house load feed mode. All four units were synchronised with the grid within 2 hours in coordination with the seamless co-operation of the grid authorities.

10.30   CSR: Further Environmental Involvement

So far, we have dealt with the active involvement of the project authorities in furthering the bio-diversity of the plant environment. This was further carried forward in the insect life of the environment when the project authorities observed a female 'Atlas' moth, considered the largest butterfly in the world with a wingspan of 25 cm, being chased by two birds in front of the office of the Director of the Kaiga stations on the 17.6.2014.

The moth, fortunately, escaped the chase and rested two days on the leaf of an ornamental plant at the above-mentioned place. The moth deposited 10 eggs on the upper portion of the leaf and more on the lower portion. The eggs on the lower part of the leaf formed caterpillars of 3 to 5 mm length each after 9 days. We look forward with great interest to the further development in the lives of this gift of nature to us.

10.31 More awards for KAGs 1 to 4

We have already discussed the awards received by KAGs 1 and 2. We will now deal with more awards for KAGs 1 to 4. KAGs 1 and 2 received the 'National level safety' award from the Directorate General, Factory Advice and Labour Institutes (DGFASLI), for the 'lowest average hazard frequency rate' performance for the year 2012. KAGs 3 and 4 were the runners-up for the year. The awards were presented by Sri. Narendra Singh Tomar, Union Minister for Labour and Employment, and Sri. Vishnu Deo Sai, Minister of State Labour Employment Department, at a function held at Vigyan Bhavan, New Delhi, in September 2014.

Sri. Suresh A. Patil, nuclear plant worker at KAGs 3 and 4, was awarded the 'Best worker' award for the year 2012, for his outstanding contribution in fuel-handling activities in KAGs 3 and 4 at the same function as stated above.

A model of a 220-MWe HWPTR was presented by Sri. M.P. Hansone, Station Director of KAGs 3 and 4, to the Goa Science Centre in Goa and it was inaugurated on 20.09.2014

10.32 Further Social Responsibility Initiatives

Apart from the CSR initiatives discussed earlier, there are several more to be dealt with. Sri. H.N. Bhat, Kaiga Station director, signed a memorandum of understanding with the 'Artificial limbs manufacturing of India', on 06.10.15, for the implementation of 'ASAN', a movement for providing artificial limbs to the local physically challenged citizens. The station carried out four preliminary surveys in the camps within a 16-km radius of the station site to identify the physically challenged citizens, during October 2015 and further into November 2015. The final distribution of limbs was planned for December 2015 at the Kaiga township.

10.33 Fire service week

The week was observed in a befitting manner in the last week of March 2010. The Kaiga township residents participated in the observance. Hands-on training on the use of fire extinguishers was imparted to them. Competitions were held during the week and prizes were distributed. The Site Director of Kaiga 1 to 4, Station Director of KAGs 1 and 2, senior officers and station employees participated in all the functions.

10.34   KAGS 1's record run

Kaiga 1 has set up another record of continuous running of 800 days as of 23.07.2018, surpassing the earlier record of RAPP unit 5 of 765 days. KAG 1 has generated 25.76 billion (2,756 million) units since the commercial operation commenced on 17.05.2016, including 4.2 billion units in the above run till 23.07.2018. The unit continued to run after the 23.07.2018 run. KAG 1 now stands second among the world PHWRs and 4th among all nuclear power reactors in the world. KAG 1 and further Indian nuclear power stations have demonstrated continuous operation for long periods exceeding 1 year, about 27 times so far.

KAG 1 beat its record of continuous run by a large margin with 895 days of continuous run as of 25.10.2018 among PHWRs. It is also second among all power reactors in the world. The unit has been operating since 13.05.2016. It may be mentioned that Indian Nuclear Power units have been continuously running 28 times so far over long periods exceeding about 2 years in each case.

10.35   Further accomplishments

KAG 1 beat its record and created a world record for continuous run with an operation of 962 days from 13.05.2016 to 31.12.2018, thus beating 940 days of UK's Daisum 2's unit 8. The unit had a capacity factor of 99.3% and generated 5 billion units of electricity during the mentioned period.

RAPP Units 3 and 4

11.1 Multiple-unit set-up in sites: siting criteria and design aspect consideration

It would be useful at this stage of the narrative to go into the criteria for setting up multiple-unit nuclear stations in India.

When the decision to set up the very first Indian nuclear power station at Tarapur (Maharashtra) was taken up, the decision was to set up two reactor units in the station. The financial sanction was already issued for both units at the beginning stage. Subsequent decisions were made to set up twin units at the RAPP and MAPP stations. However, the financial sanction was issued separately for each unit of the project. The design, procurement, planning, construction, etc. of these projects proceeded based on two units with budget provision for common services for the twin units and token budget provisions for the second unit being awaited; for projects beyond NAPP, the provision was made for the advance procurement of funds for future projects. This strategy ensured the timely availability of long-gestation critical equipment and earlier completion of projects as per the advanced schedule.

The Narora APP was the first project of the HWPT reactor type for which financial sanction was issued for the twin units together. This practice has since been followed in the case of Kakrapar, Kaiga, Tarapur units 3 and 4 (these two are of 540 MWe capacity), RAPP units 3 and 4, etc.

In line with the above policy, a decision was taken up by the Govt. of India, in 1985-1986, to set up two additional HWPT-type reactor units of 220 MWe capacity each, namely RAPP units 3 and 4.

A technical study made while the RAPP units 1 and 2 were being set up, in connection with the study of the feasibility of setting up additional reactor units at the RAPP site, had revealed that the capacity of the Rana Pratap Sagar reservoir would not be adequate for meeting the circulating water requirements of the additional units. It was hence decided to go in for natural draft hydraulic cooling towers.

11.2 Design criteria

11.2.1 Safety standards

RAPP units 3 and 4 meet all the requirements laid down in the revised safety standards adopted by the NPCIL. These features include defence in depth, redundancy, diversity and the 'feel-safe' philosophy.

11.2.2 Maximum flood level

Maximum flood level arising out of the postulated failures of the Gandhi Sagar and Rana Pratap Sagar dams situated near the site has been taken into consideration while finalising the ground level for the various structures.

11.2.3 Seismic level

The site falls under seismic zone-II as per ISI-1883 (2004), corresponding to a maximum possible earthquake having a return period of one occurrence in 10,000 years. This has been assumed while designing the various buildings and structures/systems for operating basis earthquake (OBE) and safe shut-down earthquake (SSE). A zero-period ground acceleration of 0.19 g has been assumed for SSE purposes.

11.2.4 Reactor regulatory, control, set-back and protection systems

11.2.4.1 The Regulatory System

Four regulating rods with elements containing Cobalt pellets/slugs are used for reactor regulation, set-back and flux-tilt control.

11.2.4.2 Control and set-back systems

Eight absorber rods with elements containing Cobalt pellets/slugs are used for xenon over-ride reactivity control and the addition of positive reactivity.

Two slim rods of Cadmium sandwiched in stainless steel tubes are used for the addition of negative reactivity and reactor set-back.

11.2.4.3   *The Reactor protection system*

11.2.4.3.1   Fast-acting primary shut-down system

Fourteen vertical rods of Cadmium sandwiched in SS tubes are used to trip the reactor.

Twelve liquid-poison injection shut-off tubes containing lithium penta borate solution in heavy water are used for shutting down the reactor.

11.2.4.3.2   Automatic liquid-poison addition system

The addition of Boron solution in the moderator, to augment the capacity of the slim rods, has been provided for reactor protection and control.

11.2.4.3.3   Liquid-poison injection system

The bulk addition of Boron solution to the moderator has been provided for prolonged shut-down purposes as a safeguard against unintended sub-criticality of the reactor.

11.2.5   Station electrical supervisory control and data acquisition system (SCADA)

This system was functionally developed by the NPCIL reactor control group of the BARC and the Electronics Corporation of India Ltd. (ECIL) for the first time for incorporation in RAPP units 3 and 4. The system caters to data acquisition and control of the station's electrical system and the 220-KV switchyard.

The system provides for three operator stations—two of which are located in the station control room while the third is in the switchyard; two servers are located, one each in the compressor rooms of RAPP units 3 and 4.

Five remote terminal units—one each in the control buildings of RAPP units 3 and 4, one each in the turbine building of RAPP units 3 and 4 and one in the switchyard—have been provided.

A two-layer network has been provided for the interconnection of the operator station's servers and the Remote terminal units (RTUs). The operator stations provide for the following:

(a) Mimics display

(b) Digital group display

(c) Alarm display

(d) Analog bar-graphs both pre-defined and operator-select trends

(e) History containing alarms, events and trends

(f) System display containing information about the health of the SCADA

It may be mentioned in this connection that earlier nuclear stations had provided programmable logic controllers for meeting the above-mentioned requirements. SCADA is faster and provides for better man–machine interface features. The RTUs located in the control building can be assigned emergency transfer and load-shedding functions. The RTUs can also be used for providing an expert guidance system to the plant operators during abnormal operating conditions and as a tool for in-service and shut-down maintenance.

11.2.6 Preliminary work at the site from 1986-1990

The preliminary work, such as marking levels, lay-out position, centring for various structures and related work, commenced in the middle of 1986; the construction of five work shacks, two warehouses and a concrete testing laboratory was completed as of 1989-1990.

11.2.7 Estimated cost: award of work

The estimated cost of RAPP units 3 and 4, according to the prices at 1994, was rupees 2,107 crores. From 1986-1987, the estimated cost was Rs. 1,395 crores. The escalation of Rs. 712 crores from 1986 to 1994 was on account of the following:

(a) Increase in scope of work : Rs. 137 crores

(b) Increase in cost of equipment and stores : Rs. 386 crores

(c) Incorporation of additional safety features and the increase in taxes and duties : Rs. 214 crores

(d) Interest during construction

The tender cost Rs. 657 crores for the Civil work of both the units, except for the pump-houses, stack and hydraulic cooling towers; the cost of the Civil materials supplied for free to the contractors for installation was Rs. 38.50 crores. The tender for the above work and at the above-mentioned cost was awarded to Hindustan Construction Company (HCC) in July 1988. The company stopped work on the plea that there was a delay in getting clearance from the Atomic Energy Regulation Board (AERB) for some of the awarded works, and further that there was a delay in the drawing of receipts by them for proceeding with the work. The company also held that there was a contractual–financial loss to them due to the above-mentioned reasons.

The project authorities, after studying the factors cited by HCC, concluded that there was a contractual onus on the part of the project and decided to pay a compensation of Rs. 18 lakhs to the company. The commencement of the work was delayed by 9 months and it was re-started in December 1991. An advance of Rs. 25 lakhs was also paid to the company to get the work expedited.

11.2.8   RAPP 3: main plant work during 1990-1991 and 1992-1993

The excavation for the main plant work and raft foundation (mud-mat) was completed by November 1990. Around 9500 cu. metres of concrete were poured for the RB-3 above the mud-mat during 1991-1992. About 27,000 cu. metres of concrete was scheduled to be poured during 1992-1993.

RAPP-3 in the foreground

11.2.9 Further progress from 1990-1991 and 1992-993: expenditure incurred during these periods

An expenditure of Rs. 194.05 crores was incurred as of November 1990. The above expenditure was on account of the site work incurred by the NPCIL for this work at Mumbai (Maharashtra); the site expenditure as of 1991-1992 was Rs. 24 crores and that expected for 1992-1993 was Rs. 40 crores. Overall progress at the site was 10% as of November 1990.

11.2.10 Site staff strength as of the end of 1993

A qualitative change in the type of personnel available at the site for RAPP units 3 and 4, as compared to the earlier projects, may be touched upon here. The services of professional management in the fields of administration, accounts and material management were made available at the site and this led to ease and quickness in decision-making.

RAPP units 3 and 4 had a staff strength of 770 as of 1992-1993 as against a sanctioned strength of 1,134. The total strength was expected to be less than 900 even during the peak of construction activity expected by 1994-1996.

11.2.11 Infrastructural and conventional equipment work: progress as of the end of 1993

The 650-tonne capacity crawler crane essential for the erection of major items of equipment was commissioned. Some of the technical features and cost particulars of the crane are listed below:

a) Reach : minimum 16 metres and maximum 85 metres

b) Main boom : 56 metres

c) Total weight : 1000 tonnes

d) Cost : Rs. 20 crores

Work orders for setting up infrastructural facilities and conventional piping were released. Assembly and commissioning of a 65-tonne crawler crane was completed. The contract for the main plant cable tray and lighting system was awarded.

11.2.12   Civil work of RAPP 3 and RAPP 4

11.2.12.1   Status as of the end of 1993

About 50% of the concreting work in the RB and TB of both units was completed. The quality of the concrete of the RB rafts was M-25 grade. Engineering Construction Co., a subsidiary of L&T commenced work on the natural draft cooling towers (NDCT); the excavation work on the NDCT was nearing completion. Concreting work in RAPP 3 was in progress.

11.2.12.2   Status from 1993 to 1997

The AERB put the ICD and OCD on hold as a result of the Kaiga-I dome failure on 13.05.1994. The hold was on the ICDs and OCDs of Kaiga 1, Kaiga 2, RAPP 3 and RAPP 4. The AERB set up committees to look into the causes of the Kaiga 1 dome failure. The suggestions of these committees were awaited before proceeding with RAPP 3 and RAPP 4 dome work. Other Civil work on RAPP 3 and RAPP 4 were progressing satisfactorily.

The emphasis now shifted to the expenditure work on the non-safety and conventional equipment work to compensate for the delay in the commencement of all safety-related work, of which ICD and OCD are important parts.

The above committees made their recommendations in May 1995. The improvements suggested by the committee were incorporated into the ICD and OCD work of RAPP 3 and RAPP 4. The only change incorporated in the RAPP 3 and RAPP 4 dome work was the zero-period acceleration. The required safety shut-down earthquake condition (SSE) was taken as 0.1 g as against 0.2 g for Kaiga units 1 and 2

11.2.12.3   Status as of June 1997

11.2.12.3.1   Civil work

The inner containment walls (ICW) of RAPP 3 and RAPP 4 were completed; the erection of the inner containment dome and ICD support structures was completed for both RAPP 3 and RAPP 4. All 32 trusses for the ICD staging of RAPP 3 were assembled. The laying of steel re-inforcement rods and the pre-stressing of cables for the

ICD of RAPP 3 were in good progress. The outer containment walls (OCW) of RAPP 3 had reached an elevation of 31 metres from the mean station ground level and were completed by the end of 1993.

11.2.12.3.2   Capital expenditure (CE) of RAPP 3 and 4 during 1993-1994

A sum of Rs. 192.64 crores was spent on CE.

11.2.12.3.3   Other work

All four steam generators and four PHT pump motors were installed. Nearly 70% of the nuclear equipment erection was completed. The lower and upper assemblies and drive mechanisms for reactivity control were among the major equipment erected.

11.2.12.3.4   Piping work: Status as of June 1997

Nearly 77% of nuclear piping was completed. About 500 heavy-water feeder pipes were installed. The erection of stand-pipes for the primary and secondary shut-down systems was completed.

11.2.12.4   Status as of March 1994

11.2.12.4.1   Reactor equipment

11.2.12.4.1.1   End-shields

The first and second end-shields were lowered into the Calandria vault. The SS ball filling of both the end-shields and its alignment with the Calandria, welding of the end-shield to the Calandria assemblies and routing of the assemblies were completed.

11.2.12.4.1.2   Steam generators and PHT pump motors

All four steam generators and four PHT pump motor units were erected.

11.2.12.4.1.3   Other reactor plant equipment

The FM vault coolers, pump room coolers, PHT system equipment, ion exchangers and filters for the end-shield and Calandria vault system were installed.

11.2.12.4.1.4   Overall nuclear plant equipment erection

About 45% of the erection was completed.

11.2.12.4.2 Nuclear equipment work of RAPP 3: status from the end of 1993 to early 1994

End-shields were moved to the ball-fitting enclosure in August 1992. The end-shields were the first critical nuclear components handled by the newly erected 650-tonne crawler crane at the site. The filling of balls for both the end-shields of RAPP 3 was completed and the shields were lowered into the Calandria vault in October/November 1993. The alignment, welding and routing of the end-shields were completed on RAPP units 3 and 4 with a capital expenditure of Rs. 192.64 crores.

11.2.16 Conventional plant and conventional piping erection

11.2.16.1 Status as of June 1997

11.2.16.1.1 Piping and Electrical Systems

The erection of 6.6-KV switchgear (SwGr) and 415 Motor Control Centres (MCCs) was completed to an extent of 95%. Fabrication of over 85% of conventional piping was completed. The installation of 75% cable trays was completed. All three DG sets were installed and cranked.

11.2.16.2 RAPP 4: Status as of June 1999

11.2.16.2.1 Electrical and Mechanical equipment power and control cabling

The installation of cable trays was in an advanced stage of progress. The laying of power and control cables was taken up and was in good progress. The installation of mechanical equipment was also taken in hand.

11.2.17 RAPP 3: the 220-KV Switchyard

11.2.17.1 Status as of June 1977

The 220-KV switchyard and the SUT for RAPP 3 were commissioned.

11.2.18 Common Services: status as of June 1997

The water treatment plant was commissioned.

11.2.19   RAPP 3 and RAPP 4 Civil work: status as of the 1st quarter of 1998

11.2.19.1   Inner Containment Walls (ICW)
The ICWs for both RB-3 and RB-4 were completed.

11.2.19.2   Dome work
The support structures for the domes of both RB-3 and RB-4 were almost completed. The laying of re-inforcement bars and pre-stressed cables was in progress for RB-3.

During the de-shuttering of the completed OC dome RAPP-4

11.2.20   Status as of the end of 1998
The casting of ring beams for the inner containment dome commenced on 15.05.1998 after the receipt of clearance from the AERB. Two of the three lifts for the inner containment dome of RAPP 3 were cast in the second half of 1998; the pre-stressing of the cable was taken up thereafter. The outer containment wall (OCW) of RB-4 was completed with up to 31 metres of elevation. The OCD of RB-4 was completed in 1999.

11.2.21 RAPP 3 conventional equipment: status as of the second half of 1998

11.2.21.1 *TG set and auxiliaries*

11.2.21.1.1 Condenser work

Condenser tubing work, the installation of water boxes for the condenser and hydro-testing of the condenser were completed.

11.2.21.1.2 Turbine generators

The preliminary alignment of generators with the LP turbine was completed. The preparation of HP and LP turbines for boxing up was completed by the end of 1998 and the turbine generator was inserted into position. Hydro-testing of the stator cooling water system was completed.

11.2.22 RAPP 3 Nuclear Plant work: status as of the end of 1998

11.2.22.1 *The PHT system*

All the 612 feeder pipes, including those of the secondary shut-down system, were installed and the connected PHT system piping was released for installation. The PHT piping system was completed by the end of 1998. Air-holding test and Helium-leak test were completed.

11.2.22.2 *Moderator system*

The moderator piping system was completed by the end of 1998. Air-holding test and Helium-leak test were also completed.

Hydro-test on the PHT system was taken up; the feeder/header insulation cabinet was installed; work on the emergency core cooling system (ECCS) and delayed neutron monitoring system (DNMS) was completed. The erection of the fuel-transfer equipment was completed and seal plugs were installed on the PHT tubes.

11.2.23 RAPP 4 conventional equipment: installation status as of the second half of 1998

11.2.23.1 *TG set and auxiliaries*
(1) Condenser tubing was completed

(2) The TG set generator stator was kept in position and further work on the TG set was in progress.

11.2.24   RAPP 4 Nuclear plant work: status as of the end of 1998

11.2.24.1   *The PHT system*

Around 400 out of 612 PHT system feeder pipes were installed and hydro-testing on the system was taken up.

11.2.24.2   *ECCS, shut-down coding system (SDCS) and connected systems*

The ECCS pump motor unit, bleed condenser, one FM vault atmospheric cooler and two pump-room atmospheric coolers were installed. The work on the regenerative heat exchange system, shut-down cooling system, heat-exchanger system, pump-room cooling system and ECCS equipment system were completed.

Balance equipment like the main air-lock, PHT feed pumps and shut-down cooling system pumps were received at the site. The erection of these equipment was also taken up.

11.2.24.3   *Moderator system*

The air-holding test was completed.

11.2.25   RAPP 3 and 4: overall physical progress of work

It was indicated in para 11.2.9 that the overall progress was 10% as of November 1990. The progress during subsequent years is as follows:

(1) As of 1993-1994 : 40%
(2) As of 1996-1997 : 77%
(3) As of 1997-1998 : 89%

11.2.26   RAPP 3 and RAPP 4: pre-commissioning checks and commissioning of various systems as of the end of 1998

The pre-commissioning checks of various systems were in progress. The 6.6-KV SwGr, 415-V SwGr, auxiliary transformers, 415V mccs, DG sets and control power systems were commissioned. Parts of the common service system, including the plant water system, fire water-system, and active and non-active process water cooling system, were commissioned. Third start-up transformer was charged.

11.2.27   RAPP 3 and 4: Revised cost estimate

It was stated in para 11.2.7 that the estimate, based on the prices in 1994, was Rs. 2107 crores. The revised cost estimate, based on the prices in 2000, was Rs. 2511 crores including a provision of interest during construction of Rs. 800 crores. The annual expenditure in the earlier years of construction was furnished in para 11.2.9 among other expenditure figures. Some of the latest annual expenditure figures are given below:

(1) Revised estimate for 1997-1998 : Rs. 265 crores
(2) Budget estimate for 1998-1999 : Rs. 272 crores

11.2.28   RAPP 3: status of work as of 1999

The overall progress on RAPP 3 was 90.5%. All systems are tested ready and poised for approach to first-criticality.

11.2.29   RAPP 3: First approach to criticality

The first approach to criticality started, with a count rate of 30 cycles per second (cps) with Boron concentration in the moderator at 27.8 parts per million, at 03:49 hrs on 24.12.1999. Around 20 tonnes of heavy water (D2O) with 27.8 ppm of boron oxide (B2O2) were poured into the moderator system which was already filled with 120 tonnes of heavy water. The neutron flux was simultaneously monitored. The primary shut-down system was trip tested, as per the laid down procedure, and tested at 14:20 hours with 15.1 ppm of Boron. After the 4th cycle of inserting the regulating rods, checking the valves at the ion exchanger column and withdrawing the regulating rods, the reactor became critical at 11:59 PM with 11.6 ppm of boron oxide and a count of 19,752 cycles per second (cps).

11.2.30   Milestone dates for RAPP 3

In para 11.2.11 to 11.2.27, we have seen the chronological progress of RAPP 3 and RAPP 4 work. The milestone dates in regard to the progress of RAPP 3 is listed below:

S. No.	Name of activity	Date
1	First pour of concrete	15.08.1990
2	RB raft concreting completion	30.11.1990
3	Release of the Calandria vault for erection	31.08.1993
4	End-shield grouting	30.06.1994
5	Calandria tube installation	30.03.1995
6	Coolant channel installation	05.01.1996
7	PHT feeder tube installation	18.04.1998
8	PHT system hydro test	17.05.1999
9	Hot-conditioning of the PHT system	17.08.1999
10	First date of criticality	24.12.1999
11	Date of synchronization with the NREB 10-03-2000	
12	Date of commercial operation	01.06.2000

11.2.30.1 Fuel handling simulators

A PC-based simulator was installed at the RAPP Nuclear Training Centre (NTC). The simulator was developed with the support of the NPCIL group and the Reactor control division (RCD) of BARC, both situated in Mumbai. It comprises 13 computers interconnected with the local area network (LAN). The simulator postulates normal and abnormal operations of on-power re-fuelling, spent-fuel discharge and new fuel pick-up. Facilities have been made available for the simulator instructor to introduce various malfunctions in the simulator. More experienced and licensed personnel will conduct a mandatory 2-week course for candidate control Engineers of the stations. The simulator has proved to be an excellent tool for understanding the modalities of on-power re-fuelling operations.

11.2.30.2   Energy generation: capacity factor

RAPP 3 has generated 5.5 billion units from June 2000 to June 2004. The capacity factor has varied from 74.93% to 83.98% during this period. Going further, we note that RAPP 3 has generated a total energy of 11.71 billion units from June 2000 to June 2008. Assuming the arrival rate of 2.50/unit, the total revenue for the period was 2900 crores.

11.2.31   RAPP 4: status as of mid-2000

11.2.31.1   Nuclear system work
All the work was completed; hot-conditioning of the PHT system was completed.

11.2.31.2   Moderator system work
All the work was completed and the moderator system was filled with heavy water.

11.2.31.3   Conventional system work
All the work was completed and the TG set was rolled

11.2.32   RAPP 4: pre-commissioning check and status as of the 3rd quarter of 2000
All the checks including the safety checks on the various systems were completed. All the system work was kept ready and poised for first-criticality.

11.2.33   RAPP 4: first approach to criticality
The coolant channels were loaded with nuclear fuel manually. The initial fuel loading pattern was 3637 of uranium dioxide and 35 of thorium dioxide fuel bundles. The thorium dioxide bundles were used for flux flattening and to enable the reactor to operate close to full power even during early reactor operating life. The moderator was flushed with 20 tonnes of heavy water containing about 30 ppm of Boron. Boron was then added to bring the moderator Boron value to 189 ppm. The PHT system was filled with heavy water. This was followed by the addition of 120 tonnes of heavy water to the moderator system. The final Calandria level was 94% with a Boron content of 28.9 ppm.

The first approach to criticality commenced at 13:37 hours on 2.11.2000 with the removal of Boron from the moderator. All the adjustor rods except 'UE' and 'UW' were fully kept out of the core. UE and UW were kept 80% inside the core. At 15 ppm (3 ppm away from the estimated critical Boron content), the primary shut-down system (pss) was shut down and the count rates before and after the shut-down were measured manually. Boron removal was continued, with the moderator Boron content 0.5 ppm away from the critical value. An attempt at criticality was continued with withdrawal from the core of adjustor rods NE and NW. Criticality was attained in the fourth cycle of rod withdrawal at 03:53 hrs on 3.11.2000. The critical Boron value was 11.9 ppm with rod 'UE' at 60% in and rod 'UW' 65% in.

11.2.34   RAPP 4 synchronization with the Northern Regional Grid: commercial operation

RAPP 4 was synchronised with the Northern Regional Grid at 7:23 hrs on 17.11.2000. The station was declared for commercial operation with effect from 16.12.2000.

RAPP 4 was the 4th Indian nuclear station to go into commercial operation in the year 2000. This was an international record in recent years of international nuclear history.

11.2.35   RAPP units 3 and 4: hydro test to commercial operation and comparison with Kakrapar unit 2

It took 854 days from the hydro test to commercial operation for Kakrapar Unit 2. This was reduced to 380 days for RAPP 3 and further to 161 days for RAPP 4.

11.2.36   Units generated and capacity factor

RAPP 4 generated 4.72 billion units during the period from 2000-2001, and 11 billion units during the period 2000-2001 to June 2008. The capacity factor during the period 2001-2002 to 2003-2004 varied from 74.19% to 96.64%. RAPP 4 has contributed a revenue of Rs.27.50 billion from 2001 to 2008 at an assumed chargeable rate of Rs. 2.50 per unit. RAPP units 3 and 4 contributed revenue of Rs. 56.5 billion during this period, a figure more than 2.5 times the capital cost of RAPP 3 and 4.

11.2.37   Certification as per in-service inspection requirement

Certification as per ISO-1400 for RAPP 3 and 4 was obtained during 2002-2003; the optimised in-service inspection requirements were issued as well.

11.2.38   The IAEA's 'OSART' Mission at RAPP 3 and 4

In 1982, the IAEA began an 'Operational Safety Review' of Nuclear Power Plants in various countries. Its 'Operational Safety

Review Team' (OSART) conducts this review mission. It has so far conducted 170 such missions. OSART conducted its 171st mission at RAPP units 3 and 4 from 29.10.2012 to 14.11.2012 at the request of the Govt. of India. This was the IAEA's first mission to India. Experts from Canada, Belgium, Germany, Romania, Slovakia, Sweden and the IAEA itself were involved in the review which was based on the IAEA's safety standards and good international practice. The review covered areas of management, organization, administration, operation, maintenance, technical support, radiant protection, chemistry, emergency planning and preparedness and severe accident management. The 'OSART' identified several good practices of the plant which are listed below:

1. The plant's safety culture cultivates a sense of accountability among the personnel.
2. The plant gives the staff several opportunities to expand their skills in work and participate in training programmes.
3. The plant's public-awareness program provides opportunities for the local community to learn about nuclear radiation safety.
4. The plant has an effective management training practice resulting in an effective authority system.
5. The plant uses training facilities and mock-up systems to improve maintenance quality and reduce the radiation dosages of the staff.

'OSART' has suggested the following for further enhancing the safety in the plant:

1. Enhancement of the standard of cabling conditions like laying and subsequent maintenance.
2. Enhancement of fire-door inspection and maintenance programme.
3. Enhancement of certain aspects of the plant's surveillance testing programme.
4. Enhancement of the root-causes analysis system.

OSART furnished the final report to the GOI within 3 months (by mid-February, 2013)

The authorities of the RAPP units 3 and 4 requested a follow-up 'OSART' mission within 15 months (mid-February, 2014)

11.2.39 'OSART' follow-up mission

The follow-up mission took place between the 3rd and 7th of February 2014. The mission noted the following:

1. Improvement in the condition of cable trays and control and power cable systems.
2. Improvement in the maintenance of the fire-doors.
3. Enhanced root-cause analysis of all system mishaps.

Experts from Belgium, Finland, the UK and the IAEA constituted the mission.

The IAEA conducts six 'OSART' follow-up missions annually.

11.2.40 The Ministry of Power Award

The Ministry of Power awarded its prestigious 'Gold Shield' for outstanding operational performance of the RAPP nuclear units on 22.03.2012. Sri. Sunil Kumar Shinde, Union Minister of Power handed over the shield, which was received by the Site Executive Director, Rawatbhata Rajasthan Site and station director RAPP units 3 and 4.

11.2.41 RAPP 3's record continuous run

RAPP 3 has created a record continuous run of 694 days as of 23.07.2018 and was still running thereafter. It subsequently ran up to 777 days.

Heavy-Water Projects: The Crucial Role of Heavy Water in HWPTRs

CHAPTER 12

12.1 Heavy-water management in HWPTRs

In the previous chapters, we have seen the crucial role that heavy water plays in heavy-water moderated and pressurized tube reactors. In Chapter 3, we have also seen the conscious choice that the Indian nuclear power planning Engineers have made in going in for HWPTRs as the first-generation Nuclear Power Plants. In para 5.8.1 of Chapter 5, we discussed the crucial role that the economy of heavy-water management plays in the efficient running of HWPTRs. The importance that D2O upgrading plants play in the operation of these Nuclear Power Plants was also touched upon. In this Chapter, we will look into the above-mentioned topics in detail.

12.2 Building up of the heavy-water inventory

12.2.1 Setting up of D2O production plants

As discussed earlier, an inventory of 230 tonnes of heavy water is required for the commissioning of each 235 MWe HWPTR. We have also seen in earlier chapters how the commissioning of HWPTRs in RAPP and MAPP was delayed by almost 3 to 4 years for want of the required inventory.

It was with a view of obviating such delays and making the Indian HWPTR construction and commissioning programme independent of the foreign supply of D2O that the DAE took up the programme of manufacturing indigenous D2O.

12.3 Heavy-water production methods

12.3.1 Earlier method: exchange between water and hydrogen

The most common method before the 1960s was the dual-temperature isotopic exchange of Deuterium between water and Hydrogen.

12.3.2 Later method: Ammonia–hydrogen exchange in the Baroda and Tuticorin Plants

The second method is Ammonia–Hydrogen (synthetic mixture of Hydrogen and Nitrogen) exchange for the enrichment of Deuterium, which is heavy Hydrogen (H3). A molecule of ordinary water comprises two atoms of Hydrogen and one atom of Oxygen, whereas one molecule of heavy water comprises two atoms of Deuterium (heavy hydrogen) and one atom of Oxygen. Ordinary water is H2O whereas heavy water is D2O. It may be noted that Deuterium is present to an extent of 0.016% in a Hydrogen atom.

An improvement in the process of enrichment of Deuterium, not originally contemplated by the heavy-water production designers, was the catalyst i.e. potassium oxide concentration that was used in the Baroda and Tuticorin plants. These plants operate at the maximum possible capacity. The plants are still faced with internal and external (to the plant) constraints. The external constraints relate to the concentration of Deuterium in the feed gas. This situation is not remediable any further.

12.3.3 The Thal Plant

The D2O production plant at Thal in the Raigad District of the State of Maharashtra comprises two streams, each with 55 tonnes/year capacity, and has been engineered indigenously. The plant is based on the mono-thermal Ammonia–Hydrogen process as in Tuticorin. Its feed material is synthesis gas from the Ammonia plants of Rashtriya Chemicals and Fertilisers Ltd. (RCF) at Thal. The Thal unit went into operation on 28.10.1986. The plant operates at Tuticorin at a high pressure of 250 kgs/cm sq. and at a low temperature of 27 degrees C. The cracking of Ammonia into synthesis gas is done at a pressure of 150 kgs/cm sq. and a temperature of 550 degrees C. All the major equipment, except the compressors, was procured indigenously. The heart of the process lies in the high-efficiency exchange trays. These trays too were engineered and fabricated in the country within house data generation.

12.3.4 The Talcher heavy water plant in Orissa

The plant, based on the Ammonia–Hydrogen exchange process, was set up for Rs. 50 crores and commissioned in 1979. The first quantity was produced in 1984. The feed for the plant is synthesis gas from the Ammonia plant of the Talcher fertiliser plant. The plant has, however, been facing difficulty in the assured supply of the synthesis gas from the Talcher fertilizer plant. The plant is also bedevilled by irregular power-supply from the Orissa Electricity board. Given the availability of D_2O from the other efficient D_2O plants, the DAE proposes to shut down the plant. The 400-odd workers of the Talcher D_2O plant are proposed to be absorbed into other D_2O plants of the DAE.

12.3.5 New route for the production of D_2O

The DAE, recognising the trend towards single-stream plants and their effect on stream factor and cost of production of D_2O, initiated pilot plant studies to render the D_2O plants adopting this technology independent of the fertilizer plants. A pilot plant for the extraction of Deuterium from water by the water–ammonia exchange process, set up at Baroda, has satisfactorily established the conditions for such a process, and the same is proposed to be engineered in detail with the aid of Indian consultants. This technology will be adopted not only for future plants but also for extending the life of the existing plants based on the Ammonia–Hydrogen exchange process, be it mono-thermal or dio-thermal.

12.3.6 The Kota D_2O plant: hydrogen sulphide water exchange process

While the technology for the water–ammonia exchange process was at first imported and then absorbed, that for the hydrogen sulphide water exchange process had all along been indigenous and was implemented in the Kota plant in Rajasthan. The extraction of Deuterium through the above-mentioned process was a novel technology; the toxicity of hydrogen sulphide (H_2S) and the highly corrosive environment created by the presence of H2S and water in the plant complicated the design and engineering of the plant. The plant was designed for operation at a relatively low pressure of 20 kgs/cm sq. given the stringent criteria for leak tightness of the system containing H2S gas. The permissible level of H2S in the plant environment is less than 10 ppm; the plant has an inventory of 100 tonnes of H2S. The operating experience, at a low pressure of 10-12 kgs/cm sq. of H2S, has been satisfactory.

The Kota D2O plant derives its steam supply from the Rajasthan Atomic Power Plant. The captive steam-generating plant has been sanctioned for the project by the Government to ensure the high-stream factor. With this added facility, the plant is expected to reach its design capacity of 80 tonnes of reactor-grade purity D2O per year at an early date.

12.3.7 The Manuguru (AP) plant

The plant comprises two streams each of 87 tonnes of D2O reactor-grade purity per year. The plant is based on the H2S–H2O exchange process. The steam is supplied by three boilers of 215 tonnes/hour of steam. Power for the plant is produced from two extraction turbine generators of 30 MWe capacity each.

12.3.8 The Hazira D2O Plant

Two streams of the plant based on the Ammonia–Hydrogen exchange process have an effective capacity of 110 tonnes of D2O each per year. An expenditure of Rs. 222.71 crores had been incurred as of the third quarter of 1990 as against an estimated cost of Rs. 264.39 crores. A D2O upgrading plant was commissioned at Hazira during 1992-1993. This plant would upgrade the 50-60% isotropic purity of D2O to a reactor-grade D2O of more than 99%.

12.4 Conclusion

The DAE now has eight D2O production plants in operation. Six of these are based on the Ammonia–Hydrogen exchange basis; two more plants have been on the anvil on the same basis since 1992-1997.

India is now not only self-sufficient in the production of D2O required for its PHWTRs but also has surplus capacity for export purposes.

The country was to export 100 tonnes of heavy water for use in a Nuclear Power Plant in South Korea under an agreement signed in Seoul on 1.4.1994.

The deal was worth $23 million and was the first major commercial agreement about the export of strategic nuclear material product from India.

The product was to be shipped between 1997-1998 and would be subjected to the IAEA safeguards.

India was also set to export 350 tonnes of D2O to Romania for the latter's 700 MWe heavy-water moderated reactor power plant near Bucharest. This deal was worth about $80 million.

The Role of Seismic Factors in the Design and Construction of Nuclear Power Stations: Nuclear Classes of Safety

13.1 Recent earthquakes: the causes and effects of earthquakes

13.1.1 The Uttarkashi earthquake

In 20.10.1991, an earthquake occurred in the districts of Dehradun, Chamoli, Tehri Garhwal and Uttarkashi of the State of UP. The epi-centre of this earthquake was located at Agra which is about 15 km from the town of Uttarkashi. The earthquake was rated 6.2 on the Richter scale. The earthquake resulted in the death of more than 1000 people while several hundred incurred injuries; 7,465 buildings collapsed.

13.1.2 The Latur earthquake

A more severe earthquake measuring 6.7 on the Richter scale rocked Latur and nearby areas of the Marathwada region of Maharashtra. It also affected Gulbarga, Bijapur and Raichar in the Northern region of Karnataka in the early hours of September 1993.

Nearly 7,600 people died as a result of the earthquake and 15,846 people were injured. About 19,000 houses were razed to the ground.

Some 2,17,000 houses were partially damaged. Most of these buildings appear to have been constructed with load-bearing walls raised out of stone rubble and then plastered with mud or lime plaster. No vertical or horizontal stiffener masonry seems to have been provided on the walls. Such a kind of construction would not be quake-proof.

The cost of re-building the damaged houses in Latur and other areas in the State of Maharashtra was estimated to be Rs. 1000 crores. We can thus understand the extent of injury and damages that could be caused to both persons and property by severe earthquakes.

13.1.3 The causes of earthquakes in the Indian peninsular: design of quake-proof structures

Earthquakes are caused in the Indian peninsular by the rubbing together of vast land masses (sub-plates as they are called in geological terms) along the rift or weak zones due to movement. Such movements cause high-gravity forces to be released. These forces would affect the safety of structures that have been built in these areas and may damage them severely if they are not designed to withstand such seismic (earthquake-gravity) forces. Earthquakes generate an expanding wave front starting from the focus epi-centre point at a speed of 0.1 to 1.6 metres per second and with an acceleration that could reach 0.7 to 15.34 metres/sec2. The above-mentioned figures are dependent on the intensity of the earthquake.

13.1.4 The Earth's structure and further causes for earthquakes

Let us now go into further details about the earth's structure and the probable causes of earthquakes.

It would be of interest to see a profile cross-section of the earth. The earth is spherical with comparatively flat ends at the North and South Poles. The earth is divided into six main layers marked from 1 through 6 as shown in the figure below. These represent, respectively, the inner core, outer core, lower mantle, transitional zone, upper mantle and crust. The inner core marked 1 is solid, while the outer core marked 2 is liquid. The lower mantle marked 3 is solid. The upper mantle marked 5 is the rift and may deform slowly to a plastic state. The crust marked 6 is thin and composed of Calcium and Sodium aluminosilicate minerals. The stratum is rocky and brittle. When this stratum fractures under stress, earthquakes are produced.

A study of the phenomenon of plate tectonics indicates a continuous slow movement of the plates which consist of the rigid crust and upper mantle. These plates are large slabs of 100-kilometre thickness. The motion of these electronic plates causes earthquakes and volcanoes. The study further reveals that the earth's outermost layer is broken into nearly a dozen and a half large and small fragments called plates which move relatively with a hot layer of asthenosphere below.

The details about the layers are marked in the sketch given below:

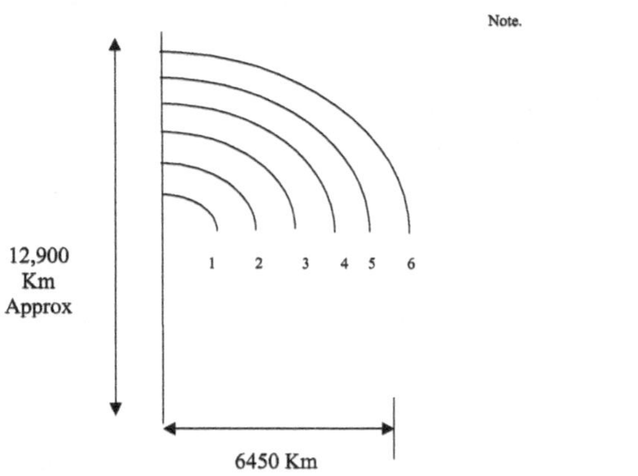

No. of layers	Depth (approx.) in Km
1	1,350
2	2,100
3	1,700
4	600
5	370
6	6.5

13.2   Normal and seismic considerations for design

Normal gravity on Earth is expressed in cm/sec2. The gravity at or near normal sea level is 981 cm. sec2. Severe earthquakes may cause an addition to normal gravity accelerations as high as 0.1 g to 0.2 g (98.1 cm to 196.2 cm per sec2) to be imposed on the structures and systems. These seismic forces act both in the horizontal and vertical directions and have to be taken into consideration while designing structures and systems. Thus, the static, motive and seismic forces have to be considered in totality while designing structures and systems.

Some very critical equipment like the electrical relays, instrumentation equipment, control equipment, etc. which perform safety functions in nuclear power stations are designed to withstand seismic forces as high as 6 g.

13.3 The seismic factor in the design of nuclear power stations

13.3.1 Seismic Design Concept

In para 9.2.1 of Chapter 9, we had seen how the Narora nuclear power station site lies in zone-4 of the Indian peninsular's seismological classification, and that the site is within a distance of 30 km from the Moradabad fault which is seismically active. Eight seismological zones in India are numbered from 1 to 8 in the ascending order of seismic severity.

We will deal with nuclear safety class requirements later in this chapter; let us now look at the definition of operating basis earthquake (OBE) and safe shut-down earthquake (SSE) requirements.

13.3.1.1 *Operating Basis Earthquake (OBE)*

OBE is the most powerful earthquake considering the regional and local geology, seismology and specific characteristics of the local surface material—which could reasonably be expected to occur in the vicinity of the plant site during the operating life of the plant. Those features and systems of the nuclear plant necessary for the continued operation of the plant without undue risk to the health and safety of the public are designed to withstand the vibrations of this class of earthquake.

13.3.1.2   *Safe shut-down earthquake (SSE)*

SSE is based on an evaluation of the maximum earthquake potential considering the local and regional geology, seismology and specific characteristics of the local sub-surface material.

It is the earthquake that produces the maximum vibratory ground motion for which certain systems and components are designed to remain functional.

13.3.1.3   *Values of OBE and SSE*

The values of OBE and SSE are site-specific and have to be arrived at by taking certain factors into account.

It may, however, be stated as an indicative value that SSE is twice that of OBE.

It was seen earlier in para 9.3.3 of Chapter 9 that the hydraulic-type cooling tower at NAPP, which is a nuclear safety Class-II structure, was designed for an OBE of peak horizontal 0.15 g and peak vertical 0.10g in this analogy. It was also stated that the towers were designed for nuclear safety Class-II requirements. The reactor building structure and systems which come under Class-I nuclear safety would have been designed for SSE of peak horizontal 0.3 g and peak vertical 0.2 g. nuclear safety Class-III and non-nuclear safety class structures and systems could be designed for lesser peak horizontal and peak vertical 'g' values.

13.3.2   The response spectrum analysis for designing nuclear structures and systems

In previous chapters, we have seen the effect of forces that may be imposed on structures and systems by seismic conditions. These conditions result in horizontal and vertical ground acceleration. The response spectrum of the structures and systems is analysed and suitable structural restraints, both rigid and flexible, are provided for these structures and systems to take care of forces caused by ground acceleration.

The ground response spectrum for a Nuclear Power Plant is derived from an ensemble of accelerograms recorded on sites with similar geological and seismo-tectonic conditions. A broad range of source and transmission path characteristics is also covered in this ensemble. A minimum of 25 accelerograms are taken to drive the response spectrum.

An analysis is also carried out by simulation, with the aid of modern computers, particularly for nuclear safety class systems and structures.

13.4 Classes of Nuclear Safety

13.4.1 General

As discussed earlier, the classes of nuclear safety are of four types namely Classes I, II, III and followed by the non-nuclear safety class. We will now see what these terms mean exactly.

There are three classes of nuclear safety for which nuclear plant structures and systems are required to be designed. They are described in the upcoming sections.

13.4.2 Nuclear Safety Class-I: structures and systems

Nuclear Safety Class-I is the classification which applies to the components-forming part which is within the reactor coolant pressure boundary; a single failure of this could cause a loss of the reactor coolant.

13.4.3 Nuclear Safety Class-II

Nuclear Safety Class-II is the classification which applies to those structures, systems and components that are not safety Class-I but which are required to fulfil a safety function.

13.4.4 Nuclear Safety Class-III

Nuclear Safety Class-III is the classification which applies to those structures, systems and components that are not safety Class-I or Class-II but whose failure would result in the release of radioactivity to the environment.

13.5 Non-nuclear Safety Class

Non-nuclear Safety Class is the classification which applies to structures, systems and components which have no direct safety function but which are connected to or influenced by equipment in nuclear safety Classes-I, II and III.

Fast-Breeder Test Reactor at Kalpakkam

CHAPTER

14

14.1 Further power reactor development

14.1.1 Choice of reactor to sustain future reactors

We have seen in para 3.1.2.2 of Chapter 3 that the second generation of reactors to be set up in India would utilize a combination of Uranium and Plutonium fuels to breed more Platinum and U233. The third-generation reactors would utilise this U233 and convert the blanket Th232 in fast-breeder reactors to produce more U233 than the equivalent of nuclear fuel they burn during power production, thus sustaining the succession of U233-fuelled reactors. It was also seen in para 3.1.3 of Chapter 3 that the burn-up rate in these FBRs would be 50,000 MWe tonnes per day of nuclear fuel as compared with the rate of 7000 to 8000 MWe tonnes per day achievable in HWPTRs. The thermal efficiency of these FBTRs would be 40-50% as against 28 to 35% of high-end HWPTRs and PWRs. It was also seen that the economics of scale could thus be attained and a self-sustaining programme of economic nuclear power production would be assured over the foreseeable future.

It would hence be necessary to set up the second-generation reactors to render the above scenario feasible. This effort has already begun.

The reactors would require Plutonium fuel, among others. The 48-MWe reactor would be one of the very few to be set up globally. The challenges to be met would be much more difficult than those faced during the scaling up of HWPTRs during the 1970s and early 1980s. The 40-MWt reactor would only be the ninth of its kind to be set up globally. The technical problems in using liquid metal in the form of Sodium as a coolant and moderator in the reactor had to be faced and solved. Liquid Sodium reacts violently with water and necessary precautions in this regard would have to be taken.

14.2   The site location

The Kalpakkam site in the State of Tamil Nadu, which houses MAPP 1 and MAPP 2, had all the infrastructural facilities for setting up the unit. Trained technicians were also ready at hand. A decision was hence taken to locate the FBTR at Kalpakkam.

14.3   Further details of the 15-MWe FBTRs

14.3.1   Fuel reactor parameters and other details

14.3.1.1   Fuel

Two kinds of fuel are used in liquid-metal fast-breeder reactors (LMFBR), namely core fuel and blanket elements.

The core elements consist of a Plutonium and U238 mixture at a ratio of 70:30 in Carbide form and are processed from the spent-fuel of operating HWPTRs. The blanket fuel consists of U238 only. The fission of the fuel core in the centre of the reactor provides almost all of the reactor power. The required fuel has been manufactured indigenously at BARC, Mumbai, and NFC, Hyderabad (AP). The design of the reactor provided for 65 fuel sub-assemblies to generate a thermal power of 42.5 MWe. Twenty-two fuel assemblies were used to attain the first-criticality. This quantity would provide for a thermal power of 10 MWe.

14.3.2   Steam generators

The reactor vessel is of 3-metre diameter while the primary circuit is of the loop type. There are two primary and secondary loops, and the number of steam generator modules per loop is 2.

The total Sodium inventory is 150 tonnes. The secondary Sodium inlet temperature is 557 degrees K while the outlet temperature to the intermediate heat-exchanger is 783 degrees K. The steam pressure at the outlet of the reactor is 134 kgs/cm2. The steam temperature is 753 degrees K.

The steam generators are once through with 7 tubes in a shell in the triple 'S' shape. Steam flow is 165 kgs/sec.

14.3.3 Feed-water heaters

Contact-type feed-water heaters have been used.

14.4 Major suppliers of equipment: scope of supply of indigenous equipment and materials

The critical components of the reactor vessel, rotating plugs, control-rod drive mechanism, steam generators (SGs), component-handling machines, turbo-generator (TG), TG control and instrumentation, reactor plant control and instrumentation packages were manufactured and supplied indigenously.

BHEL Tiruchirappalli workshop has supplied SGs, intermediate heat-exchangers and piping for the project.

Only 20% (Rs. 16 crores) of the total cost of equipment and materials of about Rs. 80 crores was on account of imports.

14.5 Project cost details

The initial cost estimate was Rs. 34.85 crores while the latest cost estimate, as of 1985, was projected at Rs. 91.98 crores.

14.6 First-criticality run up to power: initial operational experience

The reactor attained its first-criticality in 1985. Several difficulties such as the dislodgment of fuel pipes from the fuel assemblies, among others, were encountered during the raising of reactor power. Many experimental fuel pins were tried and the reactor power could be stabilised at 400 kWth. High-power operations commenced in January of 1993. The smooth enhancement of power from 400 kWth to 4000 kWth in a time of 2 hours was carried out in the first fortnight of January 1993. Operation at increasing thermal power levels was continued through the months following January 1993; successful operation at 8 MWth was established in October 1993. The reactor power was raised to 11 MWth and the station was successfully synchronised with the TN Power grid at 20:45 hrs on 29/11/93. 1 MWe of power was being fed to the grid.

The project authorities proposed to take out the experimental pins after the successful conclusion of reactor operation at 11 MWe to conduct post-irradiation tests and check other reactor Physics parameters. The reactor power would then go up to 13 MWth. The FBTR would, at that stage, have demonstrated the success of its technology. It would also be possible at that stage for FBTR to feed the electricity generated to the TNEB grid.

At the time of recording the operational experience of FBTR, the reactor was yet to attain its rated output of 15 MWe. The operational experience regarding full-power output was yet to be undergone. However, regarding fuel performance, the Carbide form of Uranium and Plutonium chosen for the FBTR was not found satisfactory. The Carbide form has a lower melting point than the Oxide form. It is hence proposed to use the Oxide forms of the two above-mentioned fuels for the 500-MWe prototype fast-breeder reactor at IGCAR.

A burn-up rate of 32,000 MWe of fuel has been achieved in FBTR against the design figure of 25,000 MWe. The fuel has operated up to a peak burn-up rate of 88,000 MWd/tonne (te) without any pin failure. Two fuel assemblies, one at 2500 MWd/te and another at 50,000 MWd/te, were taken out for examination in the hot cells at the radio metallurgical laboratory at IGCAR. The detailed investigations show that the fuel is capable of a burn-up of 1,30,000 MWd/te.

The irradiation of 30% PuO_2 and 70% UO_2 with the use of U233 as additional fissile material to achieve the required linear rating commenced in July 2003.

14.7 Re-processing of Carbide Fuel

A facility for re-processing the spent Carbide fuel has been commissioned and the re-processing has commenced. This facility closes the Carbide fuel cycle.

The Design and Development of 500-MWe HWPT Reactors

15.1 Cost economy of high-capacity reactor units

We have seen in earlier chapters that the unit capacity of commercial nuclear power reactors, so far installed and commissioned in India, has been 190 MWe at Tarapur and 235 MWe at all the other sites.

Other nuclear power-producing countries have gone in for higher-unit rated reactors to avail economic benefits of higher energy generation and lower per unit cost of generation. Unit ratings of 500, 750, 900, 1000 and 1300 MWe and even higher have been chosen by countries such as Canada, France, Japan and Russia. These countries already have very large installed capacities of nuclear power and are thus moving strongly in the path of constructing large-unit-sized stations. It is also seen that the average availability of PWR (Pressurized Water Reactor) units of 900 MWe capacity, that were commissioned in France even during the middle 1980s, has been of the order of 91%. The Canadian HWPTRs of 750 MWe and higher unit ratings have also maintained a cumulative capacity factor higher than 87% during 1989. The operating parameters of large-unit rated stations have heightened the level of confidence in utilities and hence more countries are going in for larger unit sizes. It may also be mentioned that the 500-MWe-sized thermal-generating units, that were built and commissioned in India, have achieved a plant load factor of 70% and higher.

15.2 The decision on 500 MWe HWPTR

It was under these compelling circumstances that a policy decision was taken to set up HWPTR stations of 500 MWe unit size, while at the same continuing with the installation of the then 220 MWe units which had been standardized from KAPP onwards.

15.3 A dedicated team for the 500 MWe HWPTR design

The Chairman of the Nuclear Power Board set up a dedicated team called 'The 500 MWe group' headed by an executive director to finalize the design and related parameters of the unit and prepare a detailed project report in the year 1984. In essence of this, the design would be scaled up from the existing 220 MWe units. The design concepts would require validation and this task was undertaken in close co-operation with the reactor engineering division of the BARC. The design took final shape during 1987-1988. The design basis report (DPR) was prepared and submitted to the AERB for approval.

While dealing with the above-mentioned scaling up, several problems came up. The design called for the use of more sophisticated raw materials. Larger components were required for the manufacture of large-sized critical equipment, and hence the use of more sophisticated nuclear-quality material was called for. The challenge was heightened since the country had been facing a decade-and-a-half ban on the import of nuclear materials and equipment ever since the 1974 sanctions which were inflicted after the Pokhran PNE (Peaceful nuclear explosion). The design was hence based on indigenous materials and equipment.

15.4 The reactor and the associated plant

Some of the technical data of 500 MWe HWPTR have been indicated in the table below for comparison.

Table 15.4 (t1)

540-MWe and 220-MWeReactor units:

Comparative Parameters

S. No	Name of system/structure process	540 MWe	220 MWe	Remarks
1 (a)	**Reactor Building Inner containment wall** Inside diameter, wall thickness (up to 6.9 mets height from grade) Wall thickness (general area beyond 6.9 mets height)	49.50 mets diam 1.40 mets 0.75 mets	42.56 mets diam 1.20 mets 0.61 mets	
1 (b)	**Outer containment wall** Inside diameter, wall thickness (up to 6.9 mets height from grade) Wall thickness (general area beyond 6.9 mets height)	54.72 mets diam 1.20 mets 0.61 mets	47.78 mets diam	
1 (c)	**Inner containment dome** Thickness (general area) Thickness (near steam generator openings)	0.65 mets 1.00 mets	0.47 mets 1.22 mets	
1 (d)	Openings in the ICD doc	2	4	
1 (e)	**Calandria vault** SS lines for walls and floors	10-mm thickness provided	No SS liner	
2.	**Reactor Vessel**			
2 (a)	Material of vessel	304-L SS	304-L SS	
2 (b)	Overall internal diameter	5.99 mets	13.0 mets	
2 (c)	Octagonal distance across face	4.97 mets	9.2 mets	
2 (d)	Number of coolant channels	306	392	
2 (e)	Material of coolant tube	Zircaloy-2	Zirconium-2% Niobium alloy	
2 (f)	Coolant tube inner diameter		82.6 mm	
2 (g)	Coolant tube outer diameter		90.8 mm	
2 (h)	Calandria tube inner diameter		107.6 mm	
2 (i)	Calandria tube outer diameter		110.2 mm	
3	**Reactor Fuel core**			
3 (a)	Cell array	square	square	
3 (b)	Lattice Pitch	286 mm	228.6 mm	
3 (c)	Core radius	3192 mm	2256 mm	

S. No	Name of system/structure process	540 MWe	220 MWe	Remarks
3 (d)	Core length	5944 mm	5085 mm	
4	**The Fuel System**			
4 (a)	Fuel Clad material	Zircaloy-4	Zircaloy-2	
4 (b)	Fuel bundle diameter	102.4 mm	81.7 mm	
4 (c)	Fuel element diameter	13 mm	14.24 mm	
4 (d)	No. of fuel elements per bundle	37	19	
4 (e)	No. of fuel elements per channel	13	12	
4 (f)	No. of fuel elements per bundle (approx.)	22 kgs	15.00 kgs	
5.	**The End-shield**			
5 (a)	Thickness	1.32 mets	1.12 mets	
5 (b)	Diameter	Inner assembly - 7.8 mets Outer assembly - 9.11 mets	6.325 mets	
6	**The PHT System**			
6 (a)	Number of Pressurisers	2	1	
6 (b)	Number of PHT pumps/Pump unit capacity	4 /6 MW	8 /0.8 MW 4 /2.0 MW	RAPP & MAPP All after MAPP
6 (c)	Number of PHT loops	2	1	
6 (d)	Number of hot headers	4	2	
6 (e)	Number of cold headers	4	2	
6 (f)	Coolant flow role	28.13 x 10^6 kgs/hr	10.70x10^6 kgs/hr	
6 (g)	Coolant pressure	126 kgs/cm2	96 kgs/cm2	
6 (h)	Reactor inlet temperature	260 degrees C	249 degrees C	
6 (i)	Reactor outlet temperature	304 degrees C	293 degrees C	
7	**Nuclear Steam Service System**			
7 (a)	Number of D2O/H2O heat-exchangers	4 (2 groups of 2 each)	8 (2 groups of 4 each)	

S. No	Name of system/structure process	540 MWe	220 MWe	Remarks
7 (b)	Type of heatx1 grs	Vertical U-shell and tube-type with integral steam drum and no pre-heater section	Vertical U-shell and tube-type with separate steam drum and pre-heater section	
7 (c)	Evaporation Rate	3076 tonnes/hr	1090 tonnes/hr	
7 (d)	Steam Pressure	44 kgs/cm2 gauge	42 kgs/cm2 absolute	
8	Air-locks	3 Main, emergency and FM	2 Main & emergency	

15.5 Reactor shut-down: reactor control and regulator system

15.5.1 Zonal control system: detection system

The core of the 540-MWe reactor is large compared to that of the 220-MWe units. The neutron strength is diffused because of the large core size and is loosely coupled across the cross-section. This aspect has been considered while designing the control and regulating system, thus oscillations in the axial and radial zones are possible. A study of the three-dimensional reactor kinetics was made to enable the design of the reactor control and regulating systems. A three-dimensional matrix of Cobalt, self-powered neutron detectors, was provided to monitor the core flux strength and distribution and provide the required inputs for the control and regulation of the reactor system. The ion chamber units, located on the side walls of the Calandria vault, serve to provide out-of-core detection of the neutron flux.

Adjustor rods, control absorbers, zonal control systems and in-core monitors are provided for the regulation and control of the reactors along with Xenon poison over-ride and flux monitoring. Global and local power control of the reactor during operation is achieved through the liquid zonal control system; this is the first time such a liquid zonal control system concept has been introduced into the design of the HWPTRs. Fourteen liquid zonal compartments, situated in vertical slots, are fed with light water which acts as neutron absorbers. Each compartment has three self-powered neutron detectors (SPNDs) associated with it. These detectors provide the required signals for reactor regulation; changing the height of the light-water column in zonal compartments individually or collectively depending on the operational requirements affects the required reactivity. In addition to the zonal control system, four hollow Cadmium rods are provided for the reactor power set-back function.

Seventeen stainless steel adjustable rods are Xenon over-ride functional of a total worth of 17 Mk. Poison-addition tanks having a minimum worth of 90 Mk are provided.

15.5.2   Neutron detection system

The reactor core has a network of self-powered neutron detectors (SPNDs) which are suitably located to monitor the neutron flux in the reactor. These SPNDs, 200 in number, are strategically located in the vertical and horizontal flux units. The horizontal flux units have Cobalt detectors for core-over-power protection provided by shut-down system 2. The vertical flux units have Vanadium detectors for the online flux mapping system, Cobalt detectors for shut-down system 1 and the liquid zonal control system. The various reactivity devices and associated self-powered neutron detectors of the shut-down systems necessitated 86 vertical and 13 horizontal penetrations in the Calandria.

15.5.3   Reactor shut-down system (RSS)

Two independent shut-down systems (SDS) have been provided for the first time in Indian Nuclear power projects. The SD system 1 has Cadmium SS rods that drop vertically into the under gravity and cause reactor shut-down (total worth of rods 70 Mk). The SD system 2 (SDS-2) comprises injecting Gadolinium Nitrate (gn) into the bulk of the moderator through six horizontal tubes inside the Calandria.

The net worth of Gadolinium Nitrate is 70 Mk. The SDS poison solution (Gadolinium Nitrate) is stored in 6 separate tanks, with each tank feeding one injection line connected to the poison injection unit. A single high-pressure Helium supply tank provides energy for injecting the poison. A trip signal from the triplicated instrumentation system opens the valves, making the Helium pressurize the poison tanks and push the poison into the moderator.

15.5.4 Other Reactor Safety systems

The containment system and vapour suppression systems are the same as for 235-MWe units. The shut-down systems are provided with online testing facilities by way of monitoring. These systems provide complete information on the health of all equipment. This provides safety under all the postulated conditions.

15.5.5 Fuelling machine heads: fuelling machine drives

Two fuelling machines have been provided as spare for each reactor.

Rock and pinion drive management has been provided for the FM heads.

A calibration and maintenance facility has been set up for the FM heads in the service building. This facility is capable of testing new FM heads in the first instance, and later after major over-hauls.

15.5.6 The Moderator System

The system incorporated canned motor pumps to avoid shaft seal leakages.

15.5.7 Seismic design

The system is designed for the postulated site, and the site-specific criteria based on the earlier observed data will provide for the operating basis earthquake (OBE) and safe shut-down earthquake requirements (SSE).

15.5.8 The PHT system: provision of independent loops

The system is designed for 126 kgf/cm2 pressure and a temperature of 310 degrees C.

The system consists of two independent loops, each catering to 50% of the fuel core. Each loop consists of two NSSGs (Nuclear Steam Generators), two cold headers, two hot headers and 50% of the PHT feeder pipes.

The coolant flow in the adjacent channels is in the opposite directions, as is the case of 235-MWe units to enable the equitable distribution of heat in the reactor core. The provision of independent loops reduces the quantum of loss of coolant in case of possible rupture in the PHP system piping and thus voiding of system piping.

15.5.9 Core decay heat removal: emergency core cooling system

15.5.9.1 Core heat removal

15.5.9.1.1 Under transient conditions

The system envisages steam flow through one or more specially provided valves directly to the condenser or the atmosphere, in case the turbine is unable to take in all the steam generated by the NSSGs due to the sudden drop in power demand.

15.5.9.1.2 Reactor shut-down/tripped condition under reactor tripping due to 415V Class-III power loss

If Class-III power-supply to the PHT pumps fails, the coolant circulation is maintained by the initial movement of the flywheel of the pumps. The coolant flow is continued due to thermo-siphoning from the high level of the pumps in the boiler room level to the reactor in the low level of the Calandria vault. Auxiliary feed pumps supplied by Class-III power provide 3% of the normal feed flow by pumping auxiliary feed water directly to the NSSGs. The auxiliary feed pumps take suction directly from the emergency feed-water storage tank. The coolant temperature is brought down to 150 degrees C by way of the above-mentioned process.

15.5.9.1.3 Under Shut-down conditions: Shut-down cooling system

The Shut-down cooling system is provided to bring the reactor to cold shut-down conditions. Shut-down cooling pumps powered by Class-III power circulate the primary coolant through heat-exchangers which transfer the heat to the process water system. Two shut-down cooling trains of 100% capacity are provided for each PHT loop.

15.5.9.2 *Emergency Core Cooling System (ECCS)*

15.5.9.2.1 Phase-I: Short-term cooling

The system consists of two phases namely the high-pressure light-water injection and the low-pressure long-term re-circulation. The high-pressure injection envisages the circulation of light water stored in water accumulators through the headers of the reactor PHT system.

A rupture disc is provided in the ECCS for the positive interface between heavy water and light water. The confirmation of loss of coolant before the actuation of ECCS is made as follows:

(1) Monitoring the rise in the PHT room pressure

(2) Monitoring the rise in the Calandria level

These precautions are necessary to safeguard against the down-grading of PHT heavy water due to the inadvertent initiation of ECCS. The short-term cooling lasts until the inventory of water in the accumulators is over.

15.5.9.2.2 Long-term cooling

Prolonged cooling is carried out by the pumping of light water in the suppression pool through three heat-exchangers. The circulating suppression pool water is cooled by active process water in these heat-exchangers provided in the cooling line before the water is delivered to the headers. The heavy water spilling from the PHT system break leads to the suppression pool to keep the circulation going before the full cool shut-down condition is established in the reactor system.

15.5.9.2.3 The enhancement of 500 MWe to 540 MWe HWPTRs in TAPP units 3 and 4

The 500-MWe design was enhanced to 540 MWe and the enhanced HWPTRs were constructed and commissioned satisfactorily. This has been dealt with in detail in Chapter 16.

15.5.10 Control room building: reactor auxiliary building and station auxiliary building There is a unique departure in the 500-MWe units from the earlier 235-MWe units. There

is a separate common control building to accommodate the separate control relay and metering panels and control equipment panels of the two units.

The reactor auxiliary building which houses a majority of equipment of the reactor auxiliary systems—such as the active process water system, emergency core cooling system, PHT and moderator purification system and the D_2O recovery system among others— is located very close to the reactor building.

TAPP Units 3 and 4 (540-MWe HWPTRs)

16.1 The first 540-MWe HWPTR in India: TAPP units 3 and 4

16.1.1 Site location and seismic zone layout

We have seen in Chapter 15, the need for evolving and designing higher Nuclear Power units through the 200 MW and higher units that are yet to be installed in India. We have also seen how the higher unit capacity nuclear power stations have resulted in the improvement of operational performance and continuing sustainability of nuclear power-generation in other countries. This was the reason why the 500 MWe/540 MWe twin units were considered for construction at Tarapur.

The design of the 500-MWe unit was finalized by a dedicated group of the Executive Director of the NPCIL at Mumbai from 1987 to 1988. The reasons for choosing Tarapur as the location for the twin units are listed below:

The State of Maharashtra had a large installed capacity of 12 GWe as of 1987-1988. The State of Gujarat, on the other hand, had 7 GWe at that time. These two States would avail almost all of the power to be generated by the proposed 500 MWe/540 MWe twin units. The large capacities in these two States would enable easy absorption of the power to be generated at Tarapur. We have earlier described in Chapter 7, the difficulties faced by the Rajasthan electricity board (part of the Northern regional grid) in absorbing the power generated by RAPP units 1 and 2.

There were difficulties in the operation of the generating units and re-starting of the units after tripping until arrangements for the islanding of the RAPP units, RPS dome power transmission area, were completed at a later date. On the other hand, Tarapur had a running nuclear power system with two 190 MWe units; start-up power was thus readily available. Infrastructural facilities for starting new units were also readily available at the site. These were some of the reasons for locating the first 540 MWe nuclear units at Tarapur.

We will now look at some of the site conditions and the facilities to be provided at the site for the new units.

16.1.2 Seismic zone

The site falls under seismic zone-3 of IS 1893 as of 2002. The required safety-radiated buildings, structures and systems have been designed for a zero-period acceleration of 0.2 g. We will look into more technical details of the units in the upcoming sections.

16.1.3 Flood-level grade elevation

The site is a shore-based (Arabian Sea) one. Grade elevation for the site was hence arrived at by taking into consideration extreme natural phenomena like cyclones and 'tsunamis', which occur at Oceans/River bodies following severe earthquakes that cause immense damage to life and properties. The design-basis flood-level was computed based on the postulation of the worst combination of naturally caused water-level storm-water surges and wave run-ups. The design-basis flood-level was arrived at 32 metres. The grade floor level for station structures was kept at 32.3 metres. A protection band on the seaward side of the site was provided at an elevation of 33 metres. Details of the band have been described in the upcoming sections.

TAPP-3&4 PHWR PLANT LAYOUT

1. Reactor building
2. Service building
3. Reactor aux. bldg.
4. Control building
5. T.G. building
6. Stn. aux. bldg. (a)
7. Stn. aux, bldg. (b)
8. D_2O upgrad. plant
9. CMF
10. Admin. block
11. Stack
12. Switch yard

The layout sketch of TAPP units 3 and 4 is placed above.

16.2 Reactor vessel and allied equipment: construction and physical details

16.2.1 Reactor vessel

The design and physical parameters of the vessel, boilers, shield and fuel elements have been furnished in Chapter 15 (para 15.4 and 15.5). Henceforth, other constructional features of this equipment will be explored in detail.

16.2.1.1 End-shields

These cylindrical box-type structures, integrated with the Calandria shell, are also made up of austenic stainless steel grade 304 A 240 plates. These plates are made up of 18% Chromium and 9% Nickel and are of low Carbon content. The thickness of the plates varies from 32 mm to 65 mm. The assembly quality requirement is for Nuclear Class-II. Each end-shield weighs 210 tonnes.

16.2.2 Reactor vessel

The reactor vessel (Calandria) is a horizontal cylinder shape welded to the extension of the end-shields. An integral Calandria end-shield assembly is thus ensured. The vessel is made of SS 304. It is a horizontal single-walled 32-mm thick vessel having an overall diameter of 7.864 metres (mean value), and a 7.064-diameter smaller shell of 5.282-metre length.

16.2.3 Fuelling machines

These are provided with modular features as seen in the earlier stations. A dedicated calibration and maintenance facility (CMF) for the machines has been provided.

16.2.4 Calandria tubes and coolant pressure tubes

The Calandria tubes are 392 in number and are made up of Zircaloy-4 grade. The coolant pressure tubes are also 392 in number and made up of 2.5% Zirconium and Niobium alloy.

The Calandria main shell has 85 non-radial nozzles on top of the vertical side and 13 nozzles on the west of the horizontal side for receiving and passing through reactivity control and shut-down mechanisms. It has 21 radial nozzles for receiving moderator piping and over-pressure lines.

16.2.4.1 *Annular space between pressure channels and the Calandria vessel*

The annular space between the coolant channel and the Calandria vessel is filled with dry carbon dioxide gas, to provide thermal insulation between the high-pressure high-temperature D_2O in the coolant channels and the relatively cold low-pressure D_2O in the Calandria vessel. The CO_2 gas is continuously circulated and monitored for any moisture pick-up; these checks improve the health of the coolant channels.

16.3 Electrical system details

We have seen in Chapter 15 that most of the details of a typical 500 MWe unit comprise the reactor systems/associated systems, FM system, regulatory system, safety and shut-down systems. In the upcoming sections, we will go into more detail on the electrical system.

16.3.1 System design concept of 540-MWe Tarapur units 3 and 4

The system has generally followed the state-of-the-art conditions of the prevailing nuclear units and has also incorporated the progressive improvements and modifications of the HWPTR units set up in the country.

Three-dimensional computer studies carried out during the design stage helped locate and avoid the glitches in the inter-disciplinary, mechanical, electrical, instrumentation, and nuclear interferences that had been experienced earlier during construction. This study helped avoid costly, time-consuming report work and thus speed up construction.

Specific requirements dictated by the conditions of the site and the conditions of the regional grid (WREB), to which the station would be connected, have been incorporated into the design.

16.3.2 Location of 400-KV and 220-KV SwGr environmental factor

Severe saline pollution prevails in the area. It was hence proposed to provide SF6 gas-insulated switchgear (GIS-SwGr) for the 400-KV and 220-KV systems. The two SwGr systems would be housed in separate buildings. The SwGr control systems would be housed in another separate building located in the switchyard area. The space required for housing GIS is only of the order of 15% to 20% of that required for an outdoor switchyard.

The GIS-SwGr does not fully do away with the outdoor system components. Bushings have to be installed for the connection of the GIS-SwGr to the outgoing EHV feeders, string insulators for terminating these feeders have to be air-insulated and the lightning arresters for these feeders also have to be located outdoors. All outdoor-located insulators and bushings are provided with creepage distances corresponding to very heavy pollution levels: 31 mm/KV. Creepage distances for the 400-KV and 220-KV equipment have been chosen as 13,020 mm (13.02 metres) and 7595 mm (7.595 metres), respectively. Creepage distance means the nearest distance between the EHV current-carrying parts to the nearest non-current-carrying conducting metal part; it is carried past all insulating materials. Fixed hot-line washing facilities for these insulators are provided with de-mineralized water through a remote-controlled panel located in the switchyard control room.

16.3.3 The 21-KV generator circuit breaker

A 21-KV generator circuit breaker has been provided between the generator and the generator transformer (21/400Kv GT). The generator circuit breaker (GCB) will facilitate the fast isolation of the GT in case of a fault in the GT. The provision of GCB obviates the need for fast transfer between the auxiliary supply buses in the event of a unit trip. There is a provision for by-passing the GCB. This feature enables the feed from the generator to be directed to the GT in case of emergency operational requirements.

16.3.4 The 400-KV and 220-KV bus systems

The 21 KV/400 KV GTs feed the 400-KV switchyard. The 400-KV gas-insulated switchyard would have 7 bays comprising two GT bays, four line-bays and one bus coupler bay. The 220-KV switchyard will have 5 bays comprising two SUT bays, two line-bays and a 400-KV station bus system. Start-up power is provided by one 70 MVA 220 KV/6.6 KV-6.6 KV 3-winding start-up transformer (SUT). The 220-KV system would be interconnected with the 220-KV system of TAPP units 1 and 2. A reliable on-site source for start-up power would thus also be available. The added technical advantage would be that TAPP units 3 and 4 can be islanded in case of a grid-supply failure; Class-IV supply would be available from TAPP 1 and TAPP 2 systems and the poisoning out of TAPP units 3 and 4 can thus be avoided.

Double-main bus arrangement has been provided for both the 400-KV and 220-KV systems.

The 400-KV system enables a separate off-site source of auxiliary power through the GT-UT route. Two independent 220-KV lines, one each from TAPP units 1 and 2, backed up by 220-KV lines from the MSEB's Boisar grid sub-station reinforce the start-up power for TAPP units 3 and 4.

16.3.5   Station auxiliary power-supply system: redundant supply arrangement and routing of safety and non-safety cable systems

The station auxiliary power-supply system (SAPSS) comprises the usual four categories of supply namely Classes I, II, III and IV. These are further sub-divided as follows:

A) Non-Safety related Class-IV 6.6-KV and 415-V Power-supply systems

B) Safety-related Class-III, Class-II and Class-I systems fed by 6.6 KV/415 V AC and 480/250 V DC systems.

The redundancy SAPSS consists of Division-1 and Division-2 systems and equipment. The division-1 system comprises the in-feed from one 35 MVA 21 KV/6.6 KV unit transformer (UT) and one winding of the 3-winding 70 MVA 220 KV/6.6 KV SUT (start-up transformer).

Division-2 consists of the feed from the second 35 MVA UT and the second LV winding of the SUT.

Division-1 systems and equipment are located in the Station Auxiliary Building (SAB) 'A' and those of Division-2 system are in SAB 'B'. It may be noted here by way of comparison that Kaiga units 1 and 2 and RAPP units 3 and 4 were provided with one auxiliary building for each unit. Thus, TAPP units 3 and 4 have one order of magnitude of higher safety margin than the earlier 220-MWe units.

Each of the Division-1 and Division-2 sets of systems, equipment and structures has been designed to ensure safe shut-down of the reactor station auxiliary transformers, which are of the dry-type and hence suitable for indoor installation with enhanced safety features by way of elimination of fire hazards.

The Class-II UPS is a 'state-of-the-art' system, a digitally process unit introduced in an Indian Nuclear Power Project design. The UPS was imported from Switzerland. The four Class-III 50% full capacity 2.4-MWe DG units are provided with in-line cylinders as against the earlier 'V' type models. Additional design features were provided to enable local synchronization.

16.3.6 Redundant safety-related cables: physical separation

The separation of reactor auxiliary buildings ensures that there are no physical proximate routings for redundant safety-related cables.

The conformance to the IEEE-384 clause about the physical separation of safety-related cables has been ensured. Physically separate openings have been provided in the RBS for the power, control and instrumentation cables. Power cabling through the floor opening has been avoided. Copper conductor cables have been extensively used to reduce the sizing of cables and hence of the openings in the inner containment wall (ICW) and the outer containment walls (OCW).

16.3.7 The 540-MWe turbine generator unit (TG)

The TG unit is a 3000-rpm (revolutions per minute) tandem compounded single and double-flow high-pressure cylinder exhausting into moisture separators and re-heaters which in turn feed into two double-flow low-pressure cylinders. The generator is a 21-KV 659-MVA 50-cycles per second unit. The rotor is a directly coupled hydrogen-cooled one, and the stator is de-mineralised water-cooled with hollow conductor windings.

16.4 Re-rating of TAPP units 3 and 4

The various parameters of TAPP units 3 and 4 were analysed in detail and as a result, it was decided to re-rate the units to 540 MWe each. The rated capacity should hence be read as 540 MWe, wherever the capacity of these units has been mentioned as 500 MWe in this Chapter.

16.5 Financial sanction: local-body clearances, land acquisition and rehabilitation of project-affected people (PAPs)

16.5.1 Financial sanction (FS)

The financial sanction was accorded for the two units at a 1990 cost of Rs. 2,427 crores; an escalation of 7% per year was assumed while framing the above estimate which was based on 500-MWe unit capacity. The capacity was later re-rated to 540 MWe as seen in para 16.4. The financial estimate was suitably revised to confirm to this higher capacity and was placed at Rs. 6,555 crores based on the prices in 2002.

16.5.2 Project clearances and PAP rehabilitation

Clearances for the project from statue-3 bodies—like the AERB, Maharashtra State Pollution Board (MSPB) and the Ministry of Environment, Forests and Climate Change (MoEF) of the Government of India—were obtained for the unit's 127 hectares of land acquired in the first instance of locating the main plant buildings and related facilities. An additional 215 hectares were later acquired for the 1.5-km radius exclusion zone for the units.

A package for the rehabilitation of the project-affected people (PAPs) was prepared and forwarded for the Government of Maharashtra's (GOM) approval. A sum of Rs. 65 crores was deposited with the GOM for the 1137 PAPs.

16.6 Details of major equipment suppliers: TAPP units 3 and 4

16.6.1 Calandria

Two Calandrias were ordered from Walchand Industries Ltd. The manufacture of the first unit was completed and that of the second was in progress as of 1990.

16.6.2 The NSSGs

BHEL was to supply four NSSGs for the first unit. TAPP 4 has been mentioned as the first unit as per field exigencies; progress on TAPP 4 was faster and hence has been described as the first unit. The NSSGs were being manufactured at BHEL'S heavy boilers plant at Tiruchirappalli (TN). BHEL was to supply NSSGs for TAPP 3 also. Each of these NSSGs weighs 180 metric tonnes and is 20 metres in length. The cost of these eight SSGs was Rs. 80 crores.

16.6.3 The 659-MVA Turbo-generators

The units were ordered from BHEL. The turbines were being manufactured at BHEL's Bhopal workshop and the generators at its Haridwar workshop.

16.6.4 End-shields.

These were ordered from L&T's Hazira (Gujarat) workshop.

16.6.5 The 70 MVA 220 KV/MVA 6.9 KV/ 6.9 KV suits and 35 MVA 21/6.9 KV station unit transformers

The 70 MVA 220 KV/MVA 6.9 KV/ 6.9 KV suits were ordered from Crompton Greaves while the 35 MVA 21/6.9 KV station unit transformers were ordered from TELK, Kerala, respectively.

16.7 Preliminary work at the site

16.7.1 Railway siding at the Boisar railway station

Improvements to the existing siding were carried out to receive and handle the larger and heavier equipment required for the 540 MWe units.

16.7.2 Storages, shops, offices and laboratories

The required closed and open storage for plant stores, and spaces for shops, laboratories and offices were identified. Already available storage and other facilities were being utilized, and additional facilities were under construction.

16.7.3 Construction power-supply

An already available power-supply at 415 volts was being utilized. A new 3 MVA 33/6.6 KV sub-station was established, and 6.6 KV/415-volt sub-stations were set up at the required construction load centres.

16.7.4 Construction and permanent water-supply

A 71-mm diameter, 19-km long water-supply line with a capacity of delivering 2 million litres of water per day was commissioned to meet the construction requirements.

A permanent water-supply system was being set up through the Maharashtra water-supply and sewage board. Full details about this system have been given in para 16.9.10.4.

16.7.5 Communication facilities

Wireless telephones and links to all the NPCIL plants and projects were established.

16.7.6   Residential colony

Over 156 residential units of six types were completed and occupied during the early stages of the project (1990). An additional 1,094 units, over a land area of 65 hectares in Salgaon village, were planned for construction to meet the needs of the growing workforce. Water-supply, sewage system and power-supply for the colony were commissioned.

16.8   Manpower planning

An aerial view of Rb-3, TB-3 and SB - TAPP-3
Nearly 135 personnel of various working grades and hierarchical levels were positioned at the site as of 1990. Manpower planning for a peak personnel force of 1,135 was underway.

16.9   TAPP units 3 and 4: Civil work progress during 2000-2001

A view of Reactor Building, TAPP 3&4 Rising-up

An aerial view of Rb-3, TB-3 and SB - TAPP-3

16.9.1 Tenders for cooling water pump-houses

Tenders were invited for two cooling water pump-houses, one each for TAPP units 3 and 4. The package for each unit included three concrete condenser cooling water pumps (CCW) with Closed Air Circuit Air (CACA) motors—each of 40,000 cu. metres per hour capacity at 1800-metre water column head—and three vertical turbine-type auxiliary water service pumps, each of 3,000-metre cube per hour capacity at 38-metre water column head with associated motors.

The package included the design, engineering, manufacture, erection and commissioning of all associated equipment like travelling water screens, valves, expansion joints, electrical and instrumentation systems, control panels, ventilation system, associated piping, etc.

16.9.2 Other Plant work: progress during 1999-2002

16.9.2.1 Piping and mechanical system packages and end-shield lowering

The work was awarded. The end-shields of RB-4 were lowered.

Lowering of second-end shield in second unit of Tarapur Project (3&4)

16.9.2.2 Permanent fresh water-supply
The contract was awarded.

16.9.3 TAPP units 3 and 4 progress as of June 2002

16.9.3.1 Manufacture/receipt of equipment/materials and the use of a giant crawler crane for the erection work
All main structural steel for RB-3 and RB-4 were received at the site. The erection commenced in March 2001 and was completed by June 2002.

16.9.3.2 Equipment and Materials
The following equipment and materials were received at the site:
A) The PHT pump motor units and reactor headers for TAPP 4

B)   Two primary system pressuring units

C)   One moisture separator and one steam re-heater for TAPP 4

D)   Liner tubes (part receipt) and TG equipment (part receipt) for TAPP 4

E)   1750 metric tonnes of structural steel

16.9.3.3   Tender for main plant Civil work of TAPP units 3 and 4: further progress

Tenders for the above-mentioned work at an estimated cost of Rs. 142 crores were floated during June 1995.

16.9.3.3.1   Site expenditure

An expenditure of Rs. 21.06 crores was incurred as of the end of 1993.

16.9.3.4   Site survey

The contour survey, sub-soil investigations and safe-blast studies to help excavation—both metric surveys and geotechnical investigations up to 15 metres depth of soil—were completed.

16.9.3.5   Tenders for the construction equipment

Tenders floated for a 15,000-tonne heavy-duty crawler crane were under scrutiny. The procurement of the mobile crane's materials for handling equipment, air compressors, workshop machines, etc. was in progress.

16.9.4   Civil work: progress as of 1998-1999

16.9.4.1   Commencement of excavation

The date of 10.10.1998 could be marked as a red-letter-day in the development of high-nuclear unit capacity projects in India. It was on this day that the ground-breaking ceremony for the first 540 MWe HWPTR units was held. Dr. R. Chidambaram, Chairman of the AEC, pressed the button of the remote control to trigger the first blasting of rock for the excavation at the reactor site. Sri. Ram Naik, Union Ministry, was the guest of honour at the function. Senior officials of the NPCIL based at the site and from the NPCIL corporate office, representatives of the project design consultants, the Maharashtra State Government officials, eminent public persons and industrialists from the Boisar Industrial Estate attended the function. Sri. T.S.R. Prasad, CHD of the NPCIL, presided over the function.

16.9.4.2 Capital expenditure between 1993-1994

Expenditure incurred during 1993-1994 was Rs. 63.98 crores.

16.9.4.3 Award of Civil Work

A contract for Civil work amounting to Rs. 351 crores was awarded to Larsen & Toubro on 24. 5.1999. Some of the Civil work quantities were:

1) Normal concreting : 2,60,000 cu. metres
2) High-strength concreting : 50,000 cu. metres
3) Heavy concreting (hematite) : 13,000 cu. metres
4) Re-inforcement steel : 64,000 metric tonnes
5) Structural steel : 6,200 metric tonnes

16.9.5 Progress from late 1998 to mid-1999

16.9.5.1 Progress of excavation

Nearly 1,99,750 cu. metres out of a total of 6,00,000 cu. metres of excavation in rock was completed. The estimated quantities of excavation in soil, rock-bound core and line drilling were 1,90,000 cu. metres, 1550 running metres and 1,12,000 running metres, respectively. Of the above-mentioned estimates, 1,32,000 cu. metres, 14,000 running metres and 15,420 running metres, respectively, were completed.

16.9.5.2 RB-3 and RB-4 excavation work: the construction of the site protection bund and progress as of 1999-2000

16.9.5.2.1 Excavation work

The overall excavation work on both the units was about 95% complete. The work in the RB-4 site was completed in four-and-a-half months. The completion date was 25.2.1999. The work in the RB-3 site was completed in 5 months and a quarter, the completion date was 20. 3.1999.

16.9.5.2.2   The site-protection bund

It was for the first time in the history of the construction of nuclear power stations in India, that a protection bund to contain sea erosion was provided. The protection bund involved the use of geo-textiles and geo-membranes. About 22,000 square metres of geo-textiles and 7,100 square metres of geo-membranes were used. The geo-textiles in the form of clothes were spread over the bund, and the geo-membranes in the form of packets were placed over the geo-textiles.

The geo-membrane is an impervious material that prevents the flow of sea water across its surface. The geo-textile is a pervious material which allows the sea water to seep through it and filter out the soil material, thus allowing the material to flow back to the sea. Soil erosion is thus prevented. The construction of the bund was completed quite a few weeks ahead of the schedule.

16.9.5.3   Geo-technical field tests

The tests were required for RB-4 and were carried out after the excavation work was completed. The tests were necessary to validate the design assumptions. The geological mapping was done by geologists from an independent agency, the Atomic Minerals Division of the Government of India and by consultant geologists. The following tests were carried out by these agencies:

A)   Vertical load test at 2 locations
B)   Lateral cyclic load test at one location
C)   Block vibration test at one location
D)   Rock concrete bond test in tension at 7 locations
E)   Rock concrete bond test in 1 location
The above tests were conducted on 12. 6.1999.

16.9.6   Concreting of the main plant

The first concreting in RB-4 commenced on 8.3.2000. The new high-strength M-60 grade concreting developed for the re-concreting of the failed dome of Kaiga 1 (see para 10.21.5 of Chapter 10) was used for RB-4 and would be used for RB-3 as well.

Peak consulting for RB-3 and RB-4 was conducted at the rate of 17,500 cu. metres per month. The second lift of the internals of RB-3 and RB-4 and the sixth lift of the inner containment wall up to 98 metres above ground-grade level were completed. The construction of the control building had reached the ground level. Raft concreting for RB-3 RB-4 in the CMIF area was completed.

Each lift height of the ICW was 2.54 metres and that of the OCW was 3.45 metres. The concreting of the Calandria vault of RB-4 was completed in a record period of 5-and-a-half months.

16.9.6.1 *Main plant concreting and other services: progress as of 2000-2001*

Each lift of the Calandria vault concreting was 2.665 metres.

The concreting of the TG net was also completed. The concreting and other Civil work of the condenser cooling water in-take and out-fall tunnels and trenches were taken up and were in good progress.

About 3000 metric tonnes of structural steel work was involved in RB-3 and RB-4. About 28 steel columns and 440 steel beams were to be erected. The erection was planned with the view of carrying out the work by other contractual agencies along with the steel structural work. One particular innovation of the plant concreting was the provision of left-in-place steel shuttering wherein the steel plates were used in place of wooden shuttering; this considerably reduced the time for concreting. These plates were used as embedded parts and for supporting other service work like ducts, cable trays and lighting. Thus, the time for the erection of other services was also considerably reduced. The erection work was continuous and without any interruption due to concreting.

The erection of more than 12,000 metric tonnes of structural steel was completed within 30 months.

Reactor Buildings 3 (left) and 4 (right) of TAPP 3&4

The raft of RB-3 and RB-4 was 5.5 metres in thickness and 62 metres in diameter. Its foundation was at 20.5 metres below the mean station ground level. This was the deepest foundation so far as the rafts in Indian nuclear plants are concerned. The raft foundation was completed in 6 layers and with 18 pours of concrete in all. The quantity of concrete was 16,000 cu. metres and the steel repairs used up 4150 metric tonnes. The largest single pour of concrete was 1,500 cu. metres in volume. The main steel re-inforcement bars were of 45 mm in diameter and had to be specially rolled in the steel manufacturer's factory. This re-inforcement was used in the 1st and 6th layers of the raft concreting. A total of 4,600 splice joints and 5000 build joints were used for the concreting. A total of 200 rock anchors of a total depth of 29.2 metres were provided to ensure that the raft remains in positive contact with the foundation rock under seismic conditions.

A Panoramic view of Tarapur Atomic Power Project (3&4)

16.9.6.2 *Pump house in-take and out-fall structures*

The work was completed which further enabled the commencement of related plant work.

Short-circuit test of the start-up transformers

16.9.6.3 TAPP units 3 and 4: the erection of equipment

16.9.6.3.1 The use of a giant crawler crane

Moderator heat-exchangers and re-generative coolers were received at the site. The said equipment was lowered at the sites of their installation. The use of a heavy-duty crawler crane of 15,000-metric-tonne capacity speeded up the erection work of large-size equipment.

The manufacture of turbine steam condensers, NSSGs, fuelling Machine Bridge and columns was in progress at the respective manufacturers' work as per the drawn-up schedule.

16.9.6.4 The award of electrical engineering equipment, procurement and commissioning contract (EPC)

EPC for the electricals of TAPP units 3 and 4 was given to Larsen & Toubro at an estimated cost of Rs. 445 crores. This was the first time that an EPC concept of carrying out a crucial system, such as the electricals of a Nuclear Power Project, was introduced. The concept is likely to be carried over to future Nuclear Power Projects as well. The above-mentioned contract was awarded in the middle of 2001.

16.9.6.5 Physical progress of the site erection work as of 2002

The progress of the site work as of the end of December 2002 was 38.5%.

16.9.7 Progress of work as of 2002-2003

16.9.7.1 *Civil work in RB-4*

The Inner containment wall (ICW), 75 cm thick (0.75 metres), was completed up to the ring beam of the wall. The grade of concrete for ICW was M-60, and that of the outer containment wall (OCW) was M-35. There were only two openings on the ICW and OCW as against four in KAPP units 1 and 2 through RAPP units 3 and 4.

The placement of supporting structures for the inner containment dome (ICD) had commenced. The ICD has a diameter of 60 metres and an overall height (from the ground-grade level) of 50.5 metres.

The other technical parameters of ICD and OCD are as follows:
1. Test pressure : 1.44 kg/cm^2
2. Design pressure : 1.44 kgs/cm^2
3. Seismic coefficient (zero-period ground acceleration 0.2 g)
4. Grade of concrete : ICD M-60 and OCD M-35
5. Pre-stressed concrete system : 19 KB (19 strands of 12.7-mm nominal diameter wires).
6. Ultimate tensile strength of pre-stressed cable wires : 355 tonnes.

    87% of the concreting was completed in RB-3 and RB-4.

16.9.7.2 *Progress of Civil Work from 2002 to 2003*

Around 6 lakh cu. metres of concreting was completed with a peak rate of 20,000 cu. metres per month and an average rate of 14,000 cu. metres per month.

16.9.7.2.1 Discharge channel for the circulating cooling water system

The construction of a discharge channel into the sea for the condenser circulating cooling water outlet and the excavation for the pump-house in-take structure was completed.

16.9.7.2.2   Progress on the conventional equipment as of 2002-2003

16.9.7.2.2.1   Steam condenser
The erection of the condenser was completed up to the neck. The condenser tubing was in position and ready for rolling.

16.9.7.2.2.2   Cable trays/cabling
About 26-km length of cable trays were laid. The installation of tray supports was maintained at the rate of 100 metric tonnes per month. The cabling installation rate was 150 km per month with a maximum rate of 31 km per day.

16.9.7.2.2.3   Ventilation and Air Handling System
The erection was taken up and was in good progress.

16.9.7.2.3   Fresh water-supply
The water-treatment plant for fresh water-supply to the unit was commissioned.

16.9.7.2.4   Start-up and unit transformer testing at CPRI Bengaluru
A 70 MVA 220/6.9/6.9 KV start-up transformer, manufactured by Crompton Greaves, and a 35 MVA 21 KV/6.9 KV unit transformer, manufactured by TELK in Kerala, were successfully tested for short-circuit at the Central Power Research Institute (CPRI), Bengaluru (Karnataka).

16.9.7.2.5   The RB-4 Nuclear Plant system: progress as of 2002-2003

16.9.7.2.5.1   Calandria, end-shields, moderator pumps and coolant channels
The Calandria and both end-shields were placed in position, aligned and welded in record time. About 392 Calandria tubes were rolled with the end-shields. Five moderator pump units were received at the site and installed. The installation of the coolant channels had commenced.

16.9.7.2.5.2 RB-4: progress on the fuelling machine (FM), coolant feeder tubes and nuclear piping installation

FM columns and the bridge in the North FM vault were erected. About 64% of the coolant feeder tubing was fabricated. Eddy Current testing of the heat-exchangers was performed and the stand-by coolers were installed.

16.9.7.2.6 The first Nuclear Service Steam Generators (NSSGs)

The country's first NSSG for a 540-MWe nuclear reactor (first of four) was flagged off from the BHEL Tiruchirappalli (TN) workshop on 7.12.2002, by Sri. V.K. Chaturvedi, CMD of the NPCIL, and Sri. A.K. Mathur, Executive Director of BHEL. Senior officials of the NPCIL were also present at the function. The shipped unit is for use in the TAPP unit 4. BHEL will supply all eight NSSGs for TAPP 3 and TAPP 4.

16.9.7.2.7 End-shields for TAPP 3

The first end-shield manufactured by L&T at their Hazira nuclear equipment manufacturing workshop and intended for the TAPP 3 unit was dispatched on 25.12.2002, ahead of schedule. L&T has been awarded the contract for four end-shields for TAPP 3 and 4. The second end-shield was expected to be dispatched by the end of January 2003.

16.9.8 Agreement for the construction of transmission lines for the evacuation of power: cost of power from TAPP units 3 and 4

An agreement was entered by the NPCIL and the Power Grid Corporation of India Ltd. (PGCIL) for setting up the power evacuation system of TAPP units 3 and 4. The erection work comprised the erection of 400-KV transmission lines, 400 KV/220 KV switchyard system interconnecting auto transformers; interconnection between the 400-KV system of TAPP units 3 and 4 and the 220-KV system of TAPP units 1 and 2 and associated switchgear and control system for the exchange of power being generated at TAPP units 1 and 2 and to be generated at TAPP units 3 and 4.

The agreement enables the PGCIL to line up the erection of their systems with that of the progress in the plant construction, commissioning and generation of power. The agreement indemnifies the PGCIL from payment of energy-use charges until the off-take of power by the beneficent State power utilities is finalized by these States.

The agreement was signed by Sri. K.J. Sebastian, Director of Transmission of the NPCIL, and Sri. R.B. Mishra, ED of the Western Region of the PGCIL, Mumbai, on 21.10.2002.

The NPCIL will be charging the State utilities Rs. 2.65 per unit for utilizing the power generation of TAPP units 3 and 4.

16.9.8.1　TAPP 3: progress as of 2001-2003

60% of completion was achieved.

TAPP 3&4: Coolant Channel Installation at TAPP-4

16.9.8.2　Conventional equipment installation: TAPP units 3 and 4

The 220-KV switchyard was charged on 16.4.2003. Start-up power was thus available from April 2003.

16.9.9　RB -4 Civil work: progress as of 2001-2003

The concreting of the ICD was completed in a record time of 22 days as against a target of 45 days. The concreting of the OCD was completed in 24 days. The domes had two openings for the lowering of NSSGs. The thickness of the OCD is 0.65 metres and that of the ICD is 49.5 metres; the height above the springing level is 12.5 metres and the height of the crown is 48 metres above base-ground level.

The total concreting involved in the OCD was 1600 cu. metres; the grade of concrete was M-60 and the placement temperature was 19 degrees Celsius. The peak rate of concreting was 48 cu. metres per hour and the concreting was completed in 8 pours.

Pre-stressing of all dome steel cables was completed.

16.9.9.1 TAPP units 3 and 4: further Civil work during 2003-2004

16.9.9.1.1 Infrastructural Civil Work

Civil work of the water in-take system, discharge tunnel, fresh-water/fire water-reservoir, DM plant, Chlorination plant and service water pump-house were completed ahead of the schedule.

16.9.10 RB-4: Nuclear equipment work

16.9.10.1 Coolant channel installation

All the 392 coolant channels were installed in 92 working days. There was nil rejection of the channels after installation for the first time in the history of channel installation work in the NPCIL. The contributing factors for the nil rejection were as follows:

(1) Proper pre-planning
(2) Approximate supply contract handling
(3) The deployment of trained manpower for the erection work
(4) Proper design regarding the positioning of bearing sleeves
(5) Increase in the length of bearing sleeves in the lattice tube

16.9.10.2 The RB-4 NSSGs

All the four NSSGs were lowered within one day of their receipt at the site.

16.9.10.3 RB-4 Nuclear piping: progress during 2003-2004

Nearly 100,000 (one hundred thousand) inch diameter of welding was completed. The erection of 500,000-inch metres of nuclear piping was completed. The feeder pipe erection was in progress.

16.9.10.4   TAPP units 3 and 4: conventional equipment erection during 2003-2004

The erection work was in progress. The fresh water-supply system was commissioned with a saving of Rs. 6 crores in the estimated cost. The source of water for the system was the 'Kawdas' pick-up weir of the 'Surya' irrigation project of the Govt. of Maharashtra. This project is located 45 km from the TAPP site. The work was carried out by the 'Maharashtra Jeevan Pradhikaran' for Rs. 36 crores. The system comprised the following:

(1) In-take well of 3 metres diameter
(2) Jack well of 9 metres diameter
(3) Over-head pump-house
(4) Diesel generator of 200-kVA capacity to provide everyday power for the system
(5) Raw water rising main
(6) Ground storage reservoir of 7 lakhs litre capacity at Kawdas pick-up weir

The scheme also included a 25.9 km length of raw water gravity main of the Maharashtra State from the Kawdas pick-up weir to the Gorima village conventional water treatment plant, with a branch line of 4.84 km from the gravity main to the townships of TAPP units 3 and 4.

16.9.10.5   TAPP units 3 and 4: progress as of July 2003 to the end of 2004

An overall progress of 69.3% was achieved.

16.9.10.6   TAPP 4: progress during mid-2004 to the end of 2004

16.9.10.6.1   Nuclear and Conventional Piping Work

A monthly average of 25,000-inch diameter of welding and 35,000-inch diameter of erection work was achieved to complete the major systems of piping. About 78% welding of the nuclear piping was completed.

16.9.10.6.2 Additional conventional equipment erection 2003-2004

16.9.10.6.2.1 The SUT protection system
The sprinkler system was provided for fire protection.

16.9.10.6.2.2 The 500-MWe TG and related equipment
The TG stator, weighing 260 tonnes, was kept in position. The high-pressure turbine rotor was installed and the LP turbine installation was in progress.

16.9.10.6.2.3 Control centre: Instrumentation and control power-supply
The field instrumentation work package was progressing as per schedule. The 26-volt battery banks, stand-by battery chargers and uninterrupted power-supply (UPS) systems were installed. The control centre work was in progress.

16.9.10.7 Further nuclear system work: the PHT and moderator systems

16.9.10.7.1 Piping work of the moderator system: progress from mid-September 2003 to mid-September 2004
The piping work of the moderator system was completed. The hydro test of four circuits of the moderator system was completed. The testing of the balance moderator system was in progress. A test run of the moderator system was taken up.

16.9.10.7.2 End-fitting seal plugs and reactivity system work
The installation of seal plugs on all end-fittings was completed. About 58 out of 81 vertical assemblies of the reactivity system were installed.

16.9.10.7.3 RB-4 Calandria installation and end-shield welding of TAPP 4: progress from mid-2004 to early 2005
All the work on the Calandria and the Calandria vault was completed. The filling of water and the calibration of all the Calandria parameters were completed. Calandria end-shield welding was completed in 30 days.

16.9.10.7.4 TAPP 4: further conventional equipment progress from mid-2004 to early 2005

16.9.10.7.4.1 The TG and allied equipment

The 500/540 MWe TG was boxed up after all the erection work was completed. The water filling test of the condenser was completed.

16.9.10.7.4.2 The 400-KV and 220-KV switchyard work

The installation of the gas-insulated 400-KV indoor switchyard system was completed. The 220-KV outdoor switchyard work was completed and the drawing of power from the switchyard for commissioning purposes had commenced. The clearance of the CEA (Central Electricity Authority) for the 400-KV switchyard system was received.

16.9.10.7.4.3 Cabling system, control wire termination system and the lighting system: progress from mid-2004 to early 2005

A peak cabling rate of 30.882 km per day, the highest among nuclear projects so far, was achieved. Around 80% of the cabling work was completed and 73% of the control wire terminations in the control equipment room were completed.

16.9.10.7.4.4 The reactor auxiliary unit cooling system, ventilation system, fire protection system and compressor and chiller system

The fire-hydrant system, auxiliary cooling water system, end-shield cooling water system and part of the Calandria cooling water system were commissioned and made operable from the control room. A test run of the service water pumps and the auxiliary water service pumps from the control room was completed as well. The commissioning of the 415-V Class-IV electrical system run compressors and chillers was completed. The commissioning of the 415-V Class-III electrical system run compressors was in progress. The primary containment ventilation system and control room ventilation system were commissioned.

16.9.10.8 Chronological progress as of mid-September 2004

All safety tests on the unit were completed. The Safety Review Committee (SRC) of the AERB cleared the tests and the stage was set for the first-criticality of TAPP 4.

16.9.10.9 TAPP 3: progress as of mid-2004

16.9.10.9.1 Civil work

Concreting of six inner containment walls, with 0.61 m thickness (ICW), and an outer containment dome (OCD) was completed. The concreting of the inner containment dome (ICD) was completed only in 27 days.

16.9.10.9.2: Progress on conventional equipment

The work on the turbine generator erection piping and the electrical and instrumentation system was progressing as per the drawn-up schedule.

16.9.10.10 Reactor system: TAPP 4

16.9.10.10.1 Calandria tubing and end-shield welding

The 330 Calandria tubes out of a total of 392 were installed. The work on the remaining 62 tubes was in progress. Preparations for one-end rolling of the pressure tubes were in progress. The Calandria end-shield welding was completed.

16.9.10.11 Nuclear steam service generator

Three NSSGs were lowered into position, two of them in a single day.

16.9.10.12 TAPP 4: further commissioning work

16.9.10.12.1 Piping

About 3,70,000-inch diameter (ID) and 4,90,000-inch diameter (ID) of piping work for TAPP 4 was completed in only 18 months. The monthly rates of erection were 29000 ID and 41000 ID.

16.9.10.12.2 Fuel loading

The AERB gave clearance for the fuel loading of TAPP 3 and TAPP 4 on 20.1.2005. One channel of TAPP 4 was loaded on 22.1.2005. The initial charge comprised 844 deeply depleted bundles, 1364 depleted bundles and 2888 natural Uranium bundles. The shield plugs were put in position and the moderator system was flushed after the entire fuel was loaded and the system was filled with heavy water.

16.9.10.12.3   The PHT system completion, testing and commissioning

The hydro test for the PHT system was completed. The hot-conditioning of the system was carried out in a record period of 55 months from the date of first concreting. Hot-conditioning was completed within 71 hours.

16.9.10.12.4   RB-4 leak test

The RB pressure and leak tests were conducted within 6 days. The pressure test was carried out at a value of 1.44 kg/cm2.

16.9.10.13   The TG set turning gear: TAPP 4

The set was put on turning gear, one year ahead of schedule. The gear of the previous vintage TG sets was electrical-driven. The gear for the TAPP 4 TG set was turbine-driven at 80 rpm (revolutions per minute). The turning gear for the TG sets is put on to ensure uniform thermal conditions all along the length and breadth of the set.

16.9.10.14   Safety drill at the site

The construction personnel of TAPP units 3 and 4 along with the site contractors' workforce participated in a site emergency exercise conducted by TAPPs. The exercise was conducted satisfactorily.

16.9.10.15   Evacuation of families from the exclusion zone

The evacuation of all families living within the exclusion zone of a 1.5-km radius from the site was carried out simultaneously along with their rehabilitation. The clearance from the AERB for proceeding with the first-criticality would be forthcoming only after the rehabilitation was completed satisfactorily.

16.9.10.16 *Part-commissioning simulators and licensing of personnel (2004-2005)*

The commissioning group for TAPP units 3 and 4 was set up. The part-test simulators for the electrical, reactor protector, reactor regulating and PHT systems were set up as well. Training of operators and maintainers on these simulators was completed. Licensing examinations for the various qualification batches were conducted. Full-scope simulators for training purposes were in position and commissioning was completed. The modular form of the 400-KV switchyard system was incorporated with a computerised operation system. This helped in the commissioning of the information system in just 13 months.

16.10 TAPP 4: First approach to criticality

Before the commencement of the approach to first-criticality, all the protective system fuel channel trip tests were carried out and the reactor was manually tripped through an SDS-1 trip system to ensure the healthiness of the SDSS (shut-down systems). The approach began in the last week of February 2005.

We have seen in para 16.9.10.12.2 that fuel loading had commenced. Complete loading along with creating a database for the loading took place within 10 days. The initial fuel-loading pattern was worked out in a manner that would ensure full-power operation from day one. A dual objective for the fuel-loading pattern was flux flattening and maximum saving of natural Uranium during the initial loading.

Neutron detectors were installed in designated positions inside and outside the fuel core after response testing. A Boron concentration of 155 ppm and Gadolinium concentration of 15 ppm was ensured in the 40 tonnes of moderator heavy water which were filled in. This quantity of heavy water was arrived at with a view of keeping the moderator level below the lower-most fuel channel.

The total quantity of 210 metric tonnes of heavy water was added to the primary heat transport system. The expected poison concentration was compared with the observed concentration. The values were closely matched. Inverse concentrates were plotted to monitor the extent of sub-criticality to ensure that criticality could never be attained during heavy-water addition to the moderator system.

The bulk addition of 293 tonnes of heavy water to the moderator system was carried out with a moderator level reaching 95.7% of the full Calandria tank level at 10:10 hrs on 28.2.2005. This addition resulted in a Boron concentration of 21 ppm and a Gadolinium concentration of 2 ppm at the full-tank level of the moderator. Gadolinium concentration was removed till it was below the detectable level (below 40 ppm). The Boron was then removed gradually. The reactor became critical with 9.82 ppm Boron in the moderator with adjuster rod 9 at a 50% out-position at 12:41 PM on 6.3.2005. The critical counts were about 20000 eps counts per second.

TAPP-4 racing towards completion

16.11   First-criticality of TAPP 4

It was a red-letter-day for the nation and particularly for the large community of ex- and serving NPCIL employees on 6.3.2005 when India's largest indigenously designed and constructed 540-MWe TAPP 4 unit attained criticality at 12:41 PM. The unit went critical 7 months ahead of the schedule and within 6 years of the first concreting. This was the fastest period so far in the history of Indian Nuclear Power Plant construction and commissioning. The dedicated efforts of the personnel involved in the construction and commissioning of TAPP unit 4 were rewarded on this day for this achievement.

16.12 First synchronisation with the Western Regional Grid (WREB)

The unit was successfully synchronised with the WREB on 4.6.2005, following clearance from the AERB for operation up to 50% of the rated capacity. The unit was declared for commercial operation in August 2005, about 8 months ahead of schedule.

16.13 TAPP 4: energy generation

The unit generated 6.38 billion units from August 2005 till June 2008, bringing in revenue of Rs.16 billion.

16.14 Breach of 6 years for the completion of nuclear units

There are several reasons for the breach of 6 years for the completion of nuclear projects. The goal was to attain the earliest and fastest completion period.

Steps were taken at the design stage itself to avoid repair and re-work during construction which would cause further delay. These steps have been detailed as follows:

(1) Award of mega supply and erection work packages
(2) Application of a high degree of automation in the erection work
(3) Use of high-capacity cranes for lifting and the erection of heavy equipment
(4) Open-top erection and construction
(5) Micro-level planning and monitoring
(6) Motivating and galvanising human capital towards the goal of high achievement

16.15 Chronological progress in TAPP 3 as of the end of 2004

16.15.1 Nuclear system work

16.15.1.1 Coolant channel installation and nuclear equipment piping

Coolant channels were installed and all major equipment in the RB-3 were installed. Piping work between the NSSGs and headers was in good progress. The PHT feeder erection work was completed within 5 months. The end-shield ball-filling was completed within 7 days. The TAPP 3 reactor leak test was completed. Reactor auxiliary systems were kept ready for the start of the first approach to criticality.

16.16 TAPP 3 conventional system work as of 2004-2005

Charging of SUTs (Station Unit transformers) was on hand. The units were subsequently commissioned to provide power for commissioning down-stream loads. Piping, ventilation, lighting, grounding and related systems were commissioned. The erection of the TG set and auxiliaries was completed. The TG set was put on barring gear and kept poised for commissioning.

A system known as 'pro-control' was introduced in TAPP units 3 and 4 to control and maintain the turbine speed at a pre-determined value under differing load conditions. This was a distributed digital control system with microprocessor-based intelligent-remote multiplexing capability. The system controls and monitors turbine start-up operation and shut-down. The system senses load changes and establishes set-points for changing load values.

16.17 Instrumentation and control systems: main control panels

The main control panels are physically separate from each other: a partition is provided between them. An operator console is provided for each unit. The panels are arranged in a 2-tier horse-shoe configuration with the operator console in a central position. The two tiers are in slanting horizontal and vertical configurations. The slanting horizontal portion accommodates all the controls in the form of mountable mosaic operating tiles consisting of miniature hand switches, push buttons and LED indicators. The panels in vertical configuration house the following: indicating instruments, controllers, recorders, CRT displays, window annunciators, etc.

16.18   Computer controls: the computer room

Computer controls have been extensively used in the plant. Fuel handling, safety and regulating system, the PHT system, station logic system, radiation monitoring system and operator information system are all controlled through computers. A computer room adjacent to the main control room houses the computers and computer peripherals. Redundant processor modes have been provided to enable testing and monitoring if the nodes in the main control room are not available for any reason.

16.19   Instrumentation system cabling: instrument testing and schedule of completion

A period of 17 to 22 months was given to the instrumentation and control group, starting from the time terminal points, and areas were made available to the group for the completion of its installation work. This schedule was achieved as a result of close co-ordination with the Engineering Construction Co. (ECC-a subsidiary of L&T Co.) and Electronic Corporation of India Ltd. (ECFL-a central PSU), the contractors for field mega package work and control centre instrumentation package, respectively. Details of the individual work are as stated below:

S. No.	Description	Total quantity	Max Rate per day	Remarks
1	Trunk cabling	1500 km	25.9 km	
2	Wire terminations	950000	2502	
3	Panel installation	680	17	
4	Stainless steel tube laying	70 km	3.6 km	
5	Auto tube welding joints	15000 nos	121	
6	Conductor laying per day		535 mes	
7	Branch cabling	300 km	2.25 km	
8	Testing and calibration of instruments	11000	210	

All the cable seals for the cables entering the RB-3 were completed in time. It is noteworthy that the cable leak test revealed zero leaks from the entire cable seal system.

16.20   Schedule followed for TAPP 3 work

Individual system work was completed within the targeted time. The schedule closely followed the actual time taken for TAPP 4 work which had acquired field precedence over the TAPP 3 work.

16.21   Independent verification and valuation of safety-related systems (IV and V)

Independent verification and valuation of all safety-related computer systems of TAPP 3 and TAPP 4 were carried out and review reports were submitted to the AERB.

16.22   'All systems Go' for first-criticality and the AERB approval

Individual systems were commissioned and TAPP 3 was positioned for its first approach to criticality as of May 2006, following the AERB approval.

16.23   First-criticality of TAPP 3

The first criticality was achieved on 21.5.2006.

16.24   Commercial operation of TAPP 3

TAPP 3 was declared for commercial operation on 18.8.2006.

16.25   Revenue generation of TAPP 3

The unit generated 4.64 billion units from August 2006 till June 2008, bringing in a revenue of Rs 11.6 billion.

16.26   TAPP 3 maximum continuous operation

TAPP 3 created a record about continuous operation for 522 days during 2010-2011.

16.27   Awards for TAPP units 3 and 4

The Central Electricity Authority's Shield award was given for TAPP units 3 and 4 at a ceremony held at Vigyan Bhavan, New Delhi, on 26.3.2012.

TAPP units 3 and 4 had earned the prestigious 'industrial and fire-safety award' for 2010-2011.

16.28   CSR activities of TAPP units 3 and 4: 2011 onwards

16.28.1   Construction of School

The CSR authorities of TAPP units 3 and 4 constructed a school building at a nearby Dardi village, providing complete infrastructure for the school building including classrooms and office furniture. Utility items like school bags, notebooks, etc. were also distributed to the students.

16.28.2   Primary Health Centre (PHC)

A 4-bed PHC was set up at the project-affected village, Popharan. It was inaugurated by Sri. Hemant Kumar, Site Director TAPPs on 2.1.2017. The PHC will be providing free health services to the residents in the neighbouring villages. The inaugural function was attended by 50 persons including local Maharashtra Govt. officials.

16.28.3   Environmental stewardship

TAPPs have been promoting environmental efforts to preserve local bio-diversity. As part of the promotion, an exclusive garden for butterflies was set up at the TAPP site. The garden was located inside the exclusion zone between the main gate and the guard house of the site during 2011-2012.

The beautiful garden harbours more than 2800 plants of 84 species of both nectar and host plants. Colourful flowering plants, creepers, short-sized plants, bushes, grasses and shrubs in this garden are spread over an area of 1690 mt2.

Around 30 species of butterflies are found in this garden. The peak season for butterflies is between September and February of each year. Plenty of butterflies visit and nest here during the above-mentioned season.

This garden is the second of its kind to be built in the Indian nuclear plant landscape after Kakrapar units 1 and 2.

16.28.4   The establishment of the green belt

A green belt of 1,10,000 trees belonging to 15 botanical varieties has been established on a land area of 30 hectares along the periphery of the plant-exclusion zone. A nursery capable of catering to 60,000 saplings was also established.

CHAPTER 17

The Decommissioning of Nuclear Power Stations

17

17.1 The need for decommissioning

So far, we have been dealing with the development and setting up of nuclear power stations. It is now time to look at the need for the decommissioning of Nuclear Power Plants and the modes and facilities required for the same.

17.2 The 'What' and 'Why' of decommissioning

In any field of activity, the permanent retirement of operating plants, after they have served their useful life, is inevitable. The retirement of plant and equipment in a fossil fuel-fired generating plant does not, in any major way, differ from the procedures about mechanical and electrical equipment in other industrial activities. The equipment and their parts, as the case may be, are dismantled and sold off as scrap.

17.3 The decommissioning of Nuclear equipment: the need for a special approach

The approach used for other plans, as discussed before, is not possible in a nuclear power station, particularly with regards to the nuclear portion of the plant because of the presence of radioactive elements in the equipment to be decommissioned.

Detailed guidelines and procedures for the decommissioning are hence drawn down for such a plant, much before the active implementation. The Indian and International criteria for radiation protection, transport, waste disposal and environmental safety have to be adhered to during the various stages of decommissioning.

17.4 Equipment for concern

A Nuclear Power Plant is left with some radioactive structures and worn-out equipment after say 30 to 40 years of operation. A major portion of the physical equipment say about 85% does not become radioactive during the operation and may be demolished or re-used, as the case may be, without any restriction. As we have seen in para 4.2.5.2 of Chapter 4, the nuclear fuel used in a nuclear power station contains within itself more than 90%, in fact up to 99%, of the generated radioactivity arising out of nuclear fission. This spent-fuel is taken out and shifted under sufficient radioactive protection safeguards for re-processing or suitable disposal.

The reactor vessel and piping are of major concern because of the radioactivity caused by neutron irradiation. This irradiation produces radioactive products with specific half-lives (half-life is that period required for the level of radioactivity to come down to 50% of its initial value). The deposition of corrosion products on equipment surfaces also takes place which is de-contaminated by chemical and mechanical processes.

17.5 Stages of decommissioning

There are three stages of decommissioning of Nuclear Power Plants which will be detailed in the upcoming sections.

17.5.1 Stage 1: storage with surveillance

This involves the removal of all irradiated fuel, coolant and other easily accessible radioactive materials/equipment structures and ventilation systems housed inside the containment building, under prescribed surveillance and control procedures. This process may take about two years to complete.

17.5.2 Stage 2: sealing off and restricted site release

The reactor vessel and components are sealed off for a dormant period of 30 years to enable the residual reactivity to decay to acceptable levels. Non-active areas of the plant are released for alternative use.

17.5.3 Stage 3: dismantling techniques

Major work of dismantling the reactor vessel and nuclear piping is done with specialised remote-handling techniques. The debris is taken away for permanent storage in steel-lined shallow land repositories in locations duly approved by the regulatory body. The remaining areas of the plant are decontaminated and released for unrestricted occupancy. The occupancy may include Green-field activities.

17.6 A prompt mode of decommissioning

An alternative to the 3-stage decommissioning process is the prompt mode of decommissioning which may be restored if need be, to the job in about 10 years. This comprises of Stage 1, as described in para 17.5.1, but without envisaging Stage 2, followed immediately by Stage 3, as discussed in paras 17.5.2 and 17.5.3, respectively. This method involves higher costs and higher man-rem exposures.

17.7 Global and Indian experience in the decommissioning of nuclear plant facilities

17.7.1 Reactors that are already decommissioned

More than 65 reactors have been decommissioned successfully the world over since 1960. The accumulated experience of this decommissioning has been shared by all members of the International Atomic Energy Agency (IAEA). The decommissioning charges have been estimated to be less than 2% of the cost of generating electricity, and the man-rem exposure is less than 0.1% of that caused by natural background radiation. France, an important nuclear power-producing country, had provided a sum of US \$4.9 billion for the decommissioning of one of its reactors at the end of 1999. An idea of the provisions required for decommissioning can be obtained by studying this occurrence.

17.7.2 The BARC Test Reactor

A test reactor at BARC has been decommissioned satisfactorily.

17.7.3 Shipping Port Reactor: German Reactor

The Shipping port 90-MWe PWR in the USA was the first commercial nuclear power station to safely and cost-effectively dismantle down to Stage 3 (disassembling the reactor vessel). Another reactor at the Elk River site in the USA and one reactor at Niederaichbach, Germany, have also successfully reached Stage 3 of decommissioning. The Niederaichbach NPS in Bavaria, Southern Germany, has been decommissioning over the last few years. The project was to be completed by 1996. The plant was commissioned by the Society for Nuclear Research in the early 1970s. The society is known as the Karlsruhe nuclear research centre. This heavy-water reactor cost 200 million German marks to build. The decommissioning has cost an estimated 2.80 million marks. The reactor internals have been dismantled by remotely controlled equipment. The 522 tonnes of reactor steel contain an estimated eight billion becquerels of radioactivity. The site is proposed to be converted to a green meadow.

17.4 The Berkeley Reactor

The UK's nuclear electric (NE) has completed the first stage of decommissioning of its twin magnox reactors at Berkeley at 60% of the originally estimated cost. The earlier estimates were 75 million pounds while the revised estimate is 45 million pounds. The savings have been mainly due to the cutting down of time for the decommissioning from 60 months to 42-45 months. Further savings are expected in the closing stages of the work.

17.5 The South-German Reactor

The 250-MWe Gundremmingen Nuclear Power Plant in the Southern part of Germany, which was commissioned in 1966, was taken up for decommissioning. This was the biggest unit yet to be taken up for the decommissioning and dismantling of the reactor vessel.

The 500-MWe Commercial PFBR at Kalpakkam (TN)

18.1 The Genesis of the 500-MWe PFBR

In Chapter 14, we have seen the construction and commissioning of 40-MWe FBTR at IGCAR, Kalpakkam. The reasonable success in the operation of this FBTR provided the fillip for the design, construction and commissioning of a 500-MWe Prototype fast-breeder reactor (PFBR). Only one or two other countries have ventured into fast-breeder reactors.

18.2 The constitution of working design objectives

A working group was constituted at the Indira Gandhi Centre for Atomic Research (IGCAR) for finalising the technical aspects of a commercial prototype fast-breeder reactor to ensure the feasibility of indigenous design, construction, and commissioning of the reactor. The working group examined the various aspects of options and recommended a 500-MWe pool-type liquid-metal fast-breeder reactor (MFBR). The detailed design work was taken up at IGCAR on such a reactor. The reactor was to be a much scaled-up version of the 40-MWe LMFBTR at Kalpakkam which has been operating satisfactorily since 1985. The design parameters were finalized in 1994.

It was borne in mind while designing the proposed LMFBR, that it was a vital necessity to minimize the capital expenditure along with the developed indigenous technology for the manufacture of the reactor and other critical components. This was because the reactor would be the forerunner for a few similar-sized reactors.

18.3   Vetting of the design basis report

The design basis report (DBR) for the 500-MWe PFBR was duly submitted to the AERB in early 2002. The AERB duly vetted the report and approved it in August 2002.

18.4   Financial sanction for the project

The financial sanction for the project (1 × 500 MWe) was accorded in 2003 at Rs. 3,520 crores based on the prices of 2002-2003.

18.5   Salient technical features of the Liquid Metal Fast Breeder Reactor (LMFBR)

18.5.1   Scaled-up version

We have already seen that the proposed 500-MWe LMFBR would be a much scaled-up version of the existing 40-MWe LMFBR at Kalpakkam. The design features would be similar to scaled-up ones with the exception that the nuclear fuel input would be of the uranium oxide type instead of the uranium carbide type.

18.5.2   Control and Regulating Mechanism

There are two independent fast-acting diverse drive mechanisms of the absorber rods. The mechanisms along with the absorber rods operate in a fail-safe mode. The mechanism consists of upper and lower parts. The lower part is partially immersed in a hot Sodium pool; the upper part is housed in the reactor containment building (RCB) environment. The height of each mechanism is 11 metres. Twelve mechanisms (9 CSRDMs (Control and Safety Rod Drive Mechanism) and 3 DSRDMs (Diverse Safety Rod Drive Mechanism) are housed within the restricted space available in the control plug. The head of the control and safety rod is coupled with the mobile assembly of the CSRDM by gripper fingers which are operated from the top of the CSRDM. The mobile assembly of the CSRDM is held by an electro-magnetic (EM) Motor drive assembly which rotates the screw in either direction to raise or lower the EM and hence the drive assembly. On receipt of the scram signal, the mobile assembly of the CSRDM along with the CSR is released and falls under gravity.

In the case of DSRDM, the electromagnet at the lower end of the assembly holds the head of the DSRDM. The DSR (Diverse Safety Rod) is parked above the fuel core during normal operation and always within its sheath. The Motor drive sub-assembly rotates the screw in either direction. On receipt of the scram signal, the EM is de-energized and the DSR alone is released to fall under gravity, unlike the CSRDM.

18.5.2.1   Further details of the Reactor

The reactor is of the pool-type liquid-metal fast-breeder type (LMFBR). The salient features of the LMFBR are detailed below:

S. No.	Description	Value/Quantity
1	Thermal Power	1,250 MWe
2	Electrical output	500 MWe
3	Core height	1 metre
4	Core diameter	1.9 metre
5	Fuel type	Plutonium Uranium Oxide
6	Pins per fuel assembly	217
7	Fuel pin outer diameter	6.6 mm
8	Maximum neutron flux	8× 10 raised to power of 15 neutrons per cm² sec
9	Fuel-clad material	
10	Absorber material	enr B4c
11	Primary circuit	pool type
12	Primary inlet/outlet temperature	670/c (Kelvin)
13	Steam temperature	766/c (Kelvin)
14	Steam pressure MPA	16.6
15	Containment building type	RCC rectangular shape
16	Design life	40 years

18.5.2.2   Components of the reactor

Some of the components of the reactor are detailed below:

S. No. Description

1) Main vessel
2) Core support structure
3) Core catcher
4) Core plate
5) Core
6) Safety vessel
7) Inner vessel
8) Top shield
9) Control plug
10) Control and safety rod drive mechanism
11) In-vessel transfer mechanism
12) Intermediate heat-exchanger
13) Primary Sodium pump

18.5.2.3 *Formal inauguration of the project*

The Prime Minister, Sri. Manmohan Singh, formally inaugurated work on the project on 23.10.2014. The day also marked the golden jubilee (50th Anniversary) of the Indira Gandhi Centre for Atomic Research (IGCAR) at Kalpakkam (TN).

18.5.3 Fuel Handling

Two mechanisms, by way of rotatable plugs, are provided for fuel handling. The transfer arm handles the sub-assembly positions and the in-vessel transfer positions. The sub-assemblies are vertically handled under Sodium at a depth of 8 metres below the roof slab. The inclined feed-transfer machine is used to transfer the sub-assemblies from the reactor assembly to the fuel-transfer cell.

18.5.4 Primary Sodium pumps: intermediate heat-exchangers of the reactor main vessel

18.5.4.1 *Reactor main vessel*

The vessel is of a U-type with 12.9 metres diameter and 13.2 metres height. It is made of stainless steel 316 LN grade 30-mm thick plates. It has to withstand a temperature of 547 degrees C and high radioactivity. The reactor's thermal power is 1250 MWt.

18.5.4.2 Primary pumps: NSSGs

Two primary heat pumps are used in the reactor to circulate and evacuate nuclear heat from the reactor Sodium pool. The pumps are of the vertical centrifugal type operating from a free-Sodium pool level with inert Argon cover gas. They are driven by variable-speed drive motors. The leak tightness is ensured at the top of the pump shaft between the Argon cover gas and the containment atmosphere by mechanical seals. The pumps are suspended from the reactor roof slab.

18.5.4.3 Intermediate Heat-exchangers

Four heat-exchangers are of the shell and tube type. They exchange heat from radioactive Sodium to non-radioactive Sodium. The material of construction of the IHXgs(Intermediate Heat Exchangers) is stainless steel of 316 LN grade. The tubes are of 0.8 mm wall thickness. The tubes are rolled and then welded onto the tube sheet by an automatic welding procedure.

18.5.4.4 The Steam Generators (NSSGs)

The nuclear steam service generators, four per loop (NSSGs), produce steam by transferring nuclear heat from the non-radioactive Sodium to the incoming feed water in the tube portion of the NSSGs. The NSSGs are made up of ger-1 Mo (Grade-91) ferritic steel. The steam temperature is 493 degrees C (766 degrees K) and the pressure is 16.6 absolute atmosphere, 70 kgs/cm^2. The NSSGs have Sodium in the shell side. Each tube has a single length of 23 metres. The material of construction is modified ger-1 Mo (Grade-91). Tube-to-tube sheet welding is a tri-metallic joint with an intermediate alloy 800 piece.

18.5.4.5 Decay heat removal

The reactor decay heat is removed by two types of heat-exchangers: Sodium-to-Sodium and Sodium-to-Air. The Sodium-to-Sodium heat-exchangers are similar to intermediate heat-exchangers (hexagons) whereas the Sodium-to-Air hexagons are of the finned type.

18.5.4.6 *The reactor core: design and manufacture*

The reactor core consists of the fuel assembly, control blanket and shielding sub-assemblies. The length of each fuel sub-assembly is 4.5 metres. The core diameter is twice the height. The fuel rods, 217 per fuel sub-assembly, are seamless of 6.6 mm outer diameter and 0.45 mm wall thickness. The wrapper tubes are 131.3 mm across with 302 mm wall thickness.

18.5.5 Other critical reactor components: design and construction

18.5.5.1 *The grid plate*

The grid plate is a box-type bottom structure with top end plates of 6.8 mm diameter that are inter-connected by sleeves of 1-metre height. There are 1,758 sleeves screwed onto the top and bottom plates.

18.5.5.2 *The core support structure*

This is a box-type structure of 7.8-metre diameter and 1.8-metre height which is manufactured from 25/30-mm thick SS316 LN grade material. Several stiffeners and openings are provided in the support structure.

18.5.5.3 *The roof slab*

This is also a box-type structure which is manufactured in various sectors on the shop floor and finally assembled and concreted. The concreted material (material of re-inforcement and construction) is of special Carbon steel and is chosen to obviate laminal tearing. The roof has a slew of highly restrained stiffness and T-welds. The thickness of the plate is 30 mm. The roof slab provides a top shield for the reactor structure.

18.5.5.4 *Manufacture, assembly and testing of the reactor component at site development*

The main vessel, safety vessel, inner vessel, etc., cannot be transported as single pieces from the manufacturer's shops.

Hence the components were proposed to be manufactured in parts at the manufacturer's shop and then transported and integrated at the site. The manufacture involved the die-pressing of large-diameter and thin-walled structures. The techniques were developed for carrying out dimensional inspection, welding inspection and the non-destruction of individual petals and the complete vessels.

18.5.5.5 The development of materials and components for the manufacture of large reactor equipment

18.5.5.5.1 Reactor main vessel: steam generators

About 3000 tonnes of stainless steel and 570 tonnes of modified (ger-1 Mo) Germanium-1 Molybdenum steel were required for the manufacture of nuclear steam service generators (NSSGs) for the plant. These materials were required in various forms and shapes such as plates, tubes, forgings, hollow bars, castings, welding electrodes and filler wires for the electrodes. The parameters and technical details of some of the important materials are furnished below:

a) Main vessel plates:

   8 metres in length, 3.5 metres in width and 30 mm thick SS316 LN

b) Tubes for NSSGs/other components equipment

   b.1) 23 metres of single-length ger-1 Mo material.

   b.2) 8 metres of single-length with 19 mm outer diameter and 0.8 mm wall thickness SS316 LN

18.5.5.6 The location of the PFBR

The 1×500 MWe PFBR project is located at a site about 1 km South of the existing Madras Atomic Power Station at Kalpakkam.

18.5.5.7 Period of construction

The unit is expected to be constructed and commissioned within 7 years. The construction was likely to be taken up in 2004 and the commissioning was expected by 2011.

18.5.5.8 Progress achieved so far

We have already seen that the techniques involved in the manufacture of vital components were finalized at an early stage of conception of the project. The vital components of major equipment have already been received at the site. The transfer arm and roof slab structure are in the advanced stage of manufacture. CSRDM, DSRDM and transfer arm for fuel handling will be tested at 550 degrees C at IGCAR to validate the design.

18.5.5.9 Plans for fast-breeder Reactors

Four more breeder reactors of 500-MWe capacity are proposed to be constructed. It is also proposed to take up the design and development of 1000-MWe units.

18.5.5.10 The formation of Bhartiya Nabhikiya Vidyut Nigam Ltd. (Bhavini)

'Bhavini' was incorporated as a public undertaking under the administrative control of the DAE (Department of Atomic Energy) on 22.10.2003, conforming to the Companies Act of 1956 with an authorised capital of Rs. 5,000 crores. The NPCIL took a stake of 5% in the public sector undertakings (PSU-India). The PSU had been charged with the construction, commissioning, operation and maintenance of the plant. Sri. S.K. Jain, CMD of the NPCIL, was appointed as the CMD of Bhavini. The NPCIL would provide management expertise for 'Bhavini'. The funding for the project would be 80% through equity and 20% through bonds.

18.5.5.10 1 Commencement of work

The preliminary site work and Civil work commenced during the first half of 2004.

18.5.5.11 First Concreting

Indian PM Shri M.M. Singh formally inaugurated first concreting by remote operation which triggered concreting in the Reactor Vault on 23.10.2004. The PM's visit to the site also marked the 50th anniversary of the inauguration of Indira Gandhi Centre for Atomic Research (IGCAR) at Kalpakkam, TN.

18.5.5.12 Progress as of December 2004

18.5.5.12.1 Nuclear Island buildings geo-technical investigation: mud-mat of RB

The site excavation was completed for 17 nuclear island buildings and the geo-technical investigation was completed. The work on the mud-mat for RB was completed and the raft work was in progress.

18.5.5.12.2   The site assembly shop

The building of the site shop, where large reactor components are to be integrated before the placement in the RB vault at the site was nearing completion.

18.5.5.12.3   The placing of purchase orders

The purchase orders (POs) for 9 long-delivery nuclear steam supply components were placed. The POs for 9 other components were under processing.

18.5.5.12.4   Civil work for Nuclear Island buildings

All the infrastructure like batching plants, cement silos, ice plants, concrete pumps, concrete minor transports, cranes, etc. were in position. The M-45 grade concrete design, for concreting the raft of the nuclear island and connected buildings (NICB), the quality assurance plan and the construction methodology for the raft were all approved by the AERB. The raft for the ventilation stack, involving 850 cu. metres of concreting was completed in only four days.

18.5.5.12.5   Financial outlay

The 13 purchase orders were worth Rs. 71.29 crores.

18.5.5.12.6   Physical progress

The overall progress of 4.71% was achieved as of October 2004.

18.5.5.13   Progress during 2005-2008

18.5.5.13.1   Safety vessel manufacture

The safety vessel which holds Sodium was fabricated at the site to prevent the unlikely event of a leak in the main Sodium vessel. Stringent radial and profile tolerances are required during the manufacture. The safety vessel is a thin shell with a diameter of 13.5 metres, wall thickness of 15 mm and a height of 13 metres. The material of the vessel is SS304 LN. Adherence to verticality is critical since the vessel is to be lowered during the erection within a radial gap of 160 mm. Horizontal alignment and alignment of the axis are also to be achieved within a tolerance of 13 mm.

18.5.5.13.2 Civil work

The construction of the reactor vault was completed. Other Civil package work is progressing simultaneously.

18.5.5.13.3 The EPC contracts

All major EPC contracts have been awarded and some of the work has commenced.

18.5.5.13.4 The erection of the safety vessel

The vessel was lowered successfully within the tolerances mentioned earlier.

18.5.5.13.5 Crawler Crane erection

A crawler crane of 1350-tonne capacity was procured and commissioned in record time. This will enable the erection of all the components of the nuclear steam service system (NSSGs).

18.5.5.13.6 Overall progress

About 36% of the overall progress was achieved as of August 2008.

18.5.5.14 Progress during 2008-2009

Progress was achieved on all civil, conventional and nuclear systems. The highlight of the nuclear system work was the lowering and alignment of the reactor vessel of about 296 tonnes in weight into the safety vessel in the early hours of 5.12.2009. The vessel is 12.5 metres in diameter and will hold 1150 tonnes of Sodium. The tail of the vessel is made up of SS316 LN material. The reactor vessel accommodates within itself: the grid plate, catcher, core support structure, inner vessel primary pumps, intermediate heat-exchangers, fuel and blanket assemblies, control rods and fuel-transfer machines. The core cables and core support structure were integrated with the main vessel and lowered into the safety vessel. The vessel was manufactured at the FBTR site itself under a stringent quality-control regime. The vessel was lowered using the heavy-duty crawler crane from a distance of 57 metres inside the reactor vault.

18.5.5.15 *Further progress on the project work beyond 2008-2009*

Further progress on the main vault and the auxiliary system was continued. The progress in the erection of conventional equipment continued through the years 2009-2010 and beyond. Various problems in the electro-magnetic pumps, secondary Sodium pumps and fuelling machines were encountered. These were gradually being set right with the full engagement of the suppliers of the equipment.

The erection and progress towards the commissioning of a 500-MWe PFBTR was the first of its kind and considered to be a challenge anywhere in the world.

18.5.5.16 *The target date for the commissioning of the 500-MWe PFBTR*

The IGCAR director, Dr. Arun Kumar Bhaduria, during his press conference at the site on 30.10.2020, said that the erection problems were getting sorted out and the project was in the final stages of commissioning. He also expressed his hope for the successful commissioning of the project by December 2021.

It may be pointed out that, due to unforeseen circumstances, the cost of the project has escalated. It is now projected, as of the final quarter of 2020, to be Rs. 6,840 crores (Rs. 68.4 billion) as against the initial cost of Rs. 3,490 crores (Rs 34.90 billion).

It is quite important to keep in mind that the initial experience regarding the 500-MWe PFBTR will keep costs and line schedules in proper order in the case of future projects. This will further help India to fulfil its commitment regarding Climate change: this is because the generation of electrical power through fossil fuel-powered stations results in avoidable greenhouse-gas emissions which could be avoided with the help of NPPs. At present, India's power-generation is about 65% to 70% through the above-mentioned stations. The generation of power through nuclear power stations, especially through FBTRs will significantly bring down greenhouse-gas emissions and will also contribute to meeting India's Climate-change obligations.

Global Nuclear Accidents

CHAPTER
19

19.1 General

In the foregoing chapters, we have seen the pattern of nuclear power development in India, the conscious choice of types and the generations of reactors to be set up to sustain a chain of economically viable nuclear power stations. The history of design, construction and commissioning of these stations—leading to self-sufficiency in the complete cycle of nuclear operations—starting right from the mining of nuclear ores to the permanent storage of radioactive by-products was also described.

The 'Introduction' to this book had very briefly touched upon the global nuclear accidents that have taken place so far. It is only so that we get into more detail about some of these nuclearly significant accidents. These details provide a better assessment of the following:

(1) The nature of these accidents
(2) The aftermath of the accidents
(3) The hazards of the accidents
(4) Its effect on the nearby residents of the station sites
(5) The steps taken to mitigate the after-effects of the accidents
(6) The steps taken to de-contaminate and re-start the damaged stations

19.2 The Chernobyl Reactor Accident

This is the most serious nuclear accident that has taken place globally, so far, and has been classified at level 7 of the ascending seven levels of severity of nuclear accidents formulated by the IAEA (0 to 7).

Let us first have a look at the design and construction features of the Chernobyl reactor plant.

19.2.1 Design and constructional features:

The Chernobyl Nuclear Power Plant, located in Ukraine, consists of four 1000-MWe reactor units. The 3200-MWe reactor is of the pressurised light water or RBMK type. This reactor is unique to the former USSR. Twenty-eight such reactors are in service, producing about 10% of the total installed power in the former USSR states; seven were under construction and eight more were planned as of 1988. The moderator is made up of graphite in the form of blocks which weigh a total of 2000 tonnes in the helium–nitrogen gas mixture housed inside the sealed steel enclosure. The graphite assembly is 12 metres in diameter and 7 metres in height; it operates at a temperature of 700 degrees C. The reactor is controlled by 80 manually operated rods, 21 scram rods, 36 over-compensation rods, 12 automatic control rods and 21 shortened absorber rods. The coolant pipes are made up of Zirconium and 2.5% of Niobium. There are 1693 pressure channels through which the coolant flows.

There are eight steam pipes of 200-mm-diameter per reactor. There are two primary loops of 300-mm diameter, each with its primary pump per reactor. The steam pressure is 65 atmospheres and the temperature of the steam in the steam generators (sgs) is 280 degrees C. The reactor system is designed for excess pressure of 4.5 kg/cm2, which is of about 4.25 atmospheres over and above the design pressure of 65 atmospheres. The enrichment of the Uranium fuel used is 1.5%. The overall length of the fuel is 10 metres. The active part is 7 metres in length. Two legs of fuel, each 3.5 metres in length, are held by a hollow central tube to provide spacing between the legs of the fuel. There are 18 fuel pencils of 13.5-mm diameter each. The Zircaloy cladding for the fuel is of 0.9-mm-thickness. There are 1600 fuel bundles, each of 110 kg weight. The fuelling machine carriage is capable of handling entire bundles. The re-fuelling is carried out on load by a top-mounted fuelling machine weighing more than 100 tonnes. The fuelling machine enclosure does not have any special design feature. The design has not provided any feature to improve the stability and integrity of the operation during low-power phases of the reactor.

19.2.2   Design safety features

All releases under accident conditions are local. The pressure suppression compartment pool is provided. The design takes into consideration the double-ended rupture of the 900-mm nuclear piping/vessel. The high-pressure injection of the primary coolant for 3 minutes, backed up by a less-pressure injection system, is provided to deal with abnormal conditions during operation. Emergency cooling system pipes and valves are provided. Self-powered neutron detectors (SPND) have been provided in the core in both the axial and radial directions. Around 180 Boron control rods have been provided, 59 of which are fully withdrawable for emergency control purposes. The core power distribution control is manual. The positive steam void coefficient is 20 mk. The local automatic control mode is provided for reactor operation while global control is through the ionisation chambers. Self-powered neutron detectors are provided in the individual fuel rods. The main coolant pumps are switched off to facilitate turbine unloading.

19.2.3   Reactor containment design concept

The containment is partial. The key parts of the reactor coolant pressure boundary are enclosed, based on the accident location concept. The reactor building consists of several inter-connected compartments or zones; each of these compartments or zones is designed to withstand over-pressures, depending on which part of the main coolant circuit these compartments or zones house. The leak-tight compartments designed to cope with the largest over-pressure are:

(1) The main coolant compartment (4.5 bars over-pressure)
(2) The reactor cavity compartment consists of a relatively thin steel shell with a fairly limited over-pressure capability of 0.8 bars; this is designed to handle the failure of only one of the 1661 pressure tubes.

Among the other design basis events, which the containment system is capable of handling, is the failure of the main coolant pump pressure leader of 900 mm diameter. This is considered to be the most severe design basis accident and failure of a distribution header.

In all of the above cases, steam, water and air arising from the pressure boundary breach are vented to the suppression pool situated below the reactor via non-return valves and venting channels.

There are a variety of systems for extracting heat from the containment compartments and the suppression pool. These include steam condensers in the steam corridor, a sprinkler system, a service water heat-exchanger connected to the suppression pool and jet coolers in the upper part of the compartments (these use water from the suppression pool to cool the air and to remove aerosols and steam). There is also provision in the compartment to keep the Hydrogen concentration below 0.2% by volume.

19.2.4   Experience of early operation

Power instability problems were encountered earlier in the operation. The initial power up-swing was by trial-and-error method.

19.2.5   Sequence of events leading up to the accident

The accident event occurred sometime on the night of 25.4.1986 to the morning of 26.4.1986. The 4th unit was on planned shut-down at 7% full power. The reactor was critical, but not producing any noticeable power. The engineers, from the manufacturer of the TG sets installed at Chernobyl, wanted to experiment to see if the inertia of a recently shut-down unit (unit 4) could be utilized to generate electrical power to operate the vital safety systems—such as the emergency coolant water pumps—in the event of an emergency shut-down of the reactor. This required the lowering of the power of the unit to about one-third of the normal output. The lowering was initiated through power set-back, however, the build-up of Xenon poisoning in the core resulted in the reactor thermal power dropping to 30 MWe. The operators, in an attempt to re-establish the power, withdrew the manually operated control rods, leaving only 8 rods in the core. This was a clear violation of the station operating rules which specify that not less than 30 control rods should remain in the core. The operators also shut-off the automatic shut-down mechanism. All of these actions could only be termed inexplicable.

As part of the so-called 'experiment', the step-valves supplying steam to the TG set were also closed to enable the turbine to 'free wheel'. The reactor began to overheat quickly. Even though all the control rods were subsequently driven back into the core, the reactor 'ran away' beyond control. No measures were immediately available for the control of the reactor. The fuel in the reactor rapidly overheated. The coolant water in the pressure tubes flashed into steam which was an inferior coolant to water. Hence the fuel temperature began to rise alarmingly. Within next to no time, the pressure tubes ruptured. The Zirconium/Niobium cladding of the fuel melted and split open; steam explosions and Hydrogen explosions followed, Hydrogen having been generated by zirconium–water and graphite–water reactions. This resulted in volatile fission products, fission gases and fuel fragments being released into the pressure tubes. The explosions also ruptured the reactor containment.

It was noticed by the maintenance staff, who were on duty on the fateful night, that the closure of the pressure relief device of unit 4, weighing more than 1000 tonnes, was blown off and that the fire was raging above the closure head. About 30 separate fires had resulted from the explosions. The resulting flame blazed high into the late night/early morning sky. A helicopter was immediately summoned to douse the fire. This was done by raining 1800 tonnes of sand and clay, 800 tonnes of dolomite and 40 tonnes of Boron carbide down into the path of the flame by the pilot leaning out of the cock-pit. Around 2,400 tonnes of lead were also poured to absorb the radiation that was released. This pilot picked up a sizeable radioactive dosage and died 3 years later in 1989.

Two other operating personnel also died as a direct result of the accident.

The reactor units 1, 2 and 3 were functioning satisfactorily at the time of the accident at unit 4. These three units were shut-down safely after the accident in unit 4.

Liquid Nitrogen was immediately pumped into the bottom of reactor 4. Reactor units 1, 2 and 3 were later re-started and continued to operate for varying periods following the accident.

19.2.6 Probable causes of the accident

Probable causes of the accident are enumerated below:

1) Fault in the fuel core and possible melting of fuel in the core is the most credible cause of the accident; this was assumed for purposes of safety design.
2) Steam ruptures in the monitoring system for moisture.
3) Uncontrolled power surge in the reactor.

19.2.7 Details of casualties: evacuation of the public

The worst affected personnel were the operators at the station and the connected personnel who were present at the station site during the time of the accident.

A recent estimate puts the number of deaths as a direct result of the accident to 42, including 3 in the blast itself and 28 within 3 months of the accident.

Bone-marrow transplants were carried out immediately on all those who suffered severe radiation exposure. All these recipients died of transplant rejection. This development was an eye-opener for radiologist doctors, as this implied that despite the heavy dosage of radiation received, not all bone-marrow cells of the body had been destroyed. These recipients are among the 28 who died within 3 months of the accident.

Around 48,000 residents within a 10-km zone, which included the nearest town Pripyat, and another 30,000 within a 20-km zone of the site were evacuated to safe locations. A total of 1, 35,000 people within a radius of 60 km of the station were evacuated as well. These people were examined and tended to by 1240 doctors, 3660 assistants, 920 nurses and 720 medical students.

19.2.8 Further studies and details of the accident

It was observed from further studies of the accident that vertical columns of 'shield sand' around reactor 4 had completely melted and flowed down. These two melted columns formed into a calcified mass. The supporting girder structures of the reactor had moved out and stayed put in these new positions, thus endangering the building structure. The openings of a total area of about 1500 sq. metres had been created as a result of the nuclear explosion, some of them 5-15 sq. metres in size each.

Radioactive debris from reactor 4 continued to pose a serious problem to the authorities of the Chernobyl plant, who have been engaged in the task of safeguarding the public and the environment from possible leakages of radioactivity from the damaged reactor. It has been estimated by the Ukraine authorities that about 7,40,000 TBq of radioactivity was released into the atmosphere on the first day of the accident and that the integrated total radioactivity had risen to 1.85 million TBq on the 11th day of the accident.

19.2.9 Further safety measures taken for containing radioactivity

19.2.9.1 *Containing decay heat:*

A tunnel under the damaged reactor with a cooling arrangement was constructed to contain decay heat from the reactor core.

19.2.9.2 *Encasing of Reactor 4:*

Reactor 4 was encased in walls 6 metres thick which were called the 'sarcophagus'. Round-the-clock shifts were maintained to oversee the condition of the 'sarcophagus'. The 'sarcophagus', built mostly by remote control, is to contain and prevent an estimated 400-700 kgs of Plutonium, 80 kgs of Caesium and several tonnes of radioactive waste from escaping into the outside atmosphere. The radioactive air inside the 'sarcophagus' (giant container) is exhausted through absolute filters to remove the radioactive particulates before the air is vented into the atmosphere. Stack holes have been provided in the 'sarcophagus' to enable the proper differential of temperature between the air inside and outside of the structure. In recent times, there is fear that the 'sarcophagus' is cracking.

19.2.10   Restoration of the Reactor units 1, 2 and 3 to service

Unit 1 was prepared to resume operation in late 1986, only five months after the occurrence of the disaster at Unit 4. This is to be viewed against the time of six years that it took to get TMI-1—three-mile island unit 1, USA—back online after the nuclear accident in TMI-2.

Unit 2 of Chernobyl was restored to service a little after the resumption of unit 1, but it had to be shut-down again in 1990 due to a disastrous fire in the turbine hall. Repair work on unit 2 was still going on as of 1993.

Unit 3 was also restored to service. Units 1 and 3 were used to feed power into the Russian grid.

19.2.11   Cleaning up efforts: the re-location of townships

Reactor units 1, 2, and 3 continued to operate for varying periods after the accident at unit 4 before they were shut down for detailed examination. The reactor units 1, 2, and 3 were contaminated by radioactive particles entering them through the ducts of the air-conditioning system of unit 4, which continued to operate even after the accident.

Considerable cleaning effort has been underway in these units, but their continued operation will require a stringent dose management regime, involving shift working.

About 3000 operating personnel of Chernobyl are now living in a new settlement which is connected to the main plant by a highway. Well-appointed housing was constructed for 10,000 members of the families of the power workers. A new township for 'power specialists' was also built on the banks of the Dnieper River. Some 52 new townships were built in the region alone within 6 to 8 weeks of the accident, for people evacuated from the 30-km exclusion zone around the plant.

19.2.12   The spread of radioactivity to other countries

Apart from the nearby townships of Pripyat and other inhabited areas close to the Chernobyl site, Belarus and the Bryansk region of the Russian Federation were also affected by the accident. The nearby Scandinavian countries and even the United Kingdom were reportedly affected by the leakage of radioactive products from the failed reactor. Caesium 137 and radionuclides apart from radioactive Iodine (this gets into the thyroid glands), Cerium, Zirconium, Niobium, Ruthenium, Americium, Strontium90 (this last product gets into the bones) and Plutonium (this gets into the lungs) could have leaked into the neighbouring countries.

19.2.13   Social and Health Effects: the formation of self-help groups

19.2.13.1   Social and Health Effects

The effect on the population of the affected countries, if any, remains to be documented. It may, however, be stated that the National Radiological Protection Board of the UK (NRPB) has estimated that the Chernobyl reactor accident has added about 70 uSv to the average annual effective dose—from May 1986 to April 1987—for people living in the UK and that this may eventually reach 100 uSv per annum over subsequent years. The figure of 70 uSv represents about 3% of the annual dose of 2150 uSv calculated for the year 1986, before the date of the accident. Most pessimistic assumptions indicate that 125 extra Cancer cases (not all of these need to be fatal) may occur over the next 50 years or so in the UK because of Chernobyl. This may be compared with the cases of death which are 1300 per year in the UK due to the effects of natural background radiation.

Apart from the above projection, no serious after-effects have been reported so far from other countries. One has to await further details before any firm opinion is expressed on the global fallout of the Chernobyl accident. Caesium 137 was still suspected of contaminating much of the landscape in Chernobyl as of 2011.

19.2.13.2   Formation of self-help groups:

A social consequence of the Chernobyl accident is the formation of a group of concerned mothers for children's protection nicknamed 'Mama 86'. This group, formed in 1990, seeks to draw the attention of young mothers to the link between their children's health and environmental degradation in the post-Chernobyl era. The main goal of the group is to raise awareness of environmental issues through concrete assistance to children with health problems. The group has collected enough donations to equip a medical laboratory for the medical examination of children and to provide advice to parents for their children's medical care and treatment.

The laboratory has so far reportedly examined 2000 children. There has been—according to the laboratory statistics—a 6.5% increase in childhood Pneumonia, a 4.8% increase in cases of Liver disease and a 3.5% increase in Brain Cancer among the school children of Kyiv, since the Chernobyl incident.

No incidents of Leukaemia have so far been detected among the public affected by the Chernobyl accident. The chances of the disease showing up later are not ruled out. Caesium 137 had contaminated much of the land in Chernobyl. About 20-40% of Caesium 137 and 50-60% of Iridium 137 in the total core inventory were suspected to have been released in these lands.

19.2.14   Scale of deployment of the clean-up personnel

A total number of 1,80,000 personnel were reportedly engaged in the clean-up and salvage operations of the Chernobyl reactors. One report called these personnel as 'liquidators' and further contends that only 28% of these 'liquidators' may be considered healthy as far as the after-effects of their deployment in those operations are considered.

19.2.15   Social, animal and agricultural life in and around Chernobyl after the accident

So far, as life in and around Chernobyl after the accident is concerned, agriculture and wildlife seem to be thriving even a decade after the accident. The IAEA, together with the Food and Agricultural Organization (FAO), is helping to restore arable land by growing rape seed plants on 50,000 hectares of contaminated land in Belarus, a country adjacent to Ukraine. The seed takes up and stores radionuclides from the soil in its stalks and seed coat, but not in its seed. The seeds from the plant can then be used for economically viable products such as bio-lubricants, cooking oil or high-protein cattle feed.

Rodents continue to inhabit the areas which were highly contaminated at the time of the accident. New gene-mutations among these rodents have, however, been observed. Many Ukrainian evacuees, particularly the elderly, are returning to their homes outside the 30-km population-exclusive zone. Life around the Ukraine reactor site seems to be coming back to normal, a decade or more after the accident. Even 15 years after the accident, around 2217 communities are still directly involved with the aftermath. Counter measures introduced by the authorities for these communities—such as de-contamination, radiological control on products and restrictions on the consumption of certain foodstuffs—are still enforced. These constraints, though irksome, are compensated by social and financial advantages.

The first marriage at Chernobyl, after the catastrophic accident, took place in late 2000.

19.2.16 The effect of the Chernobyl accident on global nuclear power development

As we have already seen, the Chernobyl accident falls under Class 7 of the IAEA severity which is considered to be the most severe among the seven classes. It is, however, pertinent to point out here that several incorrect inferences could be drawn from this accident. The following facts pertinent to the Chernobyl reactors may serve to present a broader picture in proper focus.

The Chernobyl reactor is of the early 1980s nuclear design vintage. Nuclear power plant design is continuously updated even to the extent of retro-fitting the reactor ancillaries that are already constructed, under construction or even under operation. Present-day PWR reactors could be said to be at least two generations, if not more, ahead of the Chernobyl reactors from the point-of-view of safety, maintainability, reliability and integrity.

Global nuclear power development has not, in any significant way, been affected by the Chernobyl accident.

19.2.17 Improvement to existing RBMK reactors

An important off-shoot of the Chernobyl accident is the decision of the Soviet nuclear station operating authorities to incorporate modifications in all existing RBMK reactors, as well as those under construction, to conform to the state-of-the-art practices to ensure that a Chernobyl-type accident does not occur again.

19.2.18 International response to the Chernobyl accident

It is in order at this stage to comment on the international response to this accident.

Very few foreign scientists have visited the site to acquaint themselves with the genesis and details of the accident. This is a matter of great surprise. With the serious nature of the accident taken into consideration, it was surprising that no IAEA team had visited the site to check up on the measures adopted to safeguard the damaged reactor and prevent further environmental hazards. The IAEA is the nodal agency for the safety of all civilian nuclear facilities and it would have been for the agency to have visited the site immediately after the accident.

One reason for the above-stated occurrence could be that Soviet Russia, despite being a member of the IAEA, had not participated in sharing any information about its Nuclear Power Plant operations to the same extent that the other members had been doing among themselves. There is a great lacuna with the then Soviet Russia and the now Russian Federation, Ukraine and other Central Asian Nations which have a very large nuclear power production programme. It would be in the interests of all members of the IAEA that this lacuna be corrected at the earliest. With the gradual political and politico-economic liberalisation programme underway in the Central Asian states, it may be expected that a more meaningful exchange of information between these States and the other member Nations of the IAEA—with regards to the operation and maintenance of nuclear power stations and other relevant areas—would contribute to the total of global operational experience.

Since the accident, the IAEA has adopted a resolution: 'Early notification and assistance conventions'. A landmark convention on nuclear safety was also held.

It is significant in this context to point out the presentation made by a 23-person Soviet delegation led by academician Valery

Legosov—a nuclear chemist and director of the prestigious Kurchalov Institute of Atomic Energy near Moscow—to an assembly of 500 nuclear experts from 45 nations when they met at the IAEA headquarters at Vienna during between 25.8.1986 to 29.8.1986 on the Chernobyl accident. A 382-page report was also submitted to the member countries of the IAEA. An out-fall of the meeting was the generation of several proposals for international collaboration in such areas as safer reactor operation, initiation of better operator/maintainer training standards, fire protection, levels of radioactive content of food articles and de-contamination methods, modelling of radioactive space, path-ways, study of long-term radiation effects and the treatment of highly irradiated victims.

19.2.19   Offer of financial aid for shutting down of the Chernobyl plant

The group of 7 industrialised nations (G7) comprising of the USA, UK, France, Germany, Canada, Japan and Italy have offered aid of $200 million to Ukraine for shutting down the Chernobyl plant at the G7 meeting which took place in July 1994 at Naples, Italy. This aid would be applied to the shutting down of the then operational unit 1 and 3 reactors as well as the unit 2 reactor on which repair was underway. Ukraine had been maintaining, all along, that it was essential to continue running the units given the acute power position in their country. It was therefore to be seen whether Ukraine would accept the offer of the G7 countries. It was learnt, as of 1992, that the Ukraine Council of Ministers had decided to de-commission reactor units 1 and 2 by the end of 2000.

The cost of shutting down reactor units 1, 2 and 3 was expected to be $4.4 billion. It was intended that two non-nuclear power stations would be built near the Chernobyl site to make up for the loss of the Chernobyl nuclear power-generation. In the G7 nuclear summit meeting held at Moscow during the third week of April 1996, the seven countries offered to fund $3 billion towards the cost of shutting down all the reactors, including the urgent repairs to the 'sarcophagus' of unit 4.

Chernobyl Reactor 1 was shut down permanently on 30.11.1996. The shutting down of unit 2 was taken up and completed after that.

The shut-down of the entire Chernobyl plant was completed when the sole surviving unit 3 was closed off on 15.12.2000.

The re-fuelling of all the reactors was to be carried out progressively from 2001, with a schedule of completion by 2008.

Ukraine and its European partners formally constructed a new metal dome enclosure for the remains of the 4th reactor. This largest moveable metal structure is 108 metres in height, weighs 36,000 tonnes and reportedly costs about $1.7 billion. The structure can withstand a tornado and can last for a century.

19.3   Stationary Light (SL-1) Reactor Accident

19.3.1   Details of the accident

The accident at SL-1, located near Idaho Falls in the USA, took place on 3.1.1961 while re-starting after a long shut-down. Three personnel on duty at the time died as a result of the accident. One of the dead was above the closure sealing of the reactor at the time of the accident. The evidence gathered from the radiological examination of the cigarette lighter of one of the casualties revealed that an 'uncontrolled chain reaction' was the reason for the accident. One of the casualties was found lying near the stairs leading to the reactor vault. Though alive at the time of the discovery, the person was pronounced dead at 11:10 PM on the day of the accident at the nearest hospital by an AEC doctor. The accident is believed to have occurred at 7:00 PM. Even though no fire resulted as a result of the accident, the activation of a fire detector (based on the latest technology) brought the first emergency response unit to the scene, which was the first to reach the site of the accident.

19.3.2   Post-accident investigation: nuclear safety of the reactor

Phase-I of the post-accident activities, which aimed at the recovery of the casualties, was completed successfully. Phase-II of the activities, aimed at determining whether the reactor was nuclear-safe, involved the use of remotely controlled and operated monitors, TV cameras, etc. for determining the condition of the fuel core, the control-rods and other internal parts of the reactor. The survey revealed that 6 nozzles of the control-rod mechanism were open to the atmosphere. It was further revealed that the closure sealing of the reactor was ripped open, such was the force of the explosion caused by the 'uncontrolled chain' of the reactor. The pressure of the explosion caused by the accident was computed to be 36 kgs/cm2.

The ambient temperature after the accident, as at the time of monitoring, was computed to be 38 degrees C above the reactor sealing and 32 degrees C on the reactor operating floor.

The SL-1 reactor containment was not designed to withstand the pressures that might be caused by an uncontrolled chain reaction. Such a reaction did take place as a result of the accident and the pressure generated by this uncontrolled reaction had caused the sealing closure to give away.

Monitoring of the atmosphere above the reactor site revealed no significant radiation above the background. The radiation at the reactor site revealed a background of approximately 2 rems per hour at a distance of 30 metres from the reactor and 0.1 rems per hour at a distance of about 300 metres from the reactor. The radiation level above the reactor sealing closure was 1000 rems per hour.

Phase-II of the post-accident activities revealed conclusively that the SL-1 reactor was nuclear-safe.

19.4   The Windscale Nuclear Accident

Britain's worst-ever nuclear accident took place on 10.10.1957 at the then Windscale (now re-named Sellafield) nuclear re-processing facility, one of the largest in the world. The Windscale accident has been classified at a severity of scale 6 (out of the 7 classes of severities in the ascending order).

It has been repeated that due to design deficiencies, a large number of spent-fuel rods fell into the air-exhaust ducts instead of the water storage pool. It was suspected that the absolute filters installed on the (exhaust gases) chimney were defective and hence failed to trap the radioactive emissions from the spent-fuel.

19.5 The Three-Mile Accident (TMI Accident)

The nuclear accident at Three-Mile reactor unit 2—located on an island in the Susquehanna River near Harrisburg in Pennsylvania, USA— occurred on 28.3.1979 and resulted in the leakage of radioactive gaseous products into the environment which caused a lot of concern. The accident which caused a partial fuel-core melt-down was attributed to mechanical failures, design flaws and operator errors. The accident was caused by a virtual stoppage of the main pumped feed-water flow to the boilers with the reactor on nearly full power. The Reactor shut down automatically. The three back-up feed-water pumps had failed to take over since the inlet valve to the steam generator (SG) had been left closed by mistake. The stoppage of feed-water flow and the loss of removal of nuclear heat from the primary coolant led to the steep rise in coolant temperature and the increase in primary coolant pressure to 171 kgs/cm^2. This led to the relief valve on the coolant pressuriser to open. The relief valve was stuck open. It was not closed after the pressure dropped and the event was not noticed by the operators for nearly 2 hours.

A considerable volume of the coolant was thus released from the primary coolant system. The cumulative effect of the above-stated chain of events was that a considerable amount of residual heat from the decaying fission products remained undissipated. The result was the exposure of the fuel core and its melt-down. The top 1.5-metre portion of the core was later found missing. Molten and solidified fuel from the portion was found in the lower portion of the reactor vessel. 50% of the fuel had melted at an early stage of the accident.

Before we proceed further on the accident, we may note that pregnant women and children within a 5-mile radius of the site were immediately evacuated. Post-accident estimates indicated that the highest possible dosage to anyone in the plant area due to leakage of radioactive gases was less than 100 mrem. The average exposure to the public of 2 million residing within 50 miles of the site was 1 mrem, which is less than an average exposure to one medical x-ray of 6 mrem. Estimates by the Dept. of Health, Education and

Welfare (USA) indicate one additional Cancer death due to lifetime exposure of the above-mentioned 2 million people living within 50 miles of the accident site as a result of the accident. It may be stated that the death, on an average lifetime basis, is due to natural causes of these 2 million people as one is a 3,25,000 population.

The operators of the Three-mile plant were fined $50,000 (dollars 50,000) by the Nuclear Regulatory Commission (NRC) of the USA for the violation of the prescribed procedures for the safe operation of the plant.

The report of the commission on the TMI accident listed one of the causes of the disaster: the undetected failure of the pilot-operated relief valves (POLV), on top of the heavy water pressurisers, to close. The accident was placed on a scale of 5 of the IAEA category of nuclear accidents (ascending from 1 to 7).

19.5.1 Corrective actions taken

The action plan initiated as a result of enquiry into the TMI accident has resulted in the following corrective actions and improvements:
1) Increase in the number of qualified personnel in the plant
2) Upgrading of operator training and licensing practices
3) Provision of detectors and instruments to permit operators to brew the status of the reactor at all times
4) Provision of Hydrogen-detecting equipment in the initial areas of operation
5) Improved monitoring of accident conditions
6) Improved inter-communication between the regulators and plant operators on an individual basis
7) Improvement in the implementation of emerging preparation plans.

19.5.2 Post-accident activities

The TMI reactor unit 2 was permanently shut down. Radioactive waste was shipped to a disposal site. The reactor fuel core debris was shipped to the Department of Energy (DAE-USA) facility. Both the reactor units would be de-commissioned with a time-frame task of the mid-1990's for completion.

19.5.3 Lessons drawn by the NPCIL of India from the TMI accident

The TMI accident led to the formation of the 'TMI accident implementation committee' by the NPCIL in India, to oversee the implementation of the TMI accident enquiry commission report as found applicable to the Indian Nuclear Power Plants. Applicable improvements were carried out by way of retro-fitting the existing plants. The recommendations were also suitably built into the design of the then-current and then-proposed nuclear plants.

19.6 The Mayak Nuclear Plant Accident

A blast took place in the Mayak nuclear plant near the Soviet city of Kyshtym on 29.9.1957. The plant was built during 1945-1948. The blast occurred when the cooling system for a tank containing thousands of tonnes of liquid nuclear waste failed. Two million curies of radioactivity were released into the atmosphere. The blast produced a radioactive cloud that spread over hundreds of miles around the plant site covering an area of 15,000 square miles and killing nearly 200 persons. As a result, 10,000 persons living near the plant were evacuated. The accident was scaled at 6 on the ascending scale of severity of the IAEA norms regarding nuclear accidents.

19.7 Blast in the nuclear plant outside Tokyo, Japan

A team of operators placed highly enriched Uranium fuel in a precipitation tank that was not designed for receiving and handling such a fuel in 1999. A critical reaction resulted from such storage. It led to the death of two workers. Dozens of workers and residents nearby were evacuated as a result of the accident.

19.8 Fire accident at the Brown Ferry Nuclear Power Station in Alabama State, USA

The Brown Ferry Nuclear Power Plant is located at Brown Ferry near the towns of Decatur and Athens on the banks of the Tennessee River in the State of Alabama. The plant houses three units of the boiling-water reactor type and is owned by the Tennessee Valley Authorities. Unit 1 is of 1,065 MWe capacity while units 2 and 3 each have a capacity of 1,113 MWe.

There was a cable seal between the reactor and the cable spreading room of Reactor Unit 1 which was commissioned on 20.1.1973. This seal was broken for installing additional cables on 25.3.1975. A temporary cable seal was set up with foamed plastic (a resilient polyurethane foam). Two coats of fire-resistant paint were applied on both sides of the temporary seal. A lighted candle was used by the maintenance staff to check leakage across the seal. It may be mentioned here that a differential pressure was maintained across the seal, the pressure on the cable spreading room being higher than that on the reactor side.

The temporary seal was highly combustible and caught fire during the test. Considerable damage was caused to the cables of RB-1 and RB-2. The barrier wall between the cable spreading room and the reactor building was 750 mm thick. Emergent measures to put out the fire did not succeed. The fire spread over hundreds of miles around the site. A condensate booster pump was used to pump water into the reactor. Normal shut-down of unit 1 was completed at 4 AM on 26.3.1975. The shut-down lasted for a year. The Nuclear Regulatory Commission (NRC) of the USA made suggestions for improving the cabling security including the provision of improved fire resistance and fire-survival cables.

The Tennessee Valley Authorities (TVA) proposed to upgrade the unit capacity to 1.28 GW (1280 MWe). The unit's license was extended up to 20.12.2033.

The fire spread through the KB side of the cable penetration at noon. The air-flow through the penetration was so strong that the fire extinguishers could not succeed in putting out the fire. The operators did not sound the fire-alarm first. The operators while reporting the fire did not contact the correct telephone number as per the emergency procedure. The two reactors in operation were not shut down and continued to operate. The ECCS had started to function. The power level in reactor unit 1 dropped suddenly. Reactor unit 1 was shut down manually by inserting the control rod while unit 2 indicated a falling power level and was put in shut-down mode. The high-pressure ECCS of unit 2 was lost. The control over the reactor relief valves was lost briefly and then restored.

The nuclear instrumentation control of unit 1 was lost and only four relief valves were under control. The reactor water level could not be maintained. Due to the non-availability of emergency pumps, a condensate booster pump was used to keep the water level at 1200 mm as against the name level of 5000 mm.

Reactor unit 2 was under partial control and similar procedures, as for reactor 1, were adopted to maintain the reactor water level.

The control room of Reactor Unit 2 was thick with smoke and fumes. Face masks and breathing apparatus were not available at the site and the fire raged for 6 hours.

The built-in site control system designed to flood the CSR with CO_2 was not operational. The electrical system required for initiating the flood had been purposely disabled earlier. When the system was finally triggered, it only succeeded in driving the smoke fumes into the control room. Manual effects for limiting the fire in the CSR did not succeed as well. On the suggestion of the Municipal fire-fighting personnel, who had reached the site by then, water was used for putting out the fire and not CO_2 (this was not an electrical fire), and as a result, the fire was put out in 20 minutes.

The reactor pressure of unit 1 increased to about 25 kgs/cm^2 by about 6.00 pm on 25.3.1995. The reactor was de-pressurised to enable. Plans were afoot in 2011 to upgrade the capacity of the three units to 1.28 GWe each.

19.9   Global core melt-down cases

There have been 11 core melt-down cases so far in global reactors, which are 437 in number with 14,000 full-power reactor-years of operation.

19.10   The Nuclear accident at Fukushima's Daiichi Nuclear plant in Japan due to 'tsunami' on 11.3.2011

19.10.1   Earthquake details

An earthquake of 8.9 on the Richter scale rocked the Sendai region in North East Japan at 14:46 hours IST (Indian Standard Time). The 'tsunami' (a word to describe an earthquake's effect on the sea/river waters) caused 7.5 metres to 15 metres high waves in the North East and South West coastal areas of Japan. The Daiichi nuclear plant at Fukushima in the North East region was severely affected.

The tsunami also caused a fire in the giant oil refinery at Ichhara. The 'tsunami' also caused major fires in Chile and South America.

The epi-centre of the 'tsunami' was located at Sendai and the affected nuclear plant was located at Fukushima which is 240 km North East of Tokyo, Japan's capital.

A 'tsunami' warning was issued for 10 nations along the coast of the Pacific Ocean. Similar warnings were issued in Australia, New Zealand, Russia, Mexico, Peru, Indonesia, Papa New Guinea, South America and the Western Coastal areas of the USA. The 'tsunami' left 4 million homes in Japan without electricity in its wake. Around 11 nuclear plants in the Fukushima region shut down automatically as per design.

19.10.2   Details of the Fukushima plant

There are 6 plants in Fukushima, out of which three were in operation at the time while the other three were under maintenance. The primary containment for the units was of the 'Mark-1' category. Three of the plants were of GE-make (USA), one was of Hitachi-make (Japan) and the remaining two were of Yoshida-make.

We will now go into more details of the consequences of the 'tsunami accident'. There was an interruption in the cooling water flow to reactors due to the 'tsunami'. All three units shut down automatically as per the design norms. The decay heat went down to 2% of the full power in 90 seconds. The Class-III DG power units started as per the design but shut down due to the effect of the 'tsunami'. The pressure in the containment structures of units 1, 2 and 3, which were in operation at the time, increased substantially due to the unavailability of cooling water which was required for the removal of decay heat. Radioactive steam leaked from the containment of these units due to built-up steam pressure. Radiation in the site boundary reached 1015 micro sievert per hour. The dose outside the site-exclusive zone was 4 to 7 micro sievert per hour. The injection of sea-water mixed with Boron was started at 20 hours, IST, on 11.3.2011. Traces of CS-137 and radiated Iodine were detected near unit 1 on 12.3.2011 at 13:30 hours, IST. This indicated the possible overheating of nuclear fuel. Explosion in units 1, 2 and 3 took place and was due to the mixing of Hydrogen and Oxygen in the containment, which was caused by the failure of the Zirconium fuel rods to react with water.

The radiation dose in unit 2 was 8217 micro s/ hour at 08:30 hours IST, on 15.3.2011. The power was restored to units 1 and 2 on 20.3.2011. This did not necessarily mean the re-starting of the primary circulating water. The integrity of the emergency-system water pumps and the entire primary water pumps would have had to be established before the re-starting of the units.

The ground self-defence forces shot, from over-head helicopters, 80 tonnes of water into the spent-fuel cooling water tank of unit 4 on 20.3.2011. A total of 2000 tonnes of water were poured into the tank as against the tank capacity of 1400 tonnes.

The operators of the Fukushima plants were trying to inject Nitrogen into unit 1 to prevent the stored-up Hydrogen from exploding. This action was needed given very high temperatures and pressures in the containment of the unit.

A 2-metre-long crack was observed in the maintenance pit next to the cooling sea-water intake of unit 2. This caused thousands of tonnes of contaminated radioactive water to flow back into the sea. Sodium silicate was used to plug the leakage since the silicate had the properties of cement.

Thousands of tonnes of low-level radioactive water, which itself was 100 times that of the safe limit, was released into the ocean to enable the storage of highly radioactive water (1000 times of limit) in containers.

Levels of radioactive water released into the sea were several million times more than the safe-limits. Radioactive Iodine 131 in sea water was 5 million times the safe limit. Caesium 137 present in the water was 1.1 million times the safe-limit.

Fuel failure was suspected in all three operating reactors.

Reactors 1 to 4 were likely to be scrapped, de-commissioned and de-mothed.

19.11 Explosion at the French Nuclear Waste-Management Site

An explosion took place on Monday, 12.9.2011, at the French nuclear management and disposal Centraco nuclear site located at Marcoule in the Languedoc-Roussillon region near the Mediterranean Sea. The explosion occurred in the oven meant for melting very low-radioactivity count metallic waste. The accident killed one person, seriously burnt another and slightly injured three others. There was no leak of radioactivity to the outside of the site and the injured were not contaminated either. A fire broke out after the explosion and was later contained.

The EDF Power Company of France operates the site which mostly handles the waste from the Nuclear Power Plant and small amounts of waste from hospitals and nuclear research laboratories. The site of the accident was sealed and ventilators were operating to isolate waste nuclear gases according to France's Nuclear and Alternative Energy Commission. The Centraco nuclear waste site is located on a 300-acre site that houses a nuclear research centre, a 'Mox' (Plutonium and Uranium) production plant and three other industries.

It may be of interest to note here that France's 58 Nuclear Power Plants meet more than 80% of the country's electrical energy. It had a $137 billion nuclear power expansion programme as of late 2011. It exports Nuclear Power Plants and treats nuclear waste from around the world.

Ecological and Environmental Issues Encountered By the Indian Nuclear Power Projects

20.1 Introduction

In para 2.2 of Chapter 2, we have seen that one of the natural safety features that must be taken into account in the design of Nuclear Power Projects is the demographic aspect. It was also discussed that an exclusion zone of 1.5 km radius from the plant site is enforced for all Nuclear Power Projects. An additional area of about 50 hectares for a 2-unit nuclear station is required for housing the plant and ancillary buildings. Quite a few hectares are also required for the township, hospital community buildings, etc. This means that the people affected by any project (PAP), as a result of the above-mentioned requirements, have to be removed from the habitat and rehabilitated elsewhere. The programme of removal and rehabilitation of PAP is one of the most difficult preliminary requirements for the construction of any Nuclear Power Project. This programme is a multi-dimensional one and embraces the human, economic and social spheres. The biggest single objective of any project is to upgrade the community life which comprises their environment, their way of life (both social and economic) and their togetherness. The compensation for property or assets lost does not really in any way make up for the real loss of life of individuals and families.

The assets can be built up overnight but the same social or ambient conditions may not be entirely reproduced elsewhere. It is on this touchstone that the success or failure of any rehabilitation programme of PAP would be judged. The rehabilitation of PAP would have to be accorded the most important part of all plant preliminary work because it calls for constant and consistent vigilance until the rehabilitation of every person is completed.

Let us now look at the specific environmental and rehabilitation problems encountered in the Indian Nuclear Power Projects.

20.2 The Tarapur Atomic Power Project (units 1 to 4)

Tarapur was a sleepy hamlet with a sparse population, a small fishing community and intermittent agricultural activity. When the momentum decision was made to set up India's first nuclear power station in the ancestral village of India's legendary Dr. H.J. Bhabha in 1960, the environmental issues in the village were not of great concern. At the time the village's only means of transportation was the bullock cart. However, when work on TAPP units 3 and 4 began in 1990, the number of PAP was much more than expected; suitable measures for their rehabilitation were taken. Environmental activities taken up in the area have been detailed in para 16.28.3 of Chapter 16.

20.3 The Rajasthan Atomic Power Project

The project is located in a semi-forested area of the State of Rajasthan. Almost all of the land required for the project, and the exclusion zone, belonged to the forest department of the State. The land came under the forest and semi-forest conservation programme of the State Govt. Alienation of the land required for the project presented no problems for the project authorities. Most of the people affected by the project were cattle farmers. The equivalent area required for the rehabilitation of PAP was readily available within proximity. Rehabilitation was done mostly without tears on any side.

20.4 The Madras Atomic Power Project

The project was located near the coast of the Bay of Bengal. There were no agricultural or other activities within the vicinity of the project area and the exclusion zone. The rehabilitation of PAP did not present any major problems.

20.5 The Narora Atomic Power Project

The project was located in the rich alluvial soil of the Ganges basin. The project site was also within the command area of the Narora barrage. The land sought to be acquired had high economic value. This factor led to a prolonged tug-of-war between the State of Uttar Pradesh and the Narora project authorities on one hand, and the land owners on the other. It is common knowledge that the rate fixed by the Govt. authorities for the acquisition of land is almost always less than the market value and hence the dispute went on until a deal was made to benefit both parties. The dispute was finally resolved. However, another dispute between the landowners and the project authorities arose when the land owners demanded that, apart from the compensation for the land acquired, at least one person from each of the families of the PAPs should be employed at the project. This was not found possible given the qualifications and skills prescribed by the project authorities for various grades of employment. Minimum educational qualification was required even for the lowest job namely that of a 'helper'. Even this post would last only until the completion of the project construction. The placement in the operation and maintenance divisions of the station would require the acquisition of a trade certificate in the relevant disciplines, apart from the completion of a training course as per the prescribed procedures. The employment dispute was finally resolved with the project authorities committing themselves to employing at least one person from the PAP family, subject to the qualification and certification clause.

20.6 The Kakrapar Atomic Power Project

The greatest problem that confronted the Indian NPPs, as of that time, was the one posed by the environmental issues at the KAPP site. The advisors were the sole inhabitants at the project site and exclusion areas. The community depended entirely on the semi-forest area that sustained and met their meagre living needs. When the political parties joined the fray, the advisory community was caught up in a fight for survival. The political parties tried to use this situation to gain electoral advantage. The Sarvodaya movement had also joined in with its stance that the project would lead to nuclear weapon proliferation and thus endanger world peace. It was really surprising that a movement saw such an instant connection between nuclear power development and nuclear proliferation.

The result of all this frenetic anti-nuclear Power Project movement was the intensive agitation of the Advisasis who were aided and abetted by the political parties and Non-Governmental organizations. The agitation led to the firing by the State police and the loss of innocent lives. There was no further impediment to the project after this agitation ended.

20.7 The Kaiga Atomic Power Project

20.7.1 Details of PAP: The extent of land acquired
About 150 families in four villages were affected by the project. Adequate compensation towards land acquisition and rehabilitation of PAP was deposited with the Karnataka Government.

Details of the land acquired, steps taken to provide green cover for the site, etc. have been detailed in para 10.1 of Chapter 10.

20.7.2 The Role of NGOs
The agitation against the Kaiga APP was spear-headed by the Committee for Alternatives to Nuclear Energy (CANE). Several other non-governmental organisations had joined the agitation as well. Kota Shivarama Karanth—the nonagenarian poet, litterateur, novelist, Sahitya Academy awardee and a doyen among the literary figures of Karnataka—had also lent his complete support to the agitation. The head of Pejavar Mutt had also taken a prominent part in the agitation. The common thread that sparked the agitation of all the above-mentioned persons and organisations was the safety hazards they perceived in the nuclear project. The project authorities had constantly pointed out the several engineered safety features that have been built into the project but to no avail.

The failure of the inner surface of the pre-stressed concrete inner dome of Kaiga Unit 1 provided another argument for the zealous opponents of the Kaiga APP. This failure has been dealt with in detail in Chapter 10 (paras 10.8.2 to 10.8.3). The failure, though serious, would not affect the safety or integrity of the dome or other nuclear class structures, since these would have to meet the stipulated internal pressure and withstand test requirements after the completion of construction/ re-construction. Further clearance of the AERB to commission these structures and systems would also be required. The AERB, on its part, grants the concurrence only after a thorough review of the test results on the dome and other nuclear class structures.

20.7.3 CANE's argument

CANE, which has been the foremost in its opposition to the Kaiga APP, has nested its stable arguments on alternatives to nuclear power. CANE would, no doubt, have been aware that the demand for power in India has always been ahead of the available supply, and that the shortages in meeting the power demands and energy requirements have been in India over the last several years and would continue to be so over the foreseeable future. Non-conventional and renewable energy sources such as solar energy, mini hydel, wind power, sea-wave power and bio-mass have their roles to play in the energy spectrum. These would, by their very nature and locations of their utilisation availability, play a marginal, though important role in the power sector. Their total potential, even in the event of their fullest utilisation, would not exceed a high single-digit percentage of the total power and energy requirement of the country. These constitute, as of now, only 0.25% of the total installed capacity of the country and may reach 2% by the end of the 20th century.

20.7.4 The impact on bio-diversity: flora and fauna

The local area of the Kaiga Atomic Power Project comes under the major forest cover of the Western Ghats. In para 10.1 of Chapter 10, we had seen that about 120 hectares of this land had been, and would be used, for housing the 6 units of Kaiga. Steps for providing a green cover for the site have been detailed in the same Chapter. The project authorities have been funding the University of Agricultural Sciences in connection with the study of flora and fauna in the district of Uttara Kannad.

20.7.5 Average permissible environmental dosages at various nuclear power sites for the operating personnel

During the period 2004-2007, the average environmental dose at the sites that are 1.6 km from the stations has varied from 32.6 micro sieverts per year in RAPPs to 1.1 ms/year at Kaiga 3. The AERB limit is 1000 ms per year.

The dosage from the natural background is 2400 ms per year. The permitted average accumulation of radioactivity per year is 5 rems. It comes to a dose of 100 milli rems per week. This dosage may be accumulated at a rate not greater than 3 rems over 13 consecutive weeks. This dosage could be received as a single one but should be avoided as far as possible.

Dosages for the general public are limited to 10% of all the accumulated single dosages.

20.7.6 India's 2nd fuel re-processing plant

India's nuclear fuel cycle set-up received a boost when the then PM inaugurated the country's second re-processing plant at Tarapur on 7.1.2011.

Installed Cost of Nuclear Power Projects: A Comparative Study of the Costs of Nuclear, Hydro-Electric and Fossil Fuel-Fired Projects

21.1 Capital cost: unit sale price

As of 2020, the total capital cost of Nuclear Power Projects completed so far and under construction—starting from RAPP 1 and ending with RAPP units 7 and 8 and Kakrapar units 3 and 4—is Rs. 61,424 crores. The average cost per MWe works out to be Rs. 6.14 crores for all the projects that are installed/expected to be installed by 2023-2024 in 10 GWe. It has to be pointed out in this connection that the above total cost and average cost per MWe includes the cost of first-time high-unit capacity projects such as the 700 MWe (KAPP units 3 and 4 and RAPP units 7 and 8), the IGWe or 1000 MWe (Kudankulam units 1 and 2) and the 540 MWe (TAPP units 3 and 4) projects. The development cost for these high-unit capacity projects has to be borne in mind.

The attached table furnishes the details of the sale cost of energy per unit for various projects such as nuclear, hydro-thermal and gas-based projects. It will be seen that the per unit generation costs at Nuclear Power Projects such as TAPP units 1 and 2, RAPP units 1 and 2 and MAPP units 1 and 2 compare favourably when viewed against hydro-electric projects such as the Loktak (48.76-54.27 paise), the gas-based unit at Kawas (183 paise-210 paise) and the

Raichur thermal unit 4 (210 paise). The Pit-based thermal projects have not been considered as they have immense advantages.

It may also be seen that the average cost of all nuclear projects from 1964 to 2020 at Rs. 6.14 crores per megawatt (based on the costs from 1964 to 2010) has to be compared with the average thermal unit cost of Rs. 3.5 crores to 4.39 crores of projects like Talcher, Raichur and Tuticorin were built between 1994 and 1999. Based on the cost of those years, the cost of materials, equipment and related items have gone from 200% to 300% between these years and the years cited above for the nuclear projects. These factors would help us to view the cost per MWe of nuclear projects so far from a proper perspective.

21.1.1   Capital Cost of Nuclear Power Projects: the NPCIL projects and project parameters

We will now present in more detail the performance of the NPCIL ever since its inauguration in 1987. A broad picture of the crucial parameters like the sales turnover, net profit, units generated, etc. called from the company's annual profit and loss statements for the period 1989-1990 and 2018-2019 has been presented in tables 21.2 t1 to t3 attached herewith. It will be seen from these tables that the annual turnover has increased 44 times; net profit: 82 times; equity: 8.5 times; reserves: 38.5 times and earnings per share: twice. These are all very impressive figures. The NPCIL has regularly been paying dividends to the GOI. The President of India holds all the shares in his official capacity as ex-officio in the NPCIL every year, right from its inception till date.

Capacity utilization of the projects has increased from 60% in 1995-1996 to 90% in 2002-2003.

21.2   The distribution of revenue

The distribution of revenue derived, as among the various sectors of charges met with for a typical 220-MWe HWPTR, is shown below.

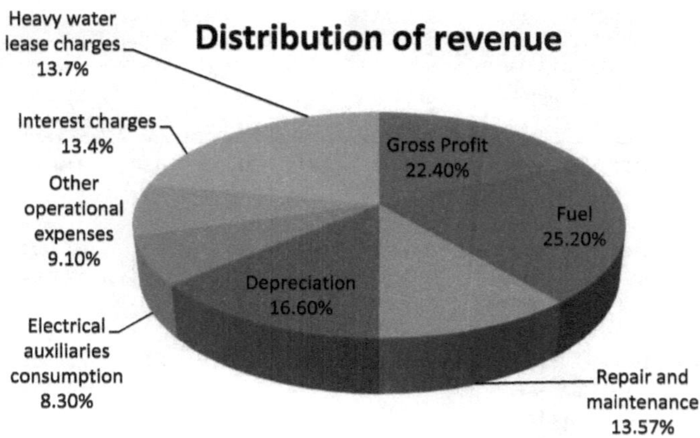

The above data pertain to the years 1990-1991 and may be taken to be typical for a 220-MWe HWPTR. It may be seen that fuel and heavy-water lease charges work out to nearly 39% of the total revenue. Gross profit at over 22% is a very satisfactory figure. To quote a fairly recent figure, for 2003-2004 the gross profit was 49% of the total revenue (extremely satisfactory).

21.1 (a) The tables showing details of costs, operating parameters and related details of the NPCIL stations

Table 21.1 (t1): It indicates the comparative cost of nuclear projects already commissioned/under construction

S. No.	Name of project	Date of first concreting	Date of first-criticality	Capital cost (crores)	Cost per MWe (cr.)	Unit capacity (MWe)	Energy cost per unit r/p	Total cost in rupees	Remark
1	TAPP 1 & 2	1964	1969	103.69	0.27	190	0.57	7460	As of 2005
2	RAPP 1	1965	1972	73.27	0.33	220	0.51	1183	As of 2010
3	RAPP 2	1968	1991	102.54	0.47	220	0.51	2770	As of 2010
4	MAPP 1	1970	1983	119.30	154	220	0.60	2200	As of June 1993

S. No.	Name of project	Date of first concreting	Date of first-criticality	Capital cost (crores)	Cost per MWe (cr.)	Unit capacity (MWe)	Energy cost per unit r/p	Total cost in rupees	Remark
5	MAPP 2	1972	1985	134.84	0.61	220	0.60	2240	As of June 1993
6	MAPP 1 & 2	1967	MAPP 1 -1989 MAPP 2 -1991	745.00	1.69	220	2.31	N1-1812 N2-2014	Up to June 2008
7	KAPP 1 & 2	1984	KAPP 1 -1992 KAPP 2-1995	1335.00	3.03	220	2.27	K1-1901 K2-1851	Up to June 2008
8	Kaiga 1 & 2	1990	K1-2000 K2-1999	3100.00	7.04	220		K1-1021 K2-1184	As of 2015
9	Kaiga 3 & 4	2002	K3-2007	3282.00	7.45	220	250	K3-038	Up to 2012
10	RAPP 3 & 4	1990	R3-1999 R4-2000	2107.00	5.27	220		R3-1126 R4-1051	Up to June 2008
11	TAPP 3 & 4	2000	T3-2006 T4-2005	6100.00	5.65	540	2.50 to 2.65	T3- 4.24 T4-5.76	Up to June 2008
12	KKNS 1 & 2	2002		14,000.00	7.00	1000	2.30		Up to June 2008
13	FBPR	2004		3520.00	7.04	500		8.52	As of 2014
14	RAPP 5 & 6	2002		3072.00	6.98	220	2.60		
15	KAPP 3 & 4	2010-2011		12000 (26.10 cost)	8.55				
16	RAPP 7 & 8	2011-2012		12000 (20.10 cost)	8.55				

Table 21.1 (t2): Comparative sale cost in terms of paise per unit for nuclear, thermal, hydro and gas-based units covering the period 1969 to 1999 have been indicated in this table.

S. No	Type of Station	Name of Station	Unit capacity	Sale cost paise/ amt.	Remarks
1	Nuclear	Tarapur units 1 & 2	190 MWe	57.00	1969 cost
2	Nuclear	Kota (Rajasthan)	220 MWe	51.00	1972 cost
3	Nuclear	Kalpakkam (Madras)	220 MWe	60.00	1982 cost
4	Nuclear	Kakrapar (Gujarat)	220 MWe	215.00	1992-1993 cost
4(a)	Nuclear	Narora (UP)	220 MWe	231.00	1991-1992 cost
5	Thermal (coal fuel)	Korba (Chhattisgarh)	210 MWe	40.52	1991-1992 cost
6	Thermal (coal fuel)	Wanakbori (Gujarat)	210 MWe	250.00	1995 cost
7	Thermal (coal fuel)	Singrauli (Madhya Pradesh)	210/500 MWe	40.46	1985 cost
8	Thermal (coal fuel)	Kota (Rajasthan)	210 MWe	76.05	1986-1987 cost
9	Thermal (coal fuel)	Talcher (Orissa)	210 MWe	180	1994 cost
10	Thermal (lignite)	Neyveli (TN)	210 MWe	53.00	
11	Thermal (lignite)	Neyveli (TN)	210/250 MWe	250.00	1999
12	Thermal (lignite)	Tuticorin (TN)	210 MWe	61.55	
13	Thermal (lignite)	Raichur (Karnataka)	210 MWe	272.00	1995
14	Hydro	Bairasiul (Himachal Pradesh)		32.94	
15	Hydro	Loktak (Manipur)		54.27	
16	Hydro	Salal (Jammu and Kashmir)		48.76	
17	Gas-based	Kawas (Gujarat)		183.00	1995-1996

Tables 21.2 (a) The distribution of revenue of the NPCIL: tables 21.2 (t1) to 21.2 (t3)

These three tables give an extract of the financial statements of the NPCIL from 1989-1990 to 2018-2019. The statements would elaborate on the immense strides made by the NPCIL in becoming one of the most profitable Central Public Sector undertakings in India.

Table 21.2 (t1): Year-wise table of the turnover, net profit, equity dividend paid, etc.

Fig. in crores   Units generated (in billions)

S. No	Description	1989-1990	1990-1991	1991-1992	1992-1993	1993-1994	1994-1995	1995-1996	1996-1997	1997-1998	1998-1999
1	Sales turnover	262	384	395	579	425	-	925	1208	1383	1902
2.	Other income								26		
3	Odd expenditure	161	228	275	353	385			625	1524	
4	Interest	39	51	52	95	110			157	152	
5	Profit before dep & tax	62	104	68	137	-70			452	420	
6	Depreciation	16	17	29	11	60			199	195	
7	Tax										
8	Net Profit before appropriation/ provisioning	46	88	39	90	-130		152	253	265	362
9	Appropriation/provisioning	13	14	18							
10	Profit after 9	33	74	21	90	-130					
11	Total units generated (in billions)	6.0	6.5	7.1	7.5	7.7	79	8.00	9.07	9.64	11.18
12	Dividend to Govt.										50.44**
13	Earning per share								622		
14	Reserves										
15	Equity				2039	2209			2924	3372	
16	Availability Factor									76	
17	Capacity Factor								60	67	71	75

S. No	Description	1999-2000	2000-2001	2001-2002	2002-2003	2003-2004	2004-2005	2005-2006	2006-2007	2007-2008	Remarks
1	Sales turnover	2111	3226	4392	4179	4035	3345	3567	3592	3334	
2.	Other income	112	181	229	242	1458	1622	619	627	933	
3	Odd expenditure	1277	1605	2298	1935	1641	1563	1793	1915	1874	
4	Interest	157	327	414	355	342	279	235	343	455	
5	Profit before dep & tax	788	1475	1909	2132	3511		2158	1961	1938	
6	Depreciation	236	353	496	472	457	283	361	664	734	
7	Tax	138	298	113	142	365	133	63	280***	127	
8	Net Profit before appropriation/ provisioning	414	825	1300	1509	2689		1734	1298	1078	
9	Appropriation/provisioning			249	8	84	5	21	7		
10	Profit after 9	414**	825	1549	1509	-2604	1705	1713	1571	1078	
11	Total units generated (in billions)	12.0	14.0	19.2	19.4	17.8	16.71	17.30	18.60	16.96	
12	Dividend to Govt.	61.50		165			3.42	514** 400*	471	324	
13	Earning per share	1601		4259	5715	7739		101.80	112.91	105.95	
14	Reserves										
15	Equity	4944		5811	7697	8932	1014.5	1014.5	1014.5	1014.5	
16	Availability Factor	86					88			83	
17	Capacity Factor	80	82.5	85	90	81	76	74		54	

** Net profit after interest payment adjusted over 1987-1988 to 1998-1999 of Rs. 329 crores.

*Maiden dividend

***Face value : Rs. 1000

****Re-imbursed by electricity utilities

*Interim dividend

**Final dividend

Table 21.2 (t2)

Figures in crores

S. No.	Particulars	Trend 31.3.2009	Trend 31.3.2010	Trend 31.3.-2011	Trend 31.3.2012
1	Net Sales	386.82 cr.	3010.56 cr.	6012.53 cr.	7913.81 cr.
2	Other optional income	3092 cr.	27.50 cr.	45.91 cr.	46.22 cr.
3	Expenditure	3260.87 cr.	2605.05 cr.	4449.05 cr.	5642.14 cr.
4	Profit from operations before other income, interest and exceptional items	576.87 cr.	433.01 cr.	1609.39 cr.	2317.89 cr.
5	Other income	641.61 cr.	763.43 cr.	828.01 cr.	748.88 cr.
6	Profit before other income and exceptional items	1218.48 cr.	1196.44 cr.	2437.40 cr.	3066.77 cr.
7	Interest	441.03 cr.	488.78 cr.	661.46 cr.	672.41 cr.
8	Exceptional items (prior period)	303.39 cr.	226.87 cr.	12.71 cr.	89.88 cr.
9	Profit before tax (6-7+8)	474.06 cr.	480.79 cr.	2381.65 cr.	1686.06 cr.
10	IT	57.64 cr.	39.49 cr.	459.78 cr.	309.74 cr.
11	Net Profit	416.41 cr.	441.30 cr.	1906.16 cr.	1376.32 cr.
12	Paid-up Equity	10145.33 cr.	10145.33 cr.	15466.57 cr.	17909.39 cr.
13	Reserves	12498.78 cr.	12040.95 cr.	12619.32 cr.	11686.23 cr.
14	EPS (FV Rs.1000)	41.04	43.50	187.61	135.66
15	Electricity Generation			26.47 bln units	32.46 bln units

Table 21.2 (t3)

S. No	Particulars	Year end 31.3.2014	Year end 31.3.2015	Year end 31.3.2016	Year end 31.3.2017	Year end 31.3.2018	Year end 31.3.2019	Year end 31.3.2020
1	Net Sales from operations	8384	8916	9626	10,003	12206	11528*	12637(1+2)
2	Other Operating income	196	41					
3	Total expenditure	5686	5964				(6)3608	6555
4	Profit from operations before other income, interest & exceptional items	2894	2993		3232	(4) 4622		6082
5	Other income	473	306					
6	Profit before interest & exceptional items	3367	3299					6082
7	Interest	482	489					
8	Exceptional items (Prior period)	1	2					
9	Profit from ordinary activities (6-7+8)	2884	2201				(9) 2819	4459
10	IT-defined tax	597				(11) 3613		120
11	Profit after tax (9-10)	2299	2201	2697	2491			4339
12	Paid-up Share Capital (Rs. 1000 per share)	10174	10174	10217	10806			
13	Paid-up debt capital	17785	12575	24308	26907	17314	17250	18200
14	Reserves	15277	16832	19392	21105	21710		
15	Debenture bond (reserve)	1288	2374				4342	
16	Earnings per share (of Rs. 1000)	225.98	216	264	241	327.31	241.19	366.23
17	Debt/Equity ratio	0.70	0.80	0.82	0.84		1.13	1.16
18	Debt service coverage ratio	1.17	0.77	0.67	1.20		1.11	1.25
19	Interest service coverage ratio	2.84	2.40	2.46	2.06		1.67	2.00
20	Net worth			29069	31911		35383	39900

21.3 Manpower strength for annual shut-down, ASD suggestion for shut-down task face, etc.

21.3.1 Typical operating stations: TAPP annual shut-down work

A typical 2 x 220-MWe HWPTR reactor's nuclear power station employs about 800 to 1000 personnel including administrative, security and support staff. A Health-Physics regimen for this staff strength is a stupendous one. The radiation protection arrangement for nuclear stations has been detailed in para 4.2.6 of Chapter 4. Apart from the regimen under normal operating conditions, it is during periods of planned or unplanned shut-down—during which routine and emergency maintenance and repair work are carried out, particularly on the nuclear plant or equipment—that this regimen acquires a more important role.

Particular mention may be made here about the re-fuelling outages in the TAPP units 1 and 2. More than 300 personnel are deployed during an outage to one unit during which other annual shut-down work is also carried out. This complement is arranged by marshalling staff from other nuclear operating stations, the Maharashtra State Electricity Board, the NPCIL office in Mumbai and the Indian Navy. Thirteen such re-fuelling outages have been carried out in respect of each unit of TAPPs as of 31.3.1994. The personnel from outside TAPP have exceeded 200 in number during the peak of such re-fuelling and annual shut-down work.

21.3.2 The permanent team for the annual shut-down

For fifteen nuclear power operating units' outages, it will take a time of about 15 months each on a schedule where the completion of all outage work in a single unit takes 1 month. The outage work in Kakrapar 2 was completed in 18 days during 2002-2003; and in 20 and 19 days, respectively, in NAPPs 1 during 2002 and 2003.

With the multiplication of generating units (mere addition may not meet our objective), the annual shut-down will be a 24 x 7,365 x 1 programme for the NPCIL. In such a situation, a 100-150 strong annual shut-down unit borne on the role of the NPCIL, under the wings of the Chief Engineer (ASD) reporting to the Director of

Operations/Executive Director, will not be an ideal outfit for the annual shut-down work. Members of such a team with its hierarchical architecture would have to be trained on the job for a few months and duly certified by the NTC before being assigned to the team. Each nuclear station would be spared the task of scouting for hands from other units for the ASD work. Maintainers and other hands, up to the senior level, from individual stations would be associated and be a part of the overall team assigned for the ASD work. The entire team would report to the station authorities at appropriate levels for the duration of the ASD. The individual stations would be billed suitably for the services of the ASD crew. This would make the special crew a self-sustaining outfit financially. This proposal, while sounding quite novel, deserves serious consideration; and unless there are other serious objectives it would certainly merit a look into.

The induction of a 100 to 150 skilled workforce is bound to take time, especially since the force would require training and licensing before induction into the force. Lateral induction from construction groups during the closing stages of construction and direct recruitment at the level of maintainers and supervisory levels may also be considered. A period of up to 2 years or longer may be required for setting up the force. Even a force in the initial stages would prove an asset from the point of review of gaining the focus that the force is bound to eventually attain. The biggest advantage of the task force is that, when formed in full, it will be a focused, skilled and ready-to-spring-to-work-at-call force that can be put to ASD work at any station. The planning, scheduling and execution of this ASD work would have to be closely relevant. The NPCIL and the station should coordinate and draw up a schedule, months in advance, as is undoubtedly done even now, to derive the maximum advantage from this force.

21.3.3 Improvement in operational efficiency

We have been looking into the various improvements in the design, construction and commissioning of Indian nuclear power stations from time to time. The steps for improving operational efficiency are on-going exercises with improvements based on operational experience. These have also been touched upon in the previous chapters.

The most important step in this direction so far is the unique emergency response centre set up at the corporate office of the NPCIL, at Anu Shakti Nagar in Mumbai, for the centralised monitoring and performance analysis of all nuclear power generating units. Voice and data communication links between the corporate office and the station sites have been strengthened in the existing dedicated VSAT (very small aperture) facility during 2006-2007.

21.3.4 Some unique operations parameters

We have seen elsewhere that TAPP 3 (540 MWe) operated continuously for 522 days during 2010-2011 and created a record for Indian nuclear power operating units. It joined a fleet of other operating units which logged a continuous operation of over 365 days.

The availability of power reactors was 91% during 2011-2012. The average capacity factor of nine reactors fuelled with imported Uranium recorded an all-time high of 97% during the year. The overall capacity factor of all 19 power reactors during that year was 79%.

21.3.5 Record in the continuous running of NPPs

The 21 NPPs in operation as of August 2019 have seen some global record-breaking continuous runs. Some 27 continuous runs of more than 1 year have been recorded, including the second global highest run of over 800 days by Kaiga 1, the highest among the HWPTRs and the 4th highest of all global power reactors. Details about the individual units have been recorded in the relevant chapters.

Overview of Nuclear Power Development in India

22.1 Milestones in the evolution of nuclear power in India: Organisational and Other Requirements

So far, we have seen in a fairly detailed manner the development of nuclear power in India. We have also discussed the details of the individual power projects

It would be of historical and other interest to set at this stage the milestone dates in the evolution of nuclear power development and management in India.

S. No.	Milestone Event	Milestone date
1	Setting up of the Dept. of Atomic Energy	1954
2	Setting up of the training school for Engineers and Scientists; their absorption	1957
3	Setting up of research reactors and nuclear power-related facilities	1956-1961
4	Start of construction of the first Indian nuclear power station	1965
5	Date of first nuclear power production	1.4.1969
6	Start of construction of the first 220-MWe HWPTR (RAPP) to be constructed and commissioned in India	1964
7	First-criticality of the first 220-MWe HWPTR (RAPP) to be constructed and commissioned in India	1972
8	Formation of the Power Projects Engineering Division (PPED)	1967
9	Start of construction of the first 220-MWe HWPTR (MAPP) to be designed, engineered, constructed and commissioned in India	1970

S. No.	Milestone Event	Milestone date
10	First-criticality of the first 220-MWe HWPTR (MAPP) to be designed, engineered, constructed and commissioned in India	1983
11	Formation of the Nuclear Power Board	1984
12	Formation of the Nuclear Power Corporation of India Ltd.	3.9.1982
13	Start of construction of India's largest unit capacity, fully indigenously designed, constructed and commissioned NPP (HWPTR 540 MWe TAPP units 3 and 4)	2000
14	First-criticality of India's largest capacity HWPTR (540 MWe)	2005
15	Start of construction of India's first 500-MWe FBPTR	2004
16	First synchronization of India's largest unit capacity	2005-2006

22.2 Formation of the Power Projects Engineering Division (PPED)

In para 8.5.1 of Chapter 8, we have seen that the PPED was formed in 1967 with Sri. H.N. Sethna as its first Director. The change vested in the PPED was also detailed in this section. The PPED was made responsible for the design and procurement of equipment for the HWPTR—India's largest unit capacity (540 MWe) generating unit to the Western Regional Grid—on 4.6.2000 and supply and provide support for the progressive construction and commissioning of RAPP, MAPP, NAPP and KAPP. We had seen how from a skeleton staff of 10, who looked after the Bombay office of the then-fledgling RAPP in 1964, this organization had grown in strength to 750 employees during the period 1970-1975. The PPED would have handled its highest tempo projects with capital costs of Rs. 2.362 billion rupees. It would perhaps be right to term an organization handling such high-cost frontier technology projects a 'Division'. It was obvious that this organization would have to be upgraded suitably to discharge adequately the responsibilities vested in it.

22.3   The Nuclear Power Board (NPB)

The Nuclear Power Board was formed in 1984 under the Dept. of Atomic Energy with Sri. M.R. Srinivasan as its first Chairman. The Board would consist of the Chairman and a specified number of members from within the Dept. of Atomic Energy as well as outside of it. The Board would be fully responsible for all the commercial Nuclear Power Projects in the country. The first challenge facing the Board would be the formidably assigned task of raising the installed nuclear power capacity to 10 GWe (10,000 MWe) by the year 2009-2010. It was getting to be obvious to one and all within the NPB, and an informed opinion outside of the NPB, that the target was too ambitious, and that it was simply impossible to approach anywhere near the hailing distance of this target. The reasons were many and the factors contributing to time and cost over-runs have been dealt with in the chapters about individual power projects. The Board would, in due course, be setting itself for more realistic targets. It is quite possible to attain the target of 10 GWe in the period 2020-2025.

22.4   The Nuclear Power Corporation of India Ltd. (NPCIL)

The scope of the Nuclear Power Board was getting wider with more Nuclear Power Projects getting sanctioned for execution. More and more nuclear power stations were also getting into commercial operation. The increasing outlays could not be met solely by budget allocations. A new set-up that would have the financial, commercial and organizational freedom required for dealing with the ever-increasing scope of work was called for. The Nuclear Power Corporation of India Ltd. (NPCIL) was formed on 17.9.1987 and took over all the responsibility that was hitherto being discharged by the Nuclear Power Board. Sri. S.L. Kate was appointed as the first managing director of the NPCIL. The secretary to the Dept. of Atomic Energy would be the ex-officio Chairman. The composition of the first board of directors was as follows:

1.   Sri. Dr. M.R. Srinivasan : Chairman
2.   Sri. S.L. Kate, Member of AEC : Managing Director

3. Sri. Bahadur Chand, Chairman of CEA : Director
4. Sri. R. Basu : Director
5. Sri. S.N. Shande, JS(F) of DAE : Director

The NPCIL completed its silver Jubilee on 17.9.2012.

The stage was thus set for the commercialization of the activities about the Nuclear Power unit construction and operation. All the then-existing nuclear power stations, except RAPP 1, were transferred to the NPCIL for operational purposes. Major decisions regarding the renovation/repair of RAPP 1 were to be taken by the DAE. The transfer of RAPP 1 to the NPCIL for operational purposes would be taken by the DAE after the reconstituted NPCIL started functioning and its recommendations were received.

22.4.1 The Reconstituted NPCIL

The NPCIL's board of directors has changed since its inception in 1987. Keeping in line with its extension of the project horizon, it was reconstituted as shown below:

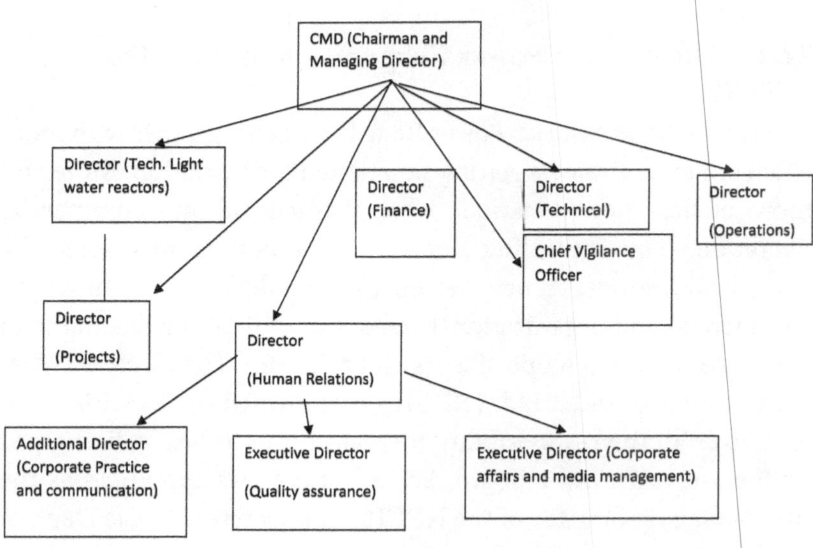

In para 22.4, we have seen that the Secretary to the DAE was designated as the Chairman ex-officio at the time of the incorporation of the NPCIL on 17.9.1987. With the increasing tempo of the NPCIL's work, it was found expeditious to make the organization truly corporate autonomous in its allotted scope of work. It was hence decided by the DAE that the Managing Director would be designated as Chairman/ Managing Director and given all the powers that the previous Chairman enjoyed.

Let us now explore the limited extracts from the Articles of Association of the NPCIL. We will also touch upon the scope of powers that the NPCIL needed to discharge its responsibilities. They are listed below:

1. Own, plan and execute an integrated programme for using nuclear fuel, a process for electricity generation and manage any type of nuclear power station.
2. To open and operate ancillary facilities for the purpose, explained in (1), such as mining and nuclear fuel production.
3. Promote the protection of the environment while constructing and operating nuclear power stations.
4. Purchasing, selling, importing, exporting and manufacturing nuclear equipment and components.
5. To co-ordinate activities of its subsidiaries.
6. Entering into arrangements and agreements with the National/State/Local authorities for furthering the interests of the NPCIL.
7. Borrow/receive money from statutory authorities for furthering its activities.
8. Construct and maintain buildings and machinery in India and other parts of the world to promote its activities.
9. Apply and obtain enactments of any statutory authorities for promoting its interests.
10. Supporting scientific, educational and charitable institutions for pursuing objectives in which the NPCIL is interested.
11. Provide for the welfare of its employees/pensioners.

22.4.2 The NPCIL personnel strength at the headquarters office

A total of 2,037 personnel—573 in administration, 249 in auxiliary to administration, 332 in technical and 883 in scientific

Cadres—were in position in the NPCIL's headquarters office in Mumbai as of early 1992.

22.4.3 Shortening of gestation period: advance procurement action

The NPCIL initiated several steps to cut down the time for the completion of the erection of NPPs. One of these was the initiation of advanced procurement of equipment needing long-delivery periods. The philosophy of advanced procurement is based on the following:

1. The Government's sanction for nuclear projects is site-specific and hence depends on the NPCIL's site selection which takes time.

2. The selection depends on a variety of factors including the most sensitive and long-drawn-out criteria of environmental and related requirements of the State/Local authorities being met.

Serialisation of orders for standard critical long-delivery equipment speeds up the site erection. We may point out in this connection that 220-MWe HWPTRs are now fully standardized and hence advance procurement is called for.

The 540-MWe HWPTRs/700-MWe HWPTRs would also be progressively standardized, paving the way for advances procurement for these higher-powered units also.

It is pertinent to point out in this connection that the DAE, set up by the Government of India, has approved the advance procurement action that would entail an expenditure of Rs. 986.75 crores for six 220-MWe units. The expenditure included an amount of Rs. 161.20 crores sanctioned earlier for the procurement of equipment for two units of 540 MWe capacity and an additional Rs. 383 crores for four units of 220-MWe capacity.

The anticipated cash flow expenditure on account of the advance procurement during the 7th five-year plan period (1987-1992) would be as follows:

540-MWe units : Rs. 369 crores
220-MWe units : Rs. 200 crores

The procurement would be for major and critical equipment like the Calandria, end-shields, coolant channel components, NSSGs, PHT/moderator pump motor units, TG units special steel, etc.

22.4.4 The NPCIL funding details

It was understood that the GOI had committed itself to a debt/equity ratio of 1:1 in the office memorandum that set up the NPCIL. The debt/equity ratio, however, had risen to 4:1 as of the third quarter of 1992. The Govt. had indicated an equity participation of Rs. 762 crores as of this period. Even then, the equity amount was not fully paid up. It would be interesting to take a look at the financial scenario for the period 1991-1992 which is stated below:

(1) Budgetary support (Govt. equity) for 1991-1992 : Rs. 130.57 crores

(2) Private placement of funds by the NPCIL : Rs. 712.18 crores

(3) Total borrowings through bonds as of 1991-1992 : Rs. 1798.26 crores

It may also be noted that the total borrowings through bonds, as of 1993-1994, was Rs. 2565 crores. Increasing interest costs as a result of the increase in debt/equity ratio is bound to erode net profits. The NPCIL would have to take steps to reduce the interest costs. See the performance schedule of projects in the table given below.

Erection/commissioning, cost-wise and time-wise data for Indian Nuclear Power Projects

S.No.	Name of Project (unit capacity)	Date of site tests enabling work	Completion cost (cr.)	Date of first concreting	Date of first-criticality	Date of commercial operation	Cost per megawatt	Total units generated		Revenue generated	
								Qty in blns	As of date	In blns (Rs.)	As of date
1.	RAPP 1 200 MWe	28.12.1964	-73.27	Early 1965	11.8.1972	6.12.1973	36 lakhs	39.56	June 2008	17.30	2003
2.	RAPP 2 200 MWe	Early 1965	102.67	Early 1965	8.10.1980	1.4.1981	51 lakhs	39.56	June 2008	17.30	2003
3	MAPP 1 200 MWe	End 1967	119.30	Early 1968	2.7.1983	27.1.1984	60 lakhs	22	June 2008	13.30	2003
4	MAPP 2 200 MWe	Mid 1969	134.84	1970-1971	13.07.2003	1.8.2003	67 lakhs	22.7	June2008	13.62	June 2008
5	NAPP 1 200 MWe	1974 end	745	Early 1975	12.3.1989	1.1.1991	186 lakhs	18.43	June 2008	43.60	June 2008
6	NAPP 2 200 MWe	1975 end	745	Early 1976	24.10.1991	01.07.1992	186 lakhs	20.14	June 2008	46.50	June 2008
7	KAPP 1 200 MWe	1981	1335	1984	3.9.1992	6.5.1993	234 lakhs	19.36	June 2008	43.94	June 2008
8	KAPP 2 200 MWe	1982	1335	1984	8.1.1995	1.9.1995	334 lakhs	18.93	June 2008	42.97	June 2008
9	RAPP 3 200 MWe	1989	2107	8/1990	24.2.1999	1.6.2000	527 lakhs	11.27	June 2008	29	June 2008
10	RAPP 4 200 MWe	1991	2107	1992	3.11.2000	16.12.2000	527 lakhs	11	June 2008	28	June 2008

S.No.	Name of Project (unit capacity)	Date of site tests enabling work	Completion cost (cr.)	Date of first concreting	Date of first-criticality	Date of commercial operation	Cost per megawatt	Total units generated		Revenue generated
11	TAPP 3 540 MWe	10.10.1998	6555	Mid 1999	21.5.2006	18.8.2006	640 lakhs	4.64	June 2008	11.60	June 2008
12	TAPP 4 540 MWe	1999	6555	8/2000	6.3.2005	8/2005	640 lakhs	6.38	June 2008	16	June 2008
13	Kaiga 1 200 MWe	1988	2896	1989	26.9.2000	16.11.2000	724 lakhs	7.63	June 2008	25	June 2008
14	Kaiga 2 200 MWe	1989	2896	1990	24.9.1999	Early 2000	724 lakhs	11.67	June 2008	27	June 2008
14(a)	Kaiga 3 200 MWe	Dec. 2001	3300	30.3.2002	26.2.2007	6.5.2007	775 lakhs	0.62	June 2008	1.55	June 2008
14 (b)	Kaiga 4 200 MWe	Middle 2002	3300	27.11.2003	27.11.2010	20.1.2011	775 lakhs	1.30	March 2012	3.20	March 2012
15	TAPP 1 220 MWe	1965	93.5	End 1965	1.2.1969	Nov 1969	25 Lakhs	75.44	June 2008	45.5	June 2008
16	TAPP 2 220 MWe	1965	93.5	Early 1966	27.2.1969	Nov. 1969	25 lakhs				
17	RAPP 5 200 MWe	Early 2001	30.72	17.10.2002	24.11.2009	4.2.2010	77 lakhs	8.52	11.8.2014	25.0 (est.)	11.8.2014
18	RAPP 6 200 MWe	Early 2002	2002 prices	Early 2003	23.1.2010	31.3.2010	77 lakhs	8.00	11.8.2014	24.50 (est)	11.8.2014
19	KAPP 3 700 MWe	Dec. 2009	12,000	Sept 2010	23.7.2020		870 lakhs				
20	KAPP 4 700 MWe	Aug. 2010	2010 prices	Early 2011			2010 prices				

S.No.	Name of Project (unit capacity)	Date of site tests enabling work	Completion cost (cr.)	Date of first concreting	Date of first-criticality	Date of commercial operation	Cost per megawatt	Total units generated	Revenue generated
21	RAPP 7 700 MWe	Aug. 2010	12,000	18.7.2011			870 lakhs		
22	RAPP 8 700 MWe	End Aug. 2010	2010 prices	6.10.2011			2010 prices		
23	KKNPP1 (1 GWe)	End 2001	20,000	31.3.2002	13.7.2013	31.12.2014	1000 lakhs	16.25 31.3.2017	63.17 31.3.2017
24	KKNPP2 (1 GWe)	Beg. 2002	2010 prices	Sept. 2002	9.7.2016	13.1.2017	1000 lakhs	63.17	
25	FBPTR 500 MWe	2003 beginning	3520 (2002-2003 prices)	2004 beg.			705 lakhs (2002-2003 prices)		

22.4.5   Financial scenarios for future projects

The 1990 costs for 6x500 MWe units (4 of these were later enhanced to 700-MWe capacity) and 4x200 MWe HWPTR units (for which advance procurement action was taken) was Rs. 90 billion (9000 crores). The as-built cost of the 2x540 MWe and 4x220 MWe units was Rs. 11,762 crores. The Kaiga units 3 and 4 and RAPP units 5 and 6, on which the erection work was progressing fast as of 2002-2003, are together expected to cost over Rs. 6354 crores. One way of sharing the project costs would be for the concerned States to contribute equity to those located in them in proportion to the amount of power allotted to them from the projects. It is understood that the states of TN, Kerala and AP have expressed a desire to participate in future projects and contribute to the equity for Kaiga units 3 and 4. One, of course, has to have one's fingers crossed for the success of such an arrangement, given the parlous state of finances of the Electricity boards of all States in India.

Well-known and successful public and private companies like BHEL, L&T, Walchandnagar Industries Ltd., ECIL and NFC Ltd. may also be roped in for the formation of joint ventures (JVs) for the construction of future projects. The legal and other requirements for such JVs would have to be looked into.

The Atomic Energy Act of India stipulates that Nuclear Power Projects would have to be carried out solely by the Government of India (NPCIL) or by a company in which the Government of India holds a minimum 5% stake. This would have to be adhered to. We may note here that the NPCIL stopped equity from the GOI from 2004-2005.

The Konkan Railway Corporation's Mankhurd—Belapur—Panvel metropolitan rail project, etc which have executed and are executing massive JVs successfully are examples which the NPCIL may follow. The NPCIL has been successfully executing equally massive projects. It may, however, open up new ways (TAPP units 3 and 4 are examples) to successfully and expeditiously complete future projects. It is pertinent to note that the GOI has authorized the NPCIL to incur capital expenditure and establish JVs and wholly owned subsidiaries both in India and abroad.

22.4.6 Past construction/commissioning performance: desirable performance for future projects

Over nearly half a century, India has constructed and commissioned 21 commercial Nuclear Power units with a total capacity of 6.78 GWe (6780 MWe). The average period for the erection and commissioning of one unit has been 8 years, with the shortest period having been TAPP units 3 and 4 each within 5 years. The above-mentioned period of 8 years may be considered reasonable for a Nuclear Power unit with all its emphasis on the safety of the public. This period has been similar to that of the advanced nuclear power-producing countries like the USA, France and Canada.

22.4.7 Funds for future projects of the NPCIL: net profit of the NPCIL

We come back now to the performance of the NPCIL during the period 1989-1990 to 2003- 2004. The net profit during this period has multiplied by 58 times. The net profit after tax depreciation and required provisions was Rs. 5662 crores. The increasing net profit has resulted in a critical profit mass which would enable the NPCIL to utilise its own generated funds to trigger the expansion of nuclear power capacity. It may be noted that these funds have increasingly been contributing to the equity of succeeding projects. The critical mass would further be strengthened after India's largest on-going project (since 2013-2014) at Kundankulam is fully commissioned.

22.5 Corporate management: quality assurance

The importance of corporate management and the need for continued fine-tuning of the management processes need no emphasis. The NPCIL has prepared, during 2005-2006, a corporate management system document outlining, among others, the quality assurance requirements for all phases of the construction and commissioning of Nuclear Power Plants as per the AERB and the IAEA DB 338 requirements.

22.6 Unique response centre of the NPCIL corporate office

A centre has been set up at the NPCIL corporate office in Mumbai, AS Nagar, for centralising the monitoring of performance analysis of all operating units. Voice and data communication links between the corporate office and the various project units have been strengthened in the existing, very small aperture terminal (VSAT), facility.

22.8 The NPCIL operating units: performance and further corporate goals

22.8.1 Nuclear units: performance

The units have so far (as of 2005) clocked 275 full reactor-years of operation.

22.8.2 Corporate goals

The main goal, among others, is the continuing adoption of the best design and construction practices—drawing upon international practices as and when necessary—for the establishment of training institutes in various nuclear power stations.

22.8.2.3 Public interaction

The NPCIL has promoted intensive interaction with the Indian general public by arranging site visits for students and professionals from various fields such as medicine, armed forces, educational institutions and farming/agriculture, among others. Mention may be made about the NPCIL's mobile vans that tour various places, particularly villages. These vans highlight the various equipment in nuclear power stations with the help of models of the equipment and components. Nuclear Science professionals, accompanying the vans, explain the role and functions of these equipment to the visitors of these vans.

It is of interest here to mention that the NPCIL was awarded the world's best public communication campaigner for its 'Atoms on wheels' van at the 'Atom Expo 2018' held in Sochi in Soviet Russia. The award was conferred by the Chief of the Atomic Energy

Corporation of Russia (Rosatom) and received by Sri. Gautam Biswas, ED (LWR) of the NPCIL.

The above-mentioned mobile exhibition vans have covered more than 6 lakhs of villagers in 1500 villages in the States of Haryana, MP, Gujarat, Maharashtra and AP over 2 years in association with the communication firm 'Media Solutions Nagpur (Mah)'. Over 22 countries participated in the above-stated competition held by 'Rosatom' in various categories.

It may also be pointed out that the 'CANDU owners Group' (COG) placed the Indian HWPTRs as the best of the global performers during the period 1999-2002. 'CANDU' is the acronym for Canadian Demonstration Unit.

Kakrapar Atomic Power Project (KAPP) Units 1 and 2

23.0 Standardised 235-MWe HWPTRs

The completion of NAPP could be said to have marked the introduction of standardised 235-MWe HWPT reactor units in future nuclear power stations. ordering of plant equipment and components was rendered possible. KAPP was the first project to benefit from this approach.

23.1 Details of the site

23.1.1 Site locational details

It was decided by the Government of India, following the recommendations of the site selection Committee, to locate the fourth HWPTR plant of 2× 235 MWe capacity at Kakrapar (21° Latitude, 72.7° E Longitude) which is about 80 km from the nearest city Surat in the State of Gujarat. The site is inhabited by tribals whose sole occupation is agriculture marked by the sparse shrubby vegetation. The forest in the area is of the Southern geographical dry deciduous variety. There is a well-developed green belt of mango, custard apple, ber and guava around the plant area. Raintrees are common. The site is located 11 km from Mandvi , the nearest town; it is also located off the Gulf of Khambhat on the Western Coast of India. The site is located 90 km from the coast of the Arabian Sea and is situated on the left bank of the Tapi River. This is the second Nuclear Power Project to be located off the upper Western Coast of India and the first to be located in the State of Gujarat.

23.1.2   Rationale for the location

The State of Gujarat does not have any coal resources. A few thermal power stations have been established in the State. It is difficult to sustain a succession of thermal power stations in the State without assured coal linkages. It may have been more feasible to establish gas-based power stations in the State with the use of the gas grid available. However, the country's first gas terminal at Dahej in the State was, at the point of this narration, only in the pipeline; no gas was available at the required scale at that time likely before 2004-2005. The gas input costs were also going up. It may be noted that the then-existing gas-based power stations such as Gandhar were finding it difficult to get assured and reliable gas supply. It may also be noted that the cost of gas per unit of electricity generation was, at that time, Rs. 105 to Rs. 200 which is a very high figure. The above-stated scenario fully endorsed the proposal for the establishment of a nuclear power station in the state.

23.1.3   Seismic Zone

The site falls under seismic zone-3 (IS 1893-2002). Required safety-related buildings, structures and systems had been designed for a zero-period ground acceleration of 0.2 g.

23.2   Sanctioned cost

The estimated cost, at 1982 prices, for the two units was Rs. 382.52 crores with a foreign exchange component of Rs. 34.77 crores. In 1994, the cost was Rs. 1,335 crores. The increase in prices was on account of the following factors:

1) Change in scope of work : Rs. 183.52 crores
2) Escalation in the cost of equipment and materials : Rs. 359.62 crores
3) Interest in debt component during construction : Rs. 310 crores
4) Other incidental causes : Rs. 99.34 crores

23.3   Circulating water requirement

The cooling water and process water requirements would be met from the Kakrapar barrage. Induced and natural draft towers would be provided to meet these requirements.

23.4 Progress of construction: preliminary work

Preliminary work commenced at the site during 1981-1982, following the issue of financial sanctions in 1981. Approach roads from Vyara to the site and from the site to the project township were laid out. Job shacks and warehouses were also built. Power-supply, required for the construction, was made available by the Electricity Board of the State of Gujarat.

23.5 Plant Civil work: the use of giant crane for the erection work

23.5.1 Award of Civil work

Civil work for the main plant and the natural draft cooling towers were awarded to Hindustan Construction Co. Work commenced simultaneously on KAPP units 1 and 2. The first concreting began in 1984.

23.5.2 ICD and OCD: RB

The inner containment domes (ICD) and outer containment domes (OCD) of the RB are identical to those in NAPP units 1 and 2. The height of the reactor building is less than that in NAPP. This is because openings are provided within the ICD and OCD at KAPP as against the opening being present only in ICD at NAPP. The provision of openings in both ICD and OCD at KAPP facilitated the erection of NSSGs and resulted in the lowering of height in the RBs of KAPP.

23.5.3 The use of a 180-metric-tonne crane

One of the important aids for the erection of heavy equipment was the use of heavy-duty crawler-mounted mobile cranes with a basic lifting capacity of 650 metric tonnes at a 12-metre radius. The capacity of the crane was 180 metric tonnes with a 57-metre main boom, 42-metre derrick and a 49-metre fly job. The crane was assembled and load-tested under the guidance of German specialists in just three weeks. The dead weight of the crane was 750 metric tonnes. The crane was manufactured by Liebherr, Germany. The crane proved to be of singular importance in the erection of heavy critical nuclear and conventional equipment.

23.6 Design changes: of both nuclear and conventional systems

23.6.1 Electrical cables

The translation into action of the concept of safety engineering for Nuclear Power Projects was carried a stage further at Kakrapar by the use of fire-survival and fire-retardant cables. The use of these cables results in several advantages, some of which are listed below:

1) The cables survive any fire for three hours at 750 degrees C; this means that the cables can be put to use even if a fire occurs and is put out within the above-mentioned period, provided the electrical characteristics have not been adversely affected by the fire.

2) These cables (where smoke-emission requirement has been specified) emit less smoke than other cables, hence firefighting and other remedial measures are more easily undertaken in the event of fire.

23.6.2 Safety-related loads: the use of FSLS cables

Fire-survival low-smoke (FSLS) cables were used in the Kakrapar Atomic Power Plant (KAPP) for fuelling safety-related loads. These cables have a low-smoke density (rated 15% approximately), a low-acid evolution (0.5% maximum), an oxygen index of 32% minimum, a temperature index of 250 minimum and can survive a fire for three hours.

23.6.3 The use of FRLS cables

Fire-retardant low-smoke (FRLS) cables were used for feeding loads other than the safety-related ones. The corresponding specifications for these cables are as stated below:

(1) Smoke density ratings : 50% (approximately)

(2) Oxygen index : 30% minimum

(3) Acid evolution : 17.5%

23.6.4 Nuclear fuel used

Thorium in the form of U233 has been used for flux flattening purposes, perhaps, for the first time in Indian HWPTRs. Thus, fuel has been used along with natural Uranium fuel to achieve more efficient fuel management in the initial fuel core.

23.6.5 Uninterruptible Power Supplies (UPS)

Two 350-KVA static UPS systems for feeding the Class-II power bus and two 60-KVA static UPS systems for feeding the Class-II control bus have been provided for the first time for each reactor unit.

23.6.6 Introduction of additional DG sets for the 415-V Class-III system

In step with the evolving safety requirements, a third 415-V Class-III DG set for each reactor unit was introduced in Indian nuclear power stations. This would provide a 100% stand-by for either of the two main 415-V Class-III buses. This is likely to become a standard design feature in future stations.

23.6.7 Improved reactor safety design features

The control room computer system for data-logging and alarm functions (CRCS) and programmed digital comparator system (PDCS) have been installed to initiate automatic reactor protective action. The Cathode ray tube (CRT) display for the electrical power system has also been introduced.

Other safety features such as the primary and secondary fast-acting shut-down systems and fast-acting person-injection system have been further refined based on the operating experience of the Indian HWPTRs.

23.6.8 The 6.6-KV and 415-V Class-III power levels

Class-III power has been made available both at the 6.6-KV and 415-V levels.

23.6.9 Provision of fire stops

All cable tray openings from one room/area to another room/area are sealed with fire-scaling material to stop fire propagation.

23.6.10 The TG set auxiliaries: the Oil system

Seal oil and flushing oil pump material are fed from the 415-V Class-II system instead of the 250-V DC system.

23.6.11   The 250-volt Class-I system

This system has been simplified and made more reliable. The details of the same will be discussed in para 23.6.12.1.

23.6.12   Changes in the equipment design: the use of improved and safer equipment

23.6.12.1   250-V batteries

Nickel–Cadmium batteries have been used in place of lead-acid batteries. These batteries are free from the 'sudden death' syndrome associated with lead-acid batteries, and their life is also longer.

23.6.12.2   The 6.6-KV/415-V auxiliary transformers

The dry type, 6.6-KV/415-V auxiliary transformers are used in place of oil-filled transformers. These transformers have been installed and form part of the switchgear itself. Long runs of cable from the auxiliary transformers to the switchgear, as discussed in earlier projects, are thus avoided.

23.6.12.3   Programmable logic controllers (PLC) provision

The PLCs have been provided for the Emergency Transfer Scheme (EMTR). The events are printed chronologically on a real-time basis. Self-diagnosis is built into the system.

23.6.12.4   Provision of an SF6 Switchgear

A sulphur hexafluoride (SF6)-filled switchgear has been provided for the 220-KV and 66-KV switchgear systems; these systems are thus rendered safer and maintenance-free. These SwGrs occupy much less space than the conventional ones.

23.6.12.5   Microprocessor-based protection relays

These relays have been introduced in many areas for electrical protection. Better accuracy, faster action time and much better reliability are thus ensured.

23.6.12.6   Coolant channel material

An improved Zirconium–Niobium alloy with 2.5% Niobium has been used for the manufacture of the coolant channels.

23.6.13 The evacuation of power from KAPP

Much thought was devoted to the feasibility of 400-KV sub-stations at KAPP for the evacuation of power from the site. The economic feasibility dictated the establishment of 220-KV DD for the purpose.

Three major 220-KV double-circuit transmission lines were decided upon as stated below:

1) 140-Km long KAPP : Vapi
2) 80-Km long KAPP : Bharuch
3) 38-Km long KAPP : Vyara

23.6.14 Construction planning: the ordering of equipment

The ordering of equipment for KAPP was taken up quite well in time and apart from the RAPP 1 unit. KAPP was the only other HWPTR project wherein no serious delays in construction were caused due to delays in the receipt of equipment at the site.

23.6.15 Details of the major suppliers

A few details of the suppliers of major equipment for KAPP are furnished below:

a) The reactor end-shields : L&T.
b) The Calandria : Walchandnagar Industries Ltd. (Part of the work was carried out at Richardson Craddus Ltd.; the finishing work was carried out at WIL, Walchandnagar)
c) Nuclear fuel Zircaloy Calandria tubes and Zirconium–Niobium alloy coolant tubes : NFC, Hyderabad.
d) The 235-MWe TG sets, auxiliaries and turbine condensers : BHEL
e) Electrics : HBB (Now ABB), NGEF, Siemens, BHEL, TELK, Voltas and WSI Podur (TN).
f) Major controls and instrumentation items : ECIL, Hyderabad
g) Large pump motor units : KSB, BHEL, Crompton Greaves, Kirloskar Electric, NGEF.

23.6.16 Automation of nuclear piping work

The extensive use of automatic welding equipment for carrying out nuclear piping work was resorted to. This has considerably reduced the time for completion of nuclear piping work.

23.6.17 Milestone dates for the manufacture and supply of a few critical equipment

23.6.17.1 *Initial development*

The Calandria for KAPP 1 was partly manufactured at Richardson & Cruddas and completed at WIL, Walchandnagar, in January 1986. The ODC left WIL during the early part of the month and reached the site after 16 days in January. The ODC route was as follows:

WIL–Bhigwan–Ahmednagar–Kopargaon–Manmad–Malegaon–Dhulia–Nevapur–Vyara–KAPP site.

23.6.17.2 *First half of 1987*

The major milestones for this period are listed below:

(1) The rotor of the HP turbine for KAPP 1 was dynamically balanced at 3000 rpm in the balancing vacuum tunnel at the BHEL workshop, Hyderabad (This test was carried out for the first time concerning an HP rotor for a nuclear duty turbine).

(2) The_250-MVA Generator Transformer for KAPP 1 was dispatched to the site by TELK, Angamaly, Kerala, in April 1987.

(3) The_250-MVA GT for KAPP 2 was in an advanced stage of manufacture at TELK and was expected to be dispatched to the site in June 1987.

The 415-V prototype switchgear panel of NGEF-make for KAPP units 1 and 2 were under type tests.

23.6.17.3 *April 1989*

The manufacture of the first steam generator unit for KAPP 1 was completed at the Tiruchirappalli workshop of BHEL. The unit was prepared for dispatch to the KAPP site.

23.6.18 Completion of KAPP 1 construction work

All the construction work of KAPP 1 was completed by the end of 1990. The project authorities were looking forward to carrying out the hot-conditioning by May 1991 and the first-criticality of KAPP 1 by the end of 1991.

23.6.19   The TG set on barring gear

The KAPP 1 TG set was successfully put on barring gear on 31.7.1991.

23.6.20   A disastrous fire in the 220-KV switchyard

A severe fire accident took place at the 220-KV switchyard in October 1991. This delayed the first-criticality of KAPP 1 to almost 1 year.

23.6.21   AERB clearance for first-criticality of KAPP 1

The AERB, during its meeting on 25.8.1992, cleared the approach to the first-criticality of KAPP 1.

23.6.22   First-criticality of KAPP 1

KAPP 1 achieved first-criticality at 12:20 AM on 3.9.1992.

23.6.23   Time of completion of KAPP 1

The time taken from fast pouring of concrete to first-criticality of KAPP 1 was 8 years. This was the second quickest time for the completion of any HWPTR unit after RAPP 1. It is to be noted that RAPP 1 was completed within 7 years and 8 months.

23.6.24   Significant milestone dates for KAPP 1

1)   Date of first-criticality : 3.9.1992
2)   Date of first synchronisation with the WREB : 24.11.1992
3)   Date of commercial operation : 6.5.1993
     With the commercial operation of KAPP 1, the installed capacity of nuclear power stations in India has gone up to 1.94 GWe.

23.6.25   Cost of generation at KAPP

The cost of power generated from KAPP 1 is expected to be Rs. 1.70 per unit according to the prices in 1989-1990. This is expected to go up to Rs. 2.28 per unit according to the prices in 1995-1996. The power to the State Electricity Boards is proposed to be sold at Rs. 2.27 per unit.

23.6.26   Early operating history of KAPP 1

KAPP 1 has been operating at a power level of 160-165 MWe after having been declared commercially operable. It had a capacity factor of 65% during the first 2 to 3 months of its commercial operation. The unit was poised to obtain the AERB's clearance for increasing the generating capacity to its rated 220 MWe. The unit had sent out 0.83 billion units of its power-generation and had earned a revenue of Rs. 178 crores. It had an unavailability factor of 100% and was operating for 97 days as of 6.1.1996. Re-fuelling of the 1,000 coolant channels was completed as of 27.1.1998. The operation of 1,000 full-power days was also completed soon thereafter.

23.6.27 KAPP 2: design changes

23.6.27.1 *Design changes as per the 'three-mile accident' implementation committee of the NPCIL*

We have seen in the earlier sections of this chapter that the changes in the design of KAPP units 1 and 2 were made as a result of the station operation experience gained in NAPPs and earlier stations and also by the recommendations of the NPCIL. We may also refer to the AERB recommendation regarding the segregation of power and control cable runs. Such segregation of cables, provision of fire barriers and installation of Hydrogen-gas leakage detectors were carried out in KAPP 2.

The above-mentioned changes were carried out in KAPP 1 as improvement work and also in other HWPT reactors in India.

23.6.27.2 *KAPP 2: progress as of the end of 1992*

All construction work was completed as of the latter half of 1992. All completed work was handed over to the station commissioning group by the end of 1992.

23.6.27.3 *The pre-commissioning test*

The integrated test of the emergency core cooling system was successfully carried out in 1993-1994.

23.6.27.4 *Flooding of the KAPP 2 site*

Heavy rains during mid-June, 1994, flooded the cable tunnel and the low-lying areas of TB 2. Necessary renovation and repair work of the affected equipment and materials were carried out successfully.

23.6.27.5   Capital expenditure during 1993-1994

An expenditure of Rs. 91.3 crores was incurred during the said period.

23.6.27.6   Authorisation for the first approach to criticality

With the successful completion of hot-conditioning of KAPP 2 on 17.11.1993, a request for the authorisation of the first approach to criticality was made to the AERB. The AERB authorised the request after conducting due safety checks at its meeting on 3.1.1995.

23.6.27.7   First-criticality of KAPP 2

The unit went critical on 8.1.1995 at 13:24 hours. It was subsequently synchronised with the WREB on 4.3.1995.

23.6.27.8   Commercial operation

The unit was declared commercial on 1.9.1995.

23.6.27.9   Setting up of the micro-earthquakes network at KAPPs

23.6.27.9.1   General details

Simultaneously, with the first-criticality of the KAPP 2 unit achieved on 4.3.1995, the micro-earthquake network at KAPPs became operational on a routine basis. The network had earlier recorded the earthquake at Koyna on 29.11.1994 (4.2 magnitude on the Richter scale) and that at Japan on 28.12.1994 (7.5 Richter scale).

KAPPs now have the capability of recording earthquakes which are too small to trigger the strong-motion instruments such as the peak-acceleration recorders, response-spectrum recorders and the accelerographs which have been provided as part of the station control and instrumentation system.

This network provides an independent source of information and supplements the station instrumentation and control system.

23.6.27.9.2   Working of the Earthquakes Sensors Network

Sensors (seismometers) are provided at remote stations. The ground vibrations picked up by these sensors are converted into electrical and subsequently to digital signals; these signals are transmitted to the Central Recording Station (CRS) located at KAPPs through ultra-high frequency (UHF) links; voice signals can also be transmitted through UHF links to the CRS. The digital signals are scanned at the CRS for the presence of any seismic signals, and data from any of the remote stations are recorded continuously on a selectable basis on paper mounted on a helical drum recorder. Whenever any seismic event is detected, digital data are collected by a programmable controller-based digital data acquisition system. The paper record provides a synoptic view of the ground motion at the selected station for 24 hours.

23.6.27.9.3 Collaboration for setting up the network: capacity of the network

The micro-earthquake instrumentation system is not yet commercially available in the country. Hence the technological capacities of Gujarat Communications and Electronics Ltd. (GCEs), BARC and the NPCIL had to be utilized to develop the network. The network has been made operational with three remote stations for Rs. 45 lakhs. The system is being augmented to provide for three additional remote stations. It is possible to augment the network to provide for a total number of 16 remote stations.

23.6.27.10   The capacity factor of KAPP units 1 and 2

For the year 2001, the capacity factors of KAPP 1 were 102.8% and 101.8% and that of 98% and 90% of KAPP 2.

23.6.27.11   Record time of annual shut-down (ASD) of KAPP 2

The ASD of KAPP 2 during 2002-2003 was completed in a record time of 18 days.

23.6.27.12   Other important operating details: highlights of the performance review of WANO

23.6.27.12.1   Revenue generated by KAPP units 1 and 2

During 2001-2002, about 3.521 billion units of electricity were generated by KAPP units 1 and 2, a record in annual power-generation. KAPP 1 had an uninterrupted run of 273 days till 15.6.2005. It generated 1.07 billion units since its last synchronisation with the WREB. From May 1993 to June 2008, the unit generated

19.36 billion units and brought in a revenue of Rs. 43.94 billion (4,394 crores). KAPP 2 generated 18.93 billion units and brought in a revenue of Rs. 42.97 billion from September 1995 to June 2008. The total revenue of KAPP units 1 and 2 was Rs. 91 billion as of 11.2.2011.

Kakrapar Atomic Power Station

23.6.27.12.2 Uninterrupted run of KAPP 2

The unit set up a new record of 219 days during 2001-2002. The unit had an uninterrupted run of 372 days during 2005-2006.

23.6.27.12.3 Accident-free days: safety award

KAPP units 1 and 2 achieved 556 accident-free days and 1,411 fire-free days, as of 31.3.2002. The station won the NPCIL industrial safety award for two consecutive years namely 2000-2001 and 2001-2002. No reportable fire accidents happened in 12 years as of January 2011. No reportable accident occurred in seven-and-a-half years as of January 2011.

23.6.27.12.4 Heavy-water economy: Re-fuelling

The station achieved the lowest D_2O loss during 2001-2002. The on-power re-fuelling of 1400 channels was carried out without any outages, during 2000-2001 and 2001-2002.

23.6.27.12.5   The 'WANO' review: other station awards

KAPP was the first Indian nuclear power station to undergo the 'WANO' review in 1998 and then again in 2008. The review had a lasting impact on the KAPPs operation.

KAPS 1 was declared No. 1 (Number one) in performance among 31 PHTRS of the world during 1992-1993, by the 'CANDU' owner group. 'CANDU' is an acronym for Canadian Demonstration Unit (for PSTRs).

Sri. R. Bhiksham, Station Director of KAPPs, was awarded the Nuclear Excellence award for the year 2002.

23.6.27.13   *Social causes activities of the station as part of corporate social responsibilities*

23.6.27.13.1   Medical aid for tribal villages

A peripheral dispensary was set up near the station site to provide medical treatment to poor tribals living in villages near the site. Free medicines have been provided for the tribal population, with active financial assistance from the medical division of KAPPs and the local voluntary health support organisations.

Free diagnostic and therapeutic facilities were provided by the KAPPs hospital for 60 ophthalmologic operations in the tribal and other nearby villages during 1999. Free spectacles, free hospitalisation and free meals for the villages were also provided.

23.6.27.13.2   Social and other activities

A 2-storey concrete school building was constructed, with a 1,160 square metre built-up area consisting of 13 rooms, blackboards, 2 fans in each room and one assembly hall, by the KAPP's CSR group as part of its corporate responsibility at Unchamala. The school began functioning in the academic year 2013-2014.

23.6.27.13.3   Base-line survey on diet, health, etc.

KAPPs has notched up another first by organising a baseline survey on diet, health and biotic environment among 14,956 persons from 17 villages around the site, who were selected based on the 'Wind rose' model. The survey was conducted through the aegis of the South Gujarat University. The survey began on 2.10.1991, a year before the first-criticality of KAPP 1 which took place on 3.9.1992. The surgery covered the dietary pattern, health profile, biotic environment and also the socio-economic conditions of the people. The health profile consisted of two parts: common diseases like goitre, cataract, mental retardation; radiation-related diseases like sterility, cancer and skin diseases and congenital deformity.

Food consumption was found to be lower than that of recommended levels. Nutrient analysis revealed a deficiency of Vitamins 'A' and 'C'. The Caloric requirement was not met and the context of Iron and Calcium in the diet was surprisingly low. Nearly 92 cases of radiation-related diseases were recorded. The biotic survey included the study of plants and animals; special attention was bestowed on aquatic flora and fauna including microscopic plants and fish. Seven locations both in the up-stream and down-stream regions of the Tapi River were chosen for the analysis of water.

The NPCIL had sanctioned Rs. 15 lakhs to the South Gujarat University for the survey covering the years 1991-1994 and an additional Rs. 56 lakhs for the years 1994-1995 and 1995-1996.

23.6.27.13.4   Green belt around KAPPs and KAPPs township award

Other awards received by KAPP include the 'Green Award' instituted by the AERB for developing a green belt around KAPP and the KAPP township.

The station received the 'Suraksha Puraskar' award for 2004 and 2005 from the National Safety Sumeet. The station received the 'Gujarat Safety Award 2004' from the Gujarat Safety Council. It also received the runners-up prize of the National Safety Award 2004 from the Ministry of Labour, Government of India.

23.6.27.13.5   KAPP 1: renovation and modernisation (R and M)

KAPP 1 was shut-down for carrying out R and M work on 1.7.2008. The unit went critical after carrying out the above-mentioned work at 07:17 hours on 31.12.2010. Re-fuelling of the unit after shut-down work was carried out with imported fuel. The unit was later synchronised with the WREB (Western Regional Electricity Board) after the mandatory approval of the AERB.

The significant works that were carried out as part of 'R and M' are as follows:
1) En-masse replacement of coolant channels
2) En-masse replacement of pressure feeder tubes
3) Planned safety upgrade works.

It may be noted that KAPP 1 and KAPP 2 units were fuelled with imported nuclear fuel and hence have been placed under the IAEA safeguards policy.

23.6.27.13.6   Certifications and CSR Award

23.6.27.13.6.1   Certifications received
For achieving significant success in the twin stations, KAPP received the following certifications:
1) ISO-14001 for environmental management systems
2) ISO-9001 for quality management systems
3) ISO-18001 for occupational health and safety management systems

The twin stations received the national 'Power line Award' which is based on the operational efficiency of power stations over the 3 years of pre-coding 2013 (2010-2013). The award was handed out by Sri. Jyoti Aditya Scindia, Minister of State for power in the Union Government, on 21.6.2013 and was received by Sri. S.L. Jain, station Director of KAPP units 1 and 2.

23.6.27.13.7   CSR: environmental safeguard work
Areas in and around the station have been rendered green as a result of planned plantation, induced bio-diversity and maintenance of water bodies in these areas. The work has attracted the White Stork, Ibis, Cormorants, Pied Kingfisher, White-breasted Kingfisher, Grey Herons, Lapwings and Purple Moorhen.

Brahminy ducks are seen on the Tapi River during the winter season. Other birds such as the Peafowl, Coucals, Watercocks, Drongos, Egrets, Bee-eaters, Red-vented Bulbuls, Oriental magpie-robins, Shrikes, Sandpipers, Fan tails, Flycatchers, Wagtails and Orioles are seen at the station site.

A garden for butterflies was established in the exclusion zone of KAPP 1 and KAPP 2 in 2009. Several species of nectar and host plants were planted to attract the winged ones. Colourful flowering plants, creepers, short plants, bushes, grass and shrubs were also planted for the same purpose. Several species of butterflies are found in the garden.

23.6.27.14   CSR (Corporate Social Responsibility) work for 2014-2015

KAPP authorities are funding a new building for the 'Dalohadiya Prathamik Shala Bhavan' school at Kakrapar. The proposed primary school building will have a two-storey RCC-frame structured unit of area 34.23 metres × 14.23 metres. The ground floor will have 5 classrooms and 1 office room. The first floor will also have the same rooms. Drinking water facilities and separate toilets for boys and girls will be provided. The estimated cost of the building was Rs. 87.21 lakhs with a completion period of 15 months. A dining place of 73.50-square-metre area and a toilet area of 24.50-square-metre area have been provided.

'Bhoomi poojan' for the building was carried out on 1.10.2014. The construction was inaugurated on the same day by Sri. D.N. Vasava, MP of the Gujarat Lok Sabha. Sri. Anandibhai Choudhari (MLA) and Sri. L.K. Jai (the KAPP site director) were present on the occasion.

Another CSR activity was the infrastructural school development in Vodkai village near the site. On 22.12.2014, 'Bhoomi Poojan' was carried out for the construction of a 2-floor RCC frame-structured school building with a 90.50 square metre area in the ground. The building was estimated to cost Rs. 43.90 lakhs. The period for completion of the building was nine months. The construction of a shed for the preparation of mid-day meals was also planned. Sri. G. Nageswar Rao, Director of the NPCIL, was the chief guest on the occasion. The Indian Chamber of Commerce and Industry conferred its 'Golden Jubilee' award on KAPP units 1 and 2 for the CSR activities it carried out during the years 2013-2014.

23.6.27.15   *Mirror operating incident in KAPP 1*

A leak in the primary coolant system of KAPP 1 resulted in the shut-down of the plant as per system planning. The cooling system and other safety systems immediately came into operation and the leakage was traced to a coolant channel. Fuel bundles from this channel were removed through remote handling and examined. No damage to the bundles was noticed. No increase in ambient radiation levels was noticed. No radiation beyond normal levels was noticed by the operating staff in the Reactor building. The plant continues to be shut down and the causes for the leakages are reportedly under investigation. Further action would be taken after the completion of the investigation in close consultation with the AERB.

23.6.27.16   *Other CSR activities*

KAPP has constructed and handed over 'Anumathak Kanya Chaatralaya', a 2-storied hostel building of 306.62 square metres of total area which is capable of accommodating 60 girl students. A toilet area of 50.12 square metres has been integrated into the building. The hostel which was handed over on 28.10.2016 is intended for the students of Shree Bhakta Sarvajanik High School at Ghata village; it cost Rs. 40.3 lakhs. KAPP had earlier constructed and handed over a boy's hostel for the same school.

23.6.27.17   *En-masse coolant channel replacement (EMCCR): KAPP units 1 and 2*

EMCCR of all 306 coolant channels in each of the KAPP units was taken up in July 2016 and was completed; both stations were brought back into operation after obtaining clearances from the AERB. The work in KAPP 2 was completed three-and-a-half months ahead of schedule in September 2018. Additional R and M work were taken up for KAPP 1 and the unit is expected to come online by May 2019.

23.6.27.18   *Additional CSR activity*

Carrying forward its CSR for 2018-2019, the authorities of Kakrapar units 1, 2, 3 and 4 handed over the 'Anumatik Haat Bazaar' to the Unchamala local authorities on 17.10.2015 to further local trade. The bazaar is a framed structure with colour-coated steel sheet roofing. The number of shops in the bazaar is 128; there are 8 blocks in the bazaar (16 in each block). The area of each shop is 4.69 square metres, the area of each block is 74.97 square metres and the total bazaar area is 752 square metres. The total constructed area of the bazaar is 1351.76 square metres, with an open circulating area of 752 square metres. The total cost of the bazaar was Rs. 40,50,000.

The 'haat' was inaugurated in the presence of Sri. Shailesh Choudhari, Sarpanch of Unchamala; Sri. Venkatachalam, director of KAPP units 3 and 4 and Sri. V.K. Sharma, director of KAPP units 1 and 2.

Sri. M.V. Parikh, Chairman of CSR, Sri B. Sridhar ACE (Eng.) and several local villagers were present at the function.

23.6.27.19 *The R and M work of KAPP 1*

The renovation and modification work—comprising the en-masse replacement of all coolant channels, en-masse replacement of heavy-water feeder tubes and other safety upgrades—were completed 3 months ahead of schedule. The reactor attained criticality on 19.5.2019 and was connected to the WREB grid on 24.5.2019.

23.6.27.20 *Health CSR work*

The KAPP's CSR team continued its CSR work by organizing an eye camp from 24.2.2019 to 26.2.2019.

The camp was conducted by the KAPP hospital in association with 'DivyaJyot Trust's' Tejas Eye Hospital Mandri. The camp was conducted at the KAPP hospital. The following treatments were offered:

a)   483 eye patients were screened.
b)   131 patients were operated for cataract and other ophthalmic surgeries.
c)   An intra-ocular lens was implanted in one patient's eye for restoration.
d)   309 corrective spectacles were given.

Food was provided for all patients and their relatives, free of cost.

Each patient was given a blanket, a napkin and a steel plate with a bowl at the time of discharge.

Kudankulam (Kk) 2 × 1 Gwe Nuclear Power Project – India's First Pressurised Advanced Light Water Reactors

24.0 Choice of advanced pressurized light water reactors (PWRS)

We have seen in chapter No. 3 the evaluation of reactor technology choice of reactor routes by India's etc. We have seen that about 15 GW of nuclear power could be generated by utilizing the present brown reserves of natural uranium the country. The choice of three steps three generation vectors to the self-sustainers breeder type reactors was also discussed.

How does the choice of PWRS fit into the above scenario? It is necessary at this stage (2002) to look at the present installed capacity of nuclear power in the country. It stands at about 2.8 GW of this total of 2.8 GW a capacity of 0.84 GW i.e. 30% of the total capacity was in the year 2000 only. Any significant increase in installed capacity would be an add on which would go.

24.0 Choice of (PWRS)

Towards building up a critical mass of capacity that would result in generation of sufficient internal financial resources to fund further capacity addition.

The Kudankulam project highest total capacity station in the country kudamkulam site is capable of 6 nos. 1 GWE units.

24.1 Location of the station.

The project is located at kudamkulam in the Tirunclwcli Kalta halans distinct of T.N. State. It is located in the Gulf of rannar, 25 kms. North East of the village of Kanya Kumari, a pilgrim centre famous for the temple of Kanyakumari and rock memorials for Swami Vivekananda and Saint Tiruvalue and a part of Radhapuram Taluk. The nearest town is Nagarkoil 35 Kms. West of the sight Tutelorien Post would be the part of disemar kation for most of the over dimensioned consignments intended for the project.

24.2 Reasons for selection of Kudamkulam site.

The site was chosen by the selection committee (SSG) of the Dept. of Atomic Energy after evaluating 3 coastal and 5 inland sites in. T.N. State during 987.88. The chosen site has the following and wattages

(1) The site provides a hard rock stratum at a reasonable depth, thus ensuring good foundation conditions

(2) The Site lies in subsonic zone no. 2 which is associated with low earthquake potential. In addition, no natural nor man mad features likely to cause include seismicity and pressure exist near the site.

(3) The site is not subject to sever cyclonic storms the minimum flood level is 6 meters above the datum level all the plant structures will be located above this flood level.

(4) Coastal conditions are conductive to formation of economic water-intake structures. Adequate depth of sea water near the site will calculate of economic length of pier and construction of economic instates water structures.

(5) Sea water in available near the site for drawal of cooling water for turbine condenser and for effluent dilution.

(6) No forest lands lay within a radius of 18 braes from the sea.

(7) No large population countries lie within a radius of 30 kms. from the site.

(8) Good accessibility to road, sea and rail lanes obtains.

(9) No hazard to safety of the plant area from man-induced events such as air craft or missile impact toxic gas release, chemical/industrial.

(10) Explosion military accident etc. is foreseen.

(11) Adequate screening value distances with respect to airports, international establishments, military installations, national highways, railway siding etc. as recommended by IAEA are available.

24.3 Power Project (T.N.)

Environmental Clearances and eco-development.

24.3.1 Details of project clearances.

Environmental impact assessment study (EIAS) of the project site was carried out and reports submitted to the various state and central authorities the state authorities concerned were the Tamil Nadu Pollution Control Boards, Share Production Committee and the Tamil Nadu State Environment Committee the Central authority was the Environmental and Forest Department. Clearances were duly received from there authorities.

24.3.2 Green-belting and Eco-development.

Kudamkulam is situated in rain shadow area. This semiarid has moderate to severe salinity and alkalinity. These factors have led to frequent crop failures and very low agricultural productivities.

A renowned NGO, the M. S. Swaminathan Research Foundation (MSSRF) initiated in 1998 a serenities research project with the objective of linking the lively hood of the coastal communities with the conservation and sanctionable use of natural resources base at the Green belling and eco-development (contd) Kudankulam site. The foundation has undertaken as part of the project green belting and demonstration of ecological options called "Nucloear and bio-technology tools for coastal scheme research" at the site.

MSSRF has planted under the above scheme a variety of specials of vegetable and other plants in a demonstration farm at the site. Some of these included radiation mutes varies of black grass granules etc. and demo plots for sustainable natured water resources management has been developed for this semiarid site. Water a scarce resource in this vegan is used economically and efficiently. Those from 7,000 samples of vary space including neem and turmeric have been planted at the site with a survival rate of 85%.

The above pioneering effort is bound to evolution formers in and around lie project site to apply the same method and thus to improve their agricultural practices.

24.4   Fresh water requirement for process water system for plant and potable for project township.

Fresh water will be drawn from prechiparai reservoir, about 65 kms. from the project site. Kudankulam 1 & 2 IGWCO project (KK 1 & 2).

24.5   Genesis of Kudankulam project.

The project was initiated as on Indo-Soviet Joint effort. It would be the first nuclear power project to be funded by Russia. A tripartite agreement between the Soviet Union and I.A.E.A. for establishment of the plant was concluded on 1409.1988. This agreement was an islanded rafter than a full scope safeguards agreement's it had however incorporated appropriate pursuit and perpetual clauses.

An inter govt. agreement (IGA) for co-operation in the design construction and commissioning of the project was signed between India and the then Soviet Russia Govts. on 30.11.1988 in New Delhi.

Soviet Russia disintegrated during December 1989. The inter govt. thus sell into a limbo. The disintegration caused a political and economic trauma in Russian Federation. The biggest unit of the then electable states of the Soviet Union. The Federation became for all practical purpose. The success for state of Soviet Russia. The Russian Federation was subsequently recognized as successor state geneses of Kudankulam 1 & 2 IGWCO of the Soviet Union Internationally and by the United Nation General Assembly. The Federation took the U.N. seat originally occupied by the Soviet Union.

The Successor Russian Federation Govt. was been to pursue the project on condition that the joint proposal be re- negotiated in hard currency terms. This condition was dictated by the severely crippled financially condition of the state.

24.6   March 1997 agreement between India and Russian Fedration Fuet and control assembly supply funding arrangement.

24.6.1   Training of Indian personally project funding repayment details.

The joint Indo-Soviet proposal for the Kudankulam project was discussed during the Indian Prime Minister Sri. Deve Gawda's visit to the Russian Federation during the last week of March 1997. It was agreed that the Russian Federation (Russia) would make available 50% of U.S. $ 3.4 billion at 1994 prices in the form of a govt. loan which would go to meet part of the cost of the 7 Soviet supplied equipment and Technical Services.

Training of Indian personnel project funding repayment details. Pre-permanent plant structural works like roads, township, domestic water supply and plant support works like offices, ware houses, construction power supply etc. would entirely be carried out by India. The above would calculate about 20% of sanctioned cost.

This loan of about $1.8 billion would serve to meet 85% of the cost of the Soviet portion of the project work which was $2.1 billion "Atom stray Export" of Russial would be responsible for the design of the project and procurement of equipment. Erection and commissioning of the plant would be carried out by Indian engineers under the supervision of Soviet experts.

The soft loan of $1.8 billion untended by Russia would carry an interest of 4% per year. It would be repaid in 14 equal yearly installments, starting a year after commercial operation of the plant.

Assuming a 2:1 loan/equity ratio, equity of `4.666 crore will have to be arranged by NPCIL the Indian Councilor part of the project through the govt of India who are the sole shareholders in all Nuclear power projects as of now.

24.6.2   Techno-Commercial Contract.

Above contract setting up KK 1 & 2 was signed o 20.07.1998by Sri. T.S.R. Prasad COD NPCEL and Mr. Kozlov, General Director Atomic ray Export of Russia.

24.6.3   MOU between Indian Russia for project implantation.

A memorandum of understanding (MOU) for implementation of KK 1 & 2 was signed between the Minister of Atomic Energy Russia and Secretary DAE/Chairman AEG in the presence of the Indian Pn. Sri. A.B. Vajpayee in Moscow on 6.11.2001. A high-land Indian delegation consisting of NPCIL officials was on laid during the ceremony. This delegation also signed, subsequently an agreement with Atomic ray Exports. Moscow for Soviet and for equipment supplies to the project amending to $1.5 billion. This agreement covered all Soviet suppliers to the project.

24.6.4   Fuel and Contract Assembly supply agreement

The agreement envisaged supply of enriched uranium fuel and critical control assembling for the entire period of operation of the reactors by Russia. The spent fuel would remain in Indian and could be reprocessed in Indian under I.A.E.A. supervision. Reprocessed fuel would also be under IAEA safeguards.

24.6.5   Training of Indian Personnel.

Indian engineers would be trained in operation and maintenance in a similar plant in operation in Russia. Russia would supply a training simulator for setting up in India.

24.7   Detailed Project Report (DPR)

Representatives of ' Atoms troy Export ' concluded their talks with represntatives of NCPIF in India in the latter half of 1997 regarding preparation of detailed project report (DPT) and preliminary safety analysis report (PSAR) for Kudankulam . DPR will deal with the following :

A. Specifications for the major componants of the reactor and balancing equipment
B. Establish scope of works falling under the role of the Russian and Indian organisations and
C. Furnish cost estimates and economic viability of the project.

The preliminary safety analysis report will furnish the plant design details and establish by analysis the safety aspects of the reactor plant.

The reports were submitted to the Indian atomic energy regulatory board (AERB) for their approval. Approval was accorded in March 2002. The board also accorded permission for commencing permanent plant construction .

Russian credit of Rs 134 crores for the preparation of above reports was received during the financial year 2001-2002.

24.8   Technical details of the plant.

24.8.1   Reactor details

24.8.1.1   *Reactor type,fuel etc.*

The plant will comprise of

2 nos. of 1000 MW (R) reactors. such water reactors have had 1000 full-power year of satisfactory performance as of 2007.

Schematic diagram of VVER 1000 and a cross section view of the reactor are shown in the sketch. VVER in Russian language parlance means 'Vo**do Vodya Noi Energetic Chisky**" Reactor Rodd V-412'. The term further means "VodaVodoenergy" water-coded water moderated reactor of 3GW thermal rating each. This type is the pre-dominate are that is on operation throughout the world with 51 units of total capacity 328G 8 nos. with a total capacity of 6.438 GW were under construction as of 1995.

Their VVER type reactors using slightly enriched uranium is quite different from the RBMK type which used graphite as moderator and boiling light water as constant.

The RBMK type of reactor was operating at Chernobyl Ukraine. The inherent physics characteristics of VVER type reactor will prevent the type of accident that accused at Chernobyl.

The VVER 392 type reactor incorporates third generation technology.

Sketch showing general arrangement of reactor.

24.8.1.2   *Reactor Details.*

RPV is housed inside the concrete pit, the reactor core, control rods and related coolant water inlet and outlet are located at the top of the RPV. The coolant is demineralised water with a pre-determined quality of boric acid added to render the water both coolant and moderator.

The reactor vessels and piping are provided with B.B. Cladding on the inside surfaces to provide necessary protection against corrosion. The outer surface of the RPV is insulated Forced Cooling of the space between the RPV and its concrete housing is provided details of the RPV and are shown in the attached sketch.

24.8.1.3   *Reactor Containment.*

Double containment concept has been adopted. The inner containment dome a 1.2 met. Thick pre-stressed concrete structure spherical in shape. ICD is provided with 6 mm thick carbon sheet steel liner. ICD has been designed with stand a maximum internal pressure of 4 kgms/em2 and the temperature conditions that may arise due to accidents. The top reaches an elevation of + 6.7 mets.

The outer containment dome (OCD) is made of 1 met thick RCC and is designed to protect the ICD from natural and man-made external hazards like tornados, hurricanes, aircraft crash etc. The outer containment is also capped with a spherical dome. The annular gap between ICD and OCD is 2.2 mets.

This annular space between the two domes is kept at a negative operating pressure to ensure that leakage of out any would be from outside about CD into ICD and not the other way.

24.8.1.3   *Reactor Primary System.*

The space ambience would be provided with cooling system and fillers. The airier the annular space would be vented through the stack after posing through fillers. Photo of RPV is attached above.

24.8.1.4   *Primary Circuit of Reactor System – Coolant Pumps NSSG – Pressurizes – Secondary Circuits.*

The coolant water in the reactor pressure vessel (RPV) acts both as coolant and moderator. A pressurizer cylindrical in shape connected to one of the few coolant circuit loops serves to keep the coolant pressurized within pre determined limits.

This presents the coolant from boiling since the coolant temperature rises to 322°c while absorbing heat from the reactor fuel care. The primary coolant circuit system compresses the RPV, primary coolant, tube side of the NSSGS associated.

Piping and Valves.

The NSSGS are horizontal shell and tube type heat exchangers. The reactor coolant water flows through a nest of U-Shaped tubes housed inside the shell of the NSSGS. The feed water occupies the shell side. The steam generated in the NSSGS is cool from the top of the NSSGS is through to the GS in the turbine boiler. There are 4 nos. coolant pumps (each of 7 MWE Capacity and Weight of 60 metric tons). Each pump has a flow of 22,000 cu mets. per hour to ensure transfer of heat from the coolant to the fuel water. The secondary circuit of the system compresses steam lines, TGS moisture separators reactor de-aerator condenser, fuel water pumps, associated pipe lines and valves.

24.8.1.5 Reactor Housing and Internals.

24.8.1.5.1 RPV internals.

The reactor core barrel accommodates the reactor core. It has perforations at the bottom. The borated coolant water enters though the upper portion of the RPV is deflected down words, enters the core barrel though the above seed perforations is pumped up through the coolant waters and enters the RPV near its middle portion.

A protective tube unit keeps the fuel core in position and guides the counted rods into the core. Core baffle is sandwiched between the core and core barrel and services as a shield for the RPV from radiation. The after internals support system for the fuel assemblies instrumentation and control and retrieval mechanism for irradiated specimens.

24.8.1.5.2 Melt Fuel Catcher reactor internal.

The most important internal in the reactor housing is the "melt fuel catcher". This is located at the bottom of the RPV housing pit. The catcher is a steel tank surrounded by water. The tank contains bricks of ferrous oxide and aluminum oxide. The main purpose of the fuel catcher is to bring down the temperature of the uranium fuel which may melt and fall down into the pit as a result of LOCA. We may mention that LOCA (Loss of Coolant Accident) is caused when there is a gross rupture in the coolant system piping, coolant is lost as a result and nuclear fuel temperature rises to melting level of the fuel. The consequences of LOCA are the same both in heavy water moderated and light water moderated reactors LOCA.

LOCA which is the maximum creditable accident in a nuclear reactor coolant pressure system has a probability rate of 704×10^{-5} per reactor year. This probability rate is only for core must down but with no release of radio activity to the public demand.

When LOCA occurs, uranium fuel which has a melting temperature of 2800°C (much higher than that of steel) will melt and in turn will melt RPV material and flow down to fuel catcher. The bricks decided above will absorb most of the heat from the melted uranium and will themselves melt in the process. The motion fuel and the bricks will form a lamp over a period of time.

Kudankulam is the second reactor in the world after the China "Lian Tun Gang" nuclear station to have this latest safety feature of a melt-fuel catcher.

24.8.1.5.3 Fuel Assembly.

The assembly is of hexagonal array. Each assembly has a weight of 705 kgms. and contains 490kgms of low enriched uranium oxide fuel (enriched to about 4% of uranium 235 isotope). The fuel in the form of pallets is encapsulated in zirconium niobium (Zr.Nb) tubes.

The assembly has 331 nos. tubes in all through which the coolant flows. The fuel is loaded in 311 nos. tubes. The balance 20 nos. tubes contain control rod and instrumentation mechanisms one central tube houses the structural element of the fuel assembly. The reactor core contains 163 nos. fuel assemblies in each tube in hexagonal array.

24.8.1.6 *Reactor Regulating System.*

The System comprises of 121 nos. control drives located on top of the RPV. These drive rods contains neutron absorbing materials and are held by electro-magnetic clutches. Reactor power is controlled by the movement of the rods in the reactor core. Insertion of the rods in the core reduces the power while their withdrawal increases the power .Failure of power supply that keeps the clutches energized results in free fall of the control rods under the force of gravitation. This leads to the sub-criticality of the reactor in period of the reactor in a period of 2 to 4 seconds.

24.8.1.7   *Boron Injection-Reactor Shutdown System.*

This is a back-up system to the control rod drive mechanism. The system injects highly concentrated boric acid (boron) into the reactor coolant circuit. One boric acid solution tank is connected to each of the four primary coolant pumps. The injection system gets actuated once it gets signal of failure of the control red drive mechanism.Even in the case of failure of power supply to the coolant pumps, the inertia energy of the large fly wheel of the pump suffices to pump sufficient boric acid into the primary coolant circuit to render the reactor sub-critical.

24.8.1.8   *Emergency Core Cooling System (ECCS).*

The system takes core of the emergency conditions that may be caused by the loss of coolant water during normal reactor operation. ECCS is of two types namely the passive ECCS and the active ECCS. It is the passive ECCS which comes into operation first to tide over the initial period of an accident like LOCA. The details of the two systems are as below.

24.8.1.8.1 Passive ECCS.

24.8.1.8.1 First Stage Hydro-accumulator Power System.

This system does not require any conventional power supply for its operation and hence will be operable during station black out conditions also. The system consists of four accumulator tanks each with a total volume of 60 met3 borated water connected to the RPV through independent nozzles. The boric acid content is of 169 ms per litre.

The accumulator tanks which are pressurized with Nitrogen occupying 10 met3 space of each tank are located above the RPV without the reactor building. The tanks have a capacity twice that of water in the RPV. Each accumulator is connected to the RPV through a check valve and associated piping. Once the pressure in the coolant drops below that in the tank, the check valve opens and borated water flows into the reactor core. This water removes the heat from the fuel core. Attached sketch indicates details of passive ECCS.

PASSIVE HEAT REMOVAL SYSTEM

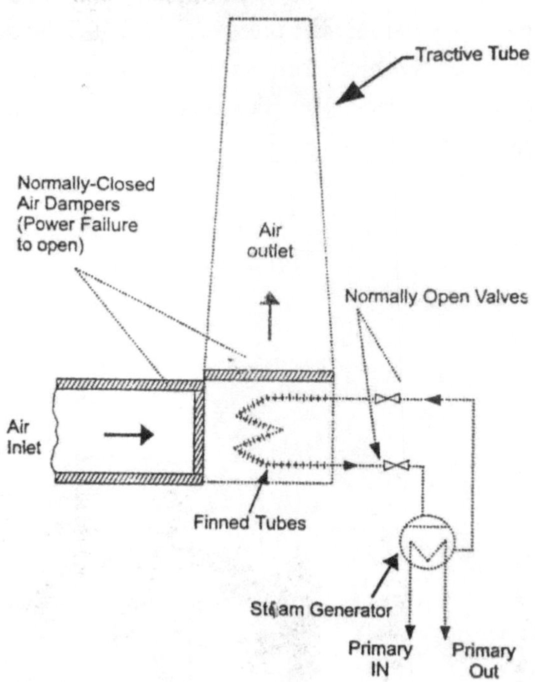

24.8.1.8.1.2   Second Stage Hydro-accumulator System.

Another passive system holding about 1000 met3 of water serves as slow discharge of water into the reactor core. This system is capable of keeping the core flooded and cooled for 24 hours in case of station black out. The second stage of hydro accumulator system consists of 8 nos. tanks above the RPV inside the R.B.

24.8.1.8.2   Active ECCS.

The passive ECCS is backed up by on active ECCS comprising of 4 nos. channels of high pressure (HP) and 4 nos. channels of low pressure (LP) sub-sect. A set of one HP channel and one LP channel is connected to a pumping system which is fed by one class III Diesel Generator Supply System. The active system is designed to remove the residual heat over a longer frame of time.

The HP system comes into operation in the event of a small rupture in the coolant system and the LP system in the event of a larger sized rapture. The passive ECCS system draws its cooling water from the spent fuel bay. The discharged water is collected from the R.B floor pump and is re-circulated by the ECCS.

24.8.1.9   Reactor primary system - hydrogen re-combination system

The system takes care of any hydrogen that may be formed by the operation of the reactor and thus may pose an explosion risk. We will now see how the hydrogen is formed.

We have seen in above paragraphs that the coolant channels are formed of zirconium - niobium. Zirconium has a high affinity for oxygen. It reacts with steam which may form in the tubes under certain conditions. The resultant chemical reaction liberates hydrogen.

Hydrogen re-combiners are located inside the upper portion of the inner containment dome. The re-combiners use "palladium" metal which acts as a catalyst to recombine hydrogen with oxygen. The effect of any explosive gas is thus neutralized.

Recombiners are of a passive nature require no power supply and thus available for action under all operating conditions.

24.8.1.10   Reactor Primary System - Purification System

The primary coolant is subject to contamination due to radio activity and chemical impurities caused by fission products. These impurities are removed by the purification system which is a closed loop comprising heat exchangers and "cationic" and "anionitic" filters. The purified coolant joins back the primary coolant system.

24.8.1.11   Venting of steam from reactor containment

The reactor containment is designed for withstanding stated internal pressure. It is hence essential to keep the pressure of the steam released in R.B within limits by a system of venting the steam particularly during accident conditions. This system releases the steam gas mixture in a pre-determined controlled manner through pipe-lines. The mixture is passed through high efficiency scrubber filters which remove the radio activity products from the mixture prior to its passage through state to upper atmosphere.

24.8.1.12   Passive Heat Removal System (PHRS)

This system is designed to meet the condition of total station black-out with the reactors and hence the system generators (SGs) in operation. It is essential in this condition to reject the heat in the secondary system of steam generation. The rejection of the heat from the primary side of the steam generators is taken care of by the fact that the SGs are horizontal vessels located above the reactor and hence thermo syphoning comes into play. The above PHRS system is the first of its kind globally.

The secondary side of the SGs (the feed water side) is connected to finned tube water-air exchangers. These heat exchangers, 12 in number are rectangular shaped and are connected to air-inlet and air-outlet dampers. The upper portion of the air-outlet duct is a 3 metre diameter chimney structure which produces a natural drought effect for the hot air-outlet to rise and flow out. The outlet duct starts at +52 meter elevation ends at +73 meters elevations on the outside of the outer containment. The heat exchangers are the located in a tertiary dome structure that is of 8 metre high over hanging cantilever box - type gallery supported on the R.B outer containment. Each heat exchanger weighs about 60 meter tons. The air-duct pipes are enclosed by a tertiary dome outside OCD. Both the air ducts are closed during normal operation. The increase in steam pressure above a pre-set value leads to the opening of both the inlet and outlet air dampers. The primary side of the heat exchangers is connected to the SGs through valves which are always kept open during the operation of the reactors.

A set of 4 such heat exchangers is connected in parallel to each other. This set is connected to one of the four SGs. These heat exchangers are located outside the OCD. There are 12 nos. heat exchangers in all.

PHRS along with emergency core cooling system (ECCS) carries out the following functions.

1.   Removal of residual heat from the reactor core for 24 hours or longer during station black out conditions.
2.   Removal of heat from the core during LOCA under station blackout conditions for 24 hours.

Tertiary dome with Passive Heat Removal System ducts during construction phase.

24.8.2   Refuelling

Refuelling is carried out once in a year, one third of the fuel in the core is replaced with fresh fuel. The refuelling is carried out as below after the reactor is shut down the coolant pressure comes down to atmospheric pressure and the coolant temperature nature to 60-70 degree C. The refuelling is expected to take about 10 days, 25 tonnes of fuel will be required to be replaced every 18 months.

All mechanical and other connections to the top cover of RPV are removed. The top cover is then readied for removal. The protective tube assembly is detached. The top cover is then lifted by a crane and shifted to its assigned place within the R.B. The spent fuel assembly from the core is examined for leak-lightness. If any of the fuel assemblies is found leaking, the same is sealed before transportation to the spent fuel bond. The leak free assemblies are also transported to the fuel pond. The remaining two-third of the fuel assemblies are rearranged in the core as per design intent. The fresh fuel would by this time have been shifted from the fresh fuel area to the holding pond. The fresh fuel is then lowered into the vacant position in the core.

The reactor pit would already have been filled with borated water to provide a height of 3 meters of water above the fuel assemblies. The entire fuel removal and re-fueling operation is carried out under the head of water. The spent fuel from the reactor core is stored within the R.B for a period of seven years. It is then shifted to the spent fuel bay outside the R.B after the expiry of this initial cooling period.

24.8.3   Waste management

Reactors would have gone through waste management facility described in 4.2.4.7 of chapter No. 4. This facility is generally for heavy water moderated and cooled Reactor Plants; some of these facilities may find application in height water moderated and cooled reactor plants. The following give in detail the management in SWRs as in Kudankulam.

The radio activity wastes in SWRs fall under solid, liquid or gaseous corresponding to low, intermediate or high level wastes. As in HWPTRs, bulk of the reactivity due to fission is contained within the spent fuel itself. The liquid waste is evaporated in a steam heated evaporator. This is compacted into a solid waste along with other solid waste generated in the plant. The vapor produced is condensed and recycled for use within the plant. Most of the solid waste comprises apparel tools.

Burnable solid waste is burnt in a specially designed incinerator. The fuel gas from the incinerator passes through high efficiency fillers which remove all the particulate activity. The gases are charged through 100 meter high stack. The non-burnable waste are treated as described in 4.2.4.2.3.

24.8.4   TGs

Each reactor provides nuclear steam for a set of turbines comprising of one no-double flow high pressure and 3 nos. double flow low pressure turbine. The turbine provide power for a directly couple 3000 RPM 24 KV 1 GWE generator. The unit is provided with a stress evaluator and an on-line monitor system that keeps a check on various parameters of the machine and its operating health.

24.8.5   Evacuation of power from the station

The power to be generated by the station is proposed to be evacuated by the following EHV transmission lives.

24.8.5.1   400 KV evacuation system

24.8.5.1.1   400 KV double circuit line between KKNPP and Tirunelveli - 100 kms length.

24.8.5.1.2   400 KV single circuit line between KKNPP and Madurai SREB grid station - 240 km length.

24.8.5.1.3   2 nos 400 KV lines/bays - spare conductor details- Quad type (4 conductors in line) - ACSR "Moose" - thermal limit - 3.2ka at 85°C.

24.8.5.1.4   220kv evacuation system

A line-in and line-out loop system has been adopted. The line-in is the 220kv single circuit line from Tulicorin - thermal power station to KKNPP - 80kms. The line-out is from KKNPP to R.S. Puclur Grid station - 17 kms.

24.8.5.1.5   One 220 KV line/bay in spare.

Conductor detail-ACSR 'Zebra' - Thermal limit - 649 amps at 75°C.

24.8.6   Allocation of power to Southern States.

The power to be generated by KK -1 and KK-2 has been allocated as follows.

Tamil Nadu State – 925 MW

Pondicherry - 67 MW

Kerala – 266 MW

Karnataka – 442 MW

Balance 300 MW remained unallocated.

24.8.7   Unit sale cost of KKNPP power.

The levellised unit cost of unexpected 30 years of operation of the station is expected to be Rs 195 per unit. This may change depending upon actual date of commissions of the two units.

24.8.8   Expected dates of commissioning

KKNPP-1 - 2006. KKNPP-2 - 2007.

24.8.9   Commercial viability of the station.

Expected annual revenue from the station at above cost and a plant load factor (plf) of 68.57 is Rs 2,022 cr. while the 14 equated yearly installments of re-payment of Russian credit commencing a year after commissioning of the plant Rs 653 cr well within year earnings.

Another indicator for the loan serviceability of the project in the "Debt coverage rates" (DSCR) which is arrived at as follows. DSCR = Gross revenue minus operating cost / principle + interest. If the above ratio is above 1 the project is economically viable. The ratio for KKNPP is 1.4 thus making it eminently viable economically.

24.8.10   Progress of KKNPP 1 & 2 as of 1992

24.8.10   Preliminary works – surveys - seismic study-set-up-land acquisition.

24.8.10.1.1   Land acquisition

Land acquisition both for plant site and losing colony was taken up. Taking over acquired land began in February 1992.

24.8.10.1.2   Preliminary works progress as of 1992 Approach road

An approach road taking off from National Highway NH-7 at Havaikinare to the site was completed.

24.8.10.1.3   Site fencing work

About 30% was completed.

24.8.10.1.4   Site surveys soil studies etc

Sub-soil investigation, background radiology survey, soil electrical resistance study, hydrographic survey off-shore hydrographic survey coasted cum near-shore dynamic studies and geo-technical investigations were completed.

24.8.10.1.5   Seismic and metrological studies.

Micro-seismic studies, metrological studies and bio fouling studies were taken up.

A digital tele-metered network for seismic study, comprising 3nos of 3-component and 4nos of single component station located in a 30km radius of the site was set up. The central receiving station was set up on a small hill at Eruvadi, about kms from the site. These stations receive data on a continuous basis from the micro-seismic station through wireless signals. The data are stored on an online computer. The network was commissioned during October 1992 and is operating satisfactorily.

24.8.10.2   KK 1 and KK 2.

The control room of KKNPP

Infrastructural works as of 1993- 2000

Bids for the falling work were invited as of 2000

Providing random rubble masonry wall and chain link fencing for the project at Chettikulam village at an estimated cost of rupees 109 lakhs.

Providing 2 nos of sea water reserve osmosis (R.O) streams of 0.25 cusec flow in each dream including intake and outfall civil structures at a point 20 kilomets. North of Kanyakumari laying of

10 kilomets. of 200 mm diameter pipe lines between the plant site and the (R.O) plant etc at an estimated cost of 19 cr.

Construction of infrastructural buildings and structures at plant site at a cost of Rs 5 crores.

24.8.10.3 *Further infrastructural work as of 2001*

24.8.10.3.1 A budget amount of rupees 100 corrode for the work to be carried out during 2001 was car-marked and would comprise the following among others.

24.8.10.3.1 Metrological laboratory

And environmental survey laboratory on which work commenced during March 2001.

24.8.10.3.1.2 Reverse osmosis (R.O) plant

Work on reverse osmosis plant (R.O) costing rupees 19 cr. was inaugurated by chairman AEC on 24.04.2001.

24.8.10.3.1.3 Site grading and levelling

Bids for the above work with a bid date of 20.03.2001 were invented. The work was expected to cost rupees 12 cr.

24.8.10.3.1.4 Infrastructural buildings

Tender for infrastructural building including internal electrification, sanitary works, domestic,fire fighting water distribution system etc were floated at an estimated cost of rupees 23 cr. with a due date of 28.03.2001.

24.8.10.3.1.5 Ground breaking Ceremony- on-site works- 3rd quarter of 2001.

The 9th of October 2001 may be looked upon as a landmark day in the story of the evolution of KK 1 and KK 2 in their construction phase; this was the day that ground breaking ceremony for the project was held.It may be no exaggeration to say that a preliminary but sure step was taken on this day in the direction of civilization of economy regenerated of this background district (Tirunelveli).

24.8.10.3.1.6   Pre-qualification bid for more civil works

P.Q bits for civil work in reactor building, reactor auxiliary buildings and related buildings under three package costing it all rupees 700 crores with varying stipulated period of completion of 49 months for one package and 35 - 44 months for the other two were invited with the due date of 12.11.2001.

Tender for construction of 540 numbers of residential buildings at an estimated cost of rupees 33 cr. were floated on 18.12.2001 with bid due date of 20.01.2002.

The above were the last major of the bids floated during 2001.

24.8.10.3.3   Staff strength

They were about 400 department personal stationed at site as of end 2001. This trend would progressively be augmented to cope with the increasing tempo of works and was expected not to exceed 266 in number.

24.8.10.4   Civil work progress as of 2002 end.

First consenting in RB-1 raft was finished on 31.03.2002 and in RB-2 during July 2002. Consenting on 8 met. thick craft of RB-1 was completed by the middle of 2002. The work on RB-2 raft was scheduled for completion by the third quarter of 2002.

24.8.10.4.2   Intake water system-temporary water system-award of work.

The work of in-take water system comprising of pump houses, tunnel chlorination of plant, temporary water system, break water dyke bridge intake structure, sea water pipeline fish protection facilities, shore protection work and related work was awarded to Hindustan construction Coy (HCC). A breakwater dyke 2 kilometer long was required to facilitate above work the structure were being constructed inside of a temporary coffer dam area (dry harbor) near the seashore on the sea sand bed. A jetty for unloading of ODCs was also constructed.

24.8.10.4.3 Award of above civil works.

The contract for the above civil work for KK 1 and KK 2 awarded to HCC was worth Rs 272 cr. The other major contract carrying out KK 1 and KK 2 works were simpler concrete pipe Limited and Larsen and Turbo subsidiary engineering construction Coy (ECC).

24.8.10.4.4   Receipt of first consignment of soviet equity-KK 1-Status as of 12/02

First consignment of 4 nos. hermetically sealed doors for RB 1 was received during December 02 and their eviction was in progress.

24.8.10.4.5   Team of Russian specialist

A team of 10 Russian specialists was present at site as of 12/02 to supervise erection work.

24.8.10.4.6   RB 1 and RB 2 Civil work status as of December 02 - Civil works - marine works

Construction above raft of RB-1 was continued. Concreting of raft of RB-2 was completed in record time during September 02. Raft foundation for reactor of auxiliary buildings was completed and work about the raft continued.

First pour of concrete for TB-1 was carried out on 31.10.2002 and concerting of raft buildings was continued.

Construction of sea-water intake and out-fall hydro-structures for condenser cooling was commenced.Special fish protection measures were taken.The marine structures word designed by the Russian organization Messrs. Gidro project, Morocco.

Construction of infrastructural at site was nearing completion.

24.8.10.4.8   Project Township work

First phase of residential quarters including guest house was completed. A 24-hour dispensary and atomic energy school were ready and in operation.

24.8.10.4.8   Sea water reverse osmosis plant.

The plant for conversion of sea water to portable use at Township was completed.Cost of water to resident would be 7 paise per litre.

24.8.10.4.9   Green activities

A green Bell along the Township boundary and other suitable plant areas were under development.

24.8.10.4.10   Fuel supply agreement

CMD NPCIL signed an agreement during early February 2003 with the president of the Russian fund manufacturing bay TVEL, Alexandra Nyago for supply of the first batch of nuclear fuel rods. The agreement was worth rupees 400 million and was part of the long-term agreement for continued supply over the 40 years expected period of operation of the KK-1 and KK-2 units.

24.8.10.10   (a) KK-1

RB-1 civil work Core-catcher

24.8.10.10   (a) .1 Progress during 2002-03

Concreting of the core catcher containment slab and walls of RB-1 was completed installation of liners for the above was nearing completion. The fabrication of core catcher tank was being carried out at St. Petersburg Russia and it was expected to be shipped to side by September.

RB-1 elevation slab was concreted during March 2003.

Concreting of the first part of the hermetically shield containment slab at +5 meters elevation was completed in July 2003.

24.8.10.10.2   RB - 1 plant works 2003-2002 period.

Erection of hermetically sealed and biological shielded doors, tanks and feelers was received at site.

24.8.10.11   KK-1 Civil works-progressas of end March 2004-RB-1

Concreting of 1.8 met.thick hermetically shield slab at+ 5.4 met. elevation was completed with a single record pour of 2900 cu mets. in 66 hours.The slab comprised 200 metric tons of rebars and 40 metric turn of embedded parts 120 templates for pre-stressing and 56 numbers hermetic - penetration embedded parts were provided for in the slab.The work was completed during September 03.

The work of erection of cylindrical structure of 44 met. diameter and 7.9 mets had reached upto 14.57 meters elevation.

24.8.10.11.2   TB-1 Progress during 2001-2004.

This rectangular concrete word structure of 96 met × 57 met and height of 43 mets. had been raised up to 14 meter elevation.

24.8.10.11.3   Under-ground tunnel works

All the above works KK-1 area were completed; Back-filling of the area was in a progress in a phased manner along with construction of permanent motorable roads.

24.8.10.11.4   Construction of other civil structure for KK-1 progress during 2001-04

Civil work construction of 10 structures including fire-fighting, water tank transformer oil emergency discharge tank and rainwater collection tank was completed.

Construction of main intake water pump house was nearing completion construction of TB operating floor at+ 7.05 met. elevation was in progress.

Nearly 60 structures were under construction at KK-1 site during the above period.

24.8.10.11.5   KK-1 Off-shore construction work during 2001-04

1.5   km. long temporary dyke construction was completed de-watering off 1, 00,000 sq. mets.of sea-bed area up to the sea-bed level involving 0.5 million met. Cube of sea water was completed with the use of 12 de-watering pumps.

24.8.10.11.6   Construction of break-water dyke intake structures, fish protection facility

Construction of 2 km. long breakwater dyke. Water in-take structure, in-take pipelines under the sea-bed, fish protection facility structures and four-bay structures was in progress.

It may be mentioned here that 4 nos. hollow - cellular concrete caisson structures were required as intake for condenser cooling water system work was in progress of these structures.

24.8.10.11.7   Reactor equipment erection -Progress during 2001-2004

Core catcher vessel weighing nearly 100 metric tons and truss canti - lever vessel weighing 140 metric tons were erected. A total of 8500 metric tons of equipment was received at site from Russia.

24.8.10.11.8   Contractor Labour strength-Further work.

With the erection of eqpt. for both KK-1 and KK-2 picking up speed or total labour strength of 6000 was available at site.

24.8.10.11.8.1   Progress as of end 2005 KK-1 civil works-RB and TB

A total quantity of 44,932 Cu. meters of concreting was carried out during August 2004, this record betraying the earlier record of 40,000 Cu. Met.per month.This performance stood as an all time record for allNPCIL projects KK-1 project had completed by end 2005 7,20,000 cu.Mets.of concreting out of the total 9,50,000 cu mets.required.

There were 7 batching plants at site with a total concerting mining capacity of 350 cu mets.per hour.concrete placing equipment consisting of placer booms, pumping units; super singers etc. were also in place to match the mixing capacity. 7noisce plants without total capacity of 140 metric tons/hour were available to facilitate controlled temperature placement TB-1 construction up to 36.5 mets.elevation including the crane beam was completed during August 2005.Construction of the primary containment wall up to 43.9 meters elevation was completed during May 2005.

24.8.11.9   Reactor plant works progress KK-1 as of end 2006.

A view of polar crane before installation of dome on the reactor building

Steam generator being erected using polar crane

Bridge assembly with main and auxiliary hoist

Crane rail being erected on the crane beam inside the containment at +43.9m

Crane girders and end girders installed at +43.9m inside reactor building

The '360degree rotating' polar crane used for handling and erecting bulky ODC equipment items in the reactor building

24.8.11.9.1 Erection works

Lower plate forming part of the core catcher system was installed. Dry shield manufactured by M/S Izhosky Zavod, St. Petersburg Russia was received in six parts and was assembled at site. Dry shield servers as thermal shield to the reactor cavity wall. It was installed in the reactor cavity shaft. The work was a precision one and completed satisfactorily.

24.8.11.9.2 NSSG-KK-1

The first nuclear service steam generator for KK-1 manufactured by M/S Zoomar Podolsk, Russia was received at site during September 2004. Three more units for KK-1 were to be received. NSSG, a horizontal shell and tube type has 10,978 tubes made of stainless steel and ways 424.5 metric tonnes. It has a diameter of 4 mets. and length of 13.85 mets. The unit was shipped from St. Petersburg to Tuticorin Port. It was transported from the above port to KKNPP mine-port through barge.

24.8.11.9.2.2 Vessel (RPV) details

The first RPV intended for KK-1 was manufactured at the plant located at St. Petersburg of Russia and was flagged off on November 8th, 2004 in the presence of high NP CIL, Department of Atomic Energy. Govt. of India, Indian consular officials and Russian representatives. The flagging ceremony was presided over by the Governor of St. Petersburg. The RPV was a letter loaded on to a barge and arrived safely at Tuticorin port on 3.01.2005.

The RPV is 5.6 mets.in diameter 11.8 mets. high and weighs about 320 tons. All heavy nuclear equipment mentioned on page 35 was erected using polar crane.

24.8.11.10 KK-1 progress during 2005-08.

Physical progress as of December 07 was 71%.

24.8.11.11.1 Civil works

The work were nearing completion concreting of inner containment dome (ICD) was completed during October 2007.A hermetically sealed steel liner of 44 meters diameter on the inside of the dome was erected. The full hemispherical steel liner ring installed was for the very first time anywhere globally. The ring was assembled in three major parts and erected in position on the dome.

Construction of the main control auxiliary building was completed up to 22.8 mets.elevation.

Concreting of outer containment dome OCD was completed during May 2008. Construction of 8 mets.high overhanging cantilever box type gallery supported on the OCD for housing of air duct pipes and forming part of the territory dome structure was also commenced during this period.

24.8.11.12   Nuclear plant works main coolant pipe welding

Erection of reactor pressure vessels and nuclear steam service equipment was completed. Erection of other nuclear auxiliary system equipment was in good progress. Welding of main coolant pipe was completed in the latter half of 2008. Welding of 32 nos. joint in 4 loops connecting the reactor pressure vessel, steam generators and coolant pumps was completed during 2nd/3rd quarter of 2008. Special induction healing technique was developed and welder qualification was carried out to meet the stringent temperature requirement for the joints welding.

24.8.11.13   Passive Heat removal system (PHRS)

All the 12 nos. heat exchangers for the system were erected.

24.8.11.14   Further conventional plant equipment

180 tonnes main turbine whole over-head travelling crane was commissioned along with 15 tons auxiliary cranes.

24.8.11.15   Nuclear fuel

The first consignment of nuclear fuel was received at site on 25.03.2008.

24.8.11.16   Polar crane

The crane mounted on circular ring frame enable travel on both horizontal and vertical axis thus enabling very heavy reactor equipment (reactor pressure vessel weighs 400 tons) to be erected at case.

24.8.11.17   IGN turbo-generator civil and mechanical works

Fly-ash concreting was for the first time adopted India for TG block.

Generator stator wing 350 meters was erected in position. The T.B. main crane has a capacity of 180 meters. An additional trolley was end erected to handle and position the stator which was the heaviest single eqpt.in station. The rotor was threaded onto status it was a very specialized operation and it was carried out with precision with the use of T.B. overhead travelling crane.

Turbo-generator hall of KKNPP-1

24.8.11.18   Station service electrical system-class III supply system control supply equipment- 220 KV Gas-Insulated (gis) supply system

Class III supply DG sets were erected. Station supply auxiliary transformers and 6.6kv. switchgear were erected control supply equipment were erected 220 kv gas-insulated supply system was commissioned during December 08.We will be delaying with the crucial electrical power.

24.8.12   KK-2 nuclear and conventional equipment system progress during 2003-08

24.8.12.1 Receipt of equipment

Reactor pressure vessel, PHT coolant pipe-line, PHT coolant pump casing coolant pressurizers, nuclear steam service generator (nssgs) H.P heaters, moisture separators, steam re-heaters, T.G. and condenser unit were received at site.

24.8.12.2 RB-1 erection of RPV

The work was carried out by contractors Larsen and Toubro under site supervision of Russian experts.The first leg of the erection operation envisaged the lifting off the 11 mets.high 4.15 mets. diameter weighing 320 metric tons onto a specially design erection trolley installed on the equipment transport at 31 metre height above the ground with the use of high capacity crane. The RPV was in the second leg of erection moved inside the RB and lowered into the R.B vault.The whole operation was carried out during December 2007.

Truss cantilever vessel being erected inside the RB-1

24.8.12.4   KK-1 and KK-2 capital expenditure as of 2005-06.

A capital expenditure of rupees 20.81 (2081 cr.) was incurred in billion the above here bringing the total so far to rupees 75.57 billion.

24.8.12.4   Independent verification and validation (IV and V)of KK-1 and KK-2

IV and V of all computer-based safety-related system was initiated during 2005-06.

24.8.13   RB-2 civil work progress during 2003-07

Construction of elevation slab was completed during June 2003. Concreting of + 5.4 mets.elevation slab was completed during December 03.

Construction of primary containment upto43.9 mets. elevation was completed during November 2005.

24.8.14   TB-2 and auxiliary building civil work-progress during 2005-2007.

Construction of TB-2 up to 36.5 mets.elevation including TB-2 over-head travelling crane beam was completed during January 2007. Construction of main control room and auxiliary building up to 22.8 mets. elevation was completed during September 2007.

24.8.15   RB-2 installation of polar crane

The crane was installed and commissioned during December 2007. This paved the way for expenditures election of major items of nuclear equipment.

24.8.16   RB-2 transport air lock-progress during 2007-08.

The huge transporter lock needed for moving ODCs into the RB and weighing 300 metric tons was installed inside the reactor containment wall and position at 31 mets.above grade level. The air lock was delivered in 5 parts. The pre assembly and welding of sections was carried out on the ground. The 350 tonnes trestle crane was used for erection of the air-lock was earlier assembled and load-tested.

The above massive operation which led to heavy reactor eqpt. being handled and positioned with precision was carried out during 2007-08.

24.8.17   RB-2 civil work ICD-Pre-stressing

Inner containment Dome (ICD) was completed during July 2008.

Pre-stressing of the dome was carried out successfully. Unbonded pre-stressing system viz Freyssinet 55 c 15 system was adopted. This is a unique feature of pre-stressing and is a model for use in the domes in future. The system consists of 128 tendons (60 u-type vertical and 68 horizontal circular in shape).Each tendon consists of 55 strands of 150 mm square cross section and is stressed at a force of 11.5 MN (mega newton).

The OCD was completed during October 2008.Territory dome of PHRS was also completed during the above period.

Dome service platform being lifted to be installed on the polar crane

24.8.18   Nuclear equipment erection -air locks, main coolant piping, auxiliary system equipment, nuclear steam system generator - progress during 2008 -2010 RB 2

Welding of main coolant piping of all thickness 70 mm commenced in early 2008 all the 32 Nos joints word scheduled for completion by early middle 2010 but carried out expeditiously by April 2009.

Direction of main and emergency air locks was taken up by middle 2009.

24.8.19   KK -1 and KK -2 use of polar crane

The twin unit project has created a record in the history of nuclear equipment erection in India by completing erection of a total of 16 numbers heavy lift pressure vessels Nssgs coolant pressure risers and coolant pumps in a period of only 12 months mid 2007 –mid 2008. The process was completed with the erection of 4 numbers NSSGS for KK 2 SG pipelines for KK-2 were elected by July 08 polar cranes was used for erection of above heavy equipment.

A glimpse of the dedication event of KKNPP-1 via video conference

24.8.20 KK 1 and KK 2 full scope simulator 4 KK 1 and KK 2

Nuclear training center of NPCIL commissioned during late 2007 early 2008 full scope simulator for VVER - 1000 reactor system at KK sight the simulator will be used for imparting training and qualifying operators and manufacturers of KK one and CTO at various levels.

A view of 1-phase, 24/400kV, 417MVA generator transformer

A view of 400kV gas-insulated switchgear(GIS)

A view of 220kV gas-insulated switchgear

A view of 400kV/220kV, 315MVA interconnecting autotransformer with 220kV gas-insulated bus inlets over supporting trestle (connected between 220kV GIS and ICT)

A view of 6kV, 6.3MW diesel generator set in emergency auxiliary power supply system building

24.9.1   Desalination plant for KK 1 and 2

The plant for catering to fresh water requirements for KK one and 2 was inaugurated the plant is based on multi vacuum compressor technology and has 4 stages in each stream the plants do not need external fresh water supply for its operation

24.9.2   Further off shoreworks floating of "Caisson" structure

The first 2400 meter caisson structure the heaviest concrete floating body of its kind made so far in the country was loaded successfully on a barge and stored to each location 1.2 kilometres from the seashore the structure will be lowered in this position the structure is 36 metres long 1.5 meters wide and 12.5 metres high this structure will be followed by the remaining 3 structures. The design of the structure was carried out by M/S Gidro project Moscow Russia under contract from NCPIL see civil contractors Hindustan construction carried out the construction and floating of the structure.

24.9.3   Project Cumulative expenditure project progress first quarter of 2008.

A cumulative expenditure of rupees 105.28 crore was incurred against sanctioned Rs 131.71 cr. This comprises of completion of various sectors as follows.

Design - 89%

Supplies - 87%

Erection - 78%

Overall project completion was as follows:

Unit - 1 - 84.6%

Unit - 2 - 75.3%

24.9.4   Conventional equipment TG stator erection KK 2.

TG unit stator one of the heaviest equipment of the project with a weight of 350 metric tons. The capacity of the turbine Hall crane is 180 metric tonnes special erection trolley was erected in order to handle the stator load. The combined crane trolley system was load tested prior to stator handling. The stator was subsequently handled successfully and placed in its position.

24.9.5 KK 1 and 2 400 KV gas insulated system (GIS) evacuation of power 2009- 10 progress.

400 KV GIS comprises of 8 nos ways (6 nos for outgoing400 KV lines one no. for inter - connecting 400 KV/ 220 KV transformer and one number for shunt reactor) was completed in a short period of 6 months. Erection of 400 KV outdoor switch yard was completed by PICIL (Power Grid Corporation of India limited).

24.9.6 KK -1 - Progress during 2009-10 - Conventional equipment

24.9.6.1 IGWe

TG set was put on barring _____ during January 2010.

24.9.6.2 Fuel Loading

Loading of 163 nos of dummy fuel each of 703 kgms of weight was begun on 23.04.2010. The loading was completed by June 2010. Loading of actual fuel was completed by September 2010. The dummy fuel was an exact replica of actual fuel except for the fact that the elements were of lead instead of uranium and the dummy fuel in used for validation of actual fuel.

24.9.6.3 Thermo - hydraulic cost - Hot conditioning - reactor checkup

Thermo - hydraulic tests were completed during the second quarter of 2010. Hot conditioning of PHT system was completed by the third quarter of 2010.

24.9.6.4 Loading of dummy fuel removed from KK - 1 into KK - 2 RPV was started by July 2010.

24.9.6.5 KK - 1 progress during 2010-11

Containment integrity (pressure boundary) test was completed satisfactory on 25.01.2011.

24.9.6.6 *KK - 1 Commissioning - Progress mid- half 2010 - Progress mid- half 2011 - Tests*

24.9.6.6.1 Functional tests

Individual functional test of equipment as part of Phase - A1 / Phase - A2 commissioning activity was completed during April 2011. Fuel loading was taken up.

24.9.6.6.2 KK - 1 Phase A3 - Commissioning primary system

24.9.6.6.2.1 Test on primary and secondary system

Hydrotest flushing and heat - run of nuclear steam service system (NSSS) was completed during August 2011. Hydrotest of NSSG was done at 24.5 MPA and of secondary circuit at 10.8 MPA during December 2011.

24.9.6.7 *Agitation against plant commissioning.*

Just as KK - 1 was getting ready for first approach to criticality local residents began an agitation against the plant. Commissioning was calling for shelving of project. One by name Uday Kumar Chairman of the "Committee against" nuclear plants was leading the agitation powerful local church authorities who had influence on the pre dominant local Christian population and fearful of losing the same due to development projects text their support to the agitation. Foreign NGOs (Non - Government Organizations) and green groups predominantly based in Scandinavian Countries and U.S also joined in the agitation. Independent third part + opinion including from inter-national professional nuclear institutions vouch safing for the integrity and inherent safety of the kudankulam plants did not deter the agitations. The agitating outfits cut off entry points to the plant. They only allowed only 50 technical staff to maintain and keep running the most essential safety systems.

The TN Government set up a 2 member committee one of which was Dr. M. R. Srinivasan former AEC Chairman and then current member of AEC to go into the safety features of the plant and present its view. The committee in its report upheld the safety and integrity of the plant.

The Tamil Nadu Government accepted on 19.03.2012 the Committee's report and authorised the commissioning of the plant, the first step of which was approach to criticality. The plant authorities then proceeded with their planned work for first approach to criticality. The 7 month long agitation had lost the plant by 30 billion (Rs. 3,000 crores) by way of re-testing and restoration of the various plant systems. A further Rs. 10 billion was also caused by lost energy generation.

The Tamil Nadu grid had a power deficit of 4GW at that time and the loss due to curtailed industrial, commercial and trade activities were immense.

24.9.7 Mock drill at plant site

A mock drill on the aspect of the KK plant was conducted at site on 09.06.2012, morning drill simulated a Fukushima (Japan) type of nuclear plant disaster. The Fukushima disaster caused by a "Tsunami" following a severe earthquake resulted in nuclear fuel core melt - down and radiation leakage in the public domain.

The public drill started off as an off site declaration of emergency at Nakkaneri Village 7kms from KK site. The District Collector of Tirunelveli was in charge of the drill. NPCIL officials including the KNPP site director Sri. R. S. Sunder acted under the Collector's orders. A target group of people was created as part of the drill. The drill was declared successful at its end. The drill was a part of and the final one of several on- site and off- site activities that are essential as mandated by AERB prior to commencing approach to first criticality of the units.

The units will now have nuclear fuel loaded in the pressure channels and would be on course of first approach to criticality. About 80 meter tonnes of fuel would be loaded initially.

24.9.8 Fuel Loading - First approach to criticality.

AERB at its 107th board meeting on 09.08.2012 approved fuel loading and approach to first criticality after review of KK - 1's safety and related aspects. Fuel loading started in the presence of IAEA officials in the third week of August 2012. IAEA had from other aspects studied the preliminary safety analysis and final safety analyses reports submitted by NPCIL. It had earlier deputed its team for physical examination of the plant prior to its safety analysis meeting. AERB team was also on land during fuel loading and first approach to criticality.

24.9.9 Last minute move by NGOs to stall first approach to criticality.

Various NGOs approached the Madras High Court to stall approach to criticality on the ground that some clearances for the plant, particularly environmental clearance were flawed. The High Court after hearing various points raised by the litigants and rejoinder by NPCIL, the defendants. The Madras High Court threw out the petitions late in August 2012 and gave the 90 ahead for fuel loading and first approach to criticality.

24.9.10.10.2.1 NGOs petition in Supreme Court

Several NGOs including Committee against Nuclear Energy (CANE) approached the Supreme Court (SC) during September - October 2012 with a plea to restrain fuel loading. Though several of the NGOs pleadings in Chennai were turned down, they put their faith in Supreme Court. Their complaint to SC was that some of the 17 recommendations of the task force set up for further scrutiny of the kudankulam nuclear plants safety features were not implemented. NPCIL in its rejoinder maintained that all these recommendation were to be implemented in their framework of 2 years. It was observed that it was concerned with people's safety and would accordingly deal the petition. It did not however restrain NPCIL from proceeding with first approach to criticality.

24.9.10.10.2.2 First approach to criticallity

NPCIL proceeded after the completion of fuel loading with first approach to criticality system integration was the first step in this approach and was accordingly taken up during January 2013.

24.9.10.3 System integration first approach to criticality

System integration forming part of first approach to criticality was commenced from mid-January 2013 with presence of AERB authorities.

24.9.10.4 AERB clearance for second heat-up and full system tests

The clearance was given during end December 2012 certain defects in equipment were found during the tests. These were attended to and clearance for second heat up run was given on 24-01-2013.

24.9.10.5 Testing of main interruption and steam discharge valves (MISV)

Testing of MISDVs was carried out satisfactorily between 29th March 2013 and 19th April 2013. The testing results revealed defects in four passive system loading valves. These were non-power operated valves. They were replaced.

Testing of instrumentation system connected with MISDV operation was taken up on 1st April 2013 and completed on 4th April 2013.

24.9.10.2.6 Staff strength as at time of first approach to criticality

Strength of 1000 personnel including departmental and contractors staff was available at KK-1 and KK-2 site at the time of first criticality of KK-1. The total strength at both KK-1 and KK-2 sites was 3000.

24.9.10.2.7 KK-1 approach to criticality - attainment of criticality

It was a red letter day in India's nuclear power development saga when KK-1 with the highest rated unit power generating capacity in the country of 1 GWe attained criticality at 11.05 pm on 13.07.2013. The process commenced on 11.07.2013 at 23.49 hours. Neutron absorber in the form of boric acid was added to the primary coolant to keep the reactor in sub-critical state. This was essential to conduct various reactor physics and reactor safety tests that were essential before the reactor attained criticality. The process was carried forward after the completion of the above test in the form of dilution of boric acid content. This resulted in a sustained fission process and attainment of first criticality.

24.9.10.2.8   Attainment of criticality KK-1

24.9.10.2.8.1   Details
The attainment of criticality of KK-1 was the culmination of a project which sought to leap from the development of large sized nuclear power plant in India. The foregoing narration in this chapter would have given even a lay reader of the book an idea of the various crucial stages involved in the design, construction and commissioning of the giant 1GWe KK-1 unit and the on - going KK-2 project.

NPCIL with the active assistance of its Soviet counter-part had finally arrived at its stated goal. It had to run an extensive public relations campaign to acquire the general public with the various stages of the project and convince it about the safety and integrity of the project. The Central Government and the concerned state government have also played their part in the final denouncement of the project.

As a former employee of NPCIL and at their ripe old age of 88 (as of August 2020) this citizen of India takes a well justified pride in this achievement.

24.9.10.2.9   Synchronization of KK-1 with Southern Regional Electricity Board Grid (SREB Grid)

KK was first synchronized with SREB grid on 22-10-13 at 2.45 hrs. The unit was generated 16 MWe at that time. The unit was within synchronized state for a period of 2 1/2 hour till 05.15 hours on 22.10.13. The unit was successfully de-synchronised for this purpose of study and evaluation of all system following the above two crucial achievements. The crust was next in the process of augmenting the generalized power in stages of 500MWe, 750Mwe and finally 1GWe (1000MWe) as per AERB clearance at each of the above three stages. The system would thoroughly be calculated and studied at each of the above stages in the following month before being declared for commercial operation.

KK is the 23rd nuclear power unit to go into operation in India.

KK after break of 2 days (on 24-10-13) resumed power generation when it was re-synchronised with SREB grid at 9.43 pm on 25-10-13. The critical power was 160Mwe and later 240MWe. The power would be scaled upto 1GWe.

24.9.10.9 *Corporate Social responsibility (CSR)*

As part of CSR, KK project authorities have constructed a school known as MuthuRama middle school in the vicinity of the project.

24.9.10.10 *Full capacity operation of KK-1*

KK-1 attained full 1GWe capacity operation on the 7th June 2014. It was shut down on 30.07.2014 for more tests required to declare the unit for commercial operation. It is likely to be declared for commercial operation within 2 to 3 months of its functioning on full 1GWe capacity.

During the operation of the above plan and during trial run of turbine as part of proving tests a vital component of the turbine got unscrewed and hit the turbine assembly causing significant damage to the turbine. The faulty component was replaced from KK-2 turbine assembly and the rectified turbine brought back to full operational capacity.

BHEL is looking for the possibility of manufacture of the component at its work (for future maintenance use as required.

24.9.10.11 *KK-1's Shutdown for maintenance.*

KK-1 had operated for 4701 hours (nearly 200 days) as of latter fort night of October 2014 since it's synchronization to SREB (Southern Regional Electricity Board) during October 2013. It had generated 2.83 billion units of electricity. The revenue generated during this period was Rs. 7 billion.

Its annual shut down was due and taken up. A fuller checkup of the turbine was later up after the repair was carried out as mentioned earlier. The unit was expected to be back on line by December 2014.

24.9.10.12 *KK-1 was re-commissioned at 09.59 pm on 07.12.2014. It is operating at 70 MWe immediately after the above. It was proposed to be gradually increased to its full operational capacity of 1 GWe.*

24.9.10.13 *KK-1 is commercial operation*

KK-1 was declared for commercial operational from the midnight of 20.12.2014.

24.9.14   KK – 2 – Hot run test

Hot run test began on 28.02.15 steam flowpaths andsteam relief devices on the steam main supply line to TG of KK – 2 would be tested.

24.9.15   Maintenance shut down of KKR – 1

The unit was shut down on 24 -6 -2015 for annual maintenance work which would last for 60 days. The unit was generating 600MW when shutdown 50 fuel bundles out of the entire 163 bundles would be replaced during the shut down .The unit had generated 6.873 billion unit during the 9,267 hours of operation since 31 -12- 2013 when the unit was declared commercially operable and has earned the revenue of nearly Rs 20 billion.

24.9.16   KK 1 and 2 – CSR during 2015

KK 1 and 2 Station authorities built as part of CSR during 2015 a classroom block at NVC. Government school, Radhapuram near the station site. The block was named after the late APJ Abdul Kalam on 10-08-2015.

24.9.16.2   KK 1 and 2 built classroom building at Panchayat Union Primary School in K.C ------- village was inaugurated on 26-08- 15.

24.9.16.3   KK 1 and 2 built classroom constructed at Government High School , Vaddakunkulam village,----- Kinara post was inaugurated during September 2015.

24.9.16.4   KK 1 and 2 - Biodiversity as part of CSR

The station has been promoting biodiversity in nearby areas.Rose ringed parakeet was noticed at Anu – Vijay

24.9.17   Mishap in KK 1 township at Chettihulam

KK 1 was shut down on 13-9-14 when a vital component in the turbine blade got unscrewed and hit the turbine casing causing significant damage. The faulty component was replaced from the turbine of KK -2 and the reactor unit recommissioned.

24.9.18   KK 1 and 2 's continuing CSR activities

KK 1 and 2 units have as a part of CSR activity during 2014-15 have ------over Rs 75 cr. for constructing concrete houses under their neighbourhood development programme to the Tiruneveli district Collector Sri M. KarunaKaran during December 14 .The whole housing plan envisages construction of 10,000 houses and improvement of infrastructure needed for housing activity at the total cost of Rs.500 crore. Continuing environmental promotion activity saw rose-ringed parakeet at Anuvijay township at Chettikulam.

24.9.19   Recommissioning of KK -1 after refueling and maintenance outage

The outage was extended to attend to all required maintenance work in addition to refueling. All the works completed and the process of criticality approach was started at 23.49 Hrs on 20-01-16. The unit attained criticality at 16.19 hours on 21-01-16. Power would be increased in stages to attain rated 1GW.

24.9.20   KK – 1 Plant load factor (PLF) – Tariff as of August 16

The PLF of KK NPS – 1 as of above date was 100 %.The station has generated 11.269 billion units so far.Tariff as of August 16 was Rs. 3.89 per unit.The unit generated 16.242 billion units of electricity as of 31-03-2017 and saved 13.69 million tonnes of CO_2 emission. KK – 1and 2 generated 44.167 billion units power output as of May 20 and saved CO_2 emission of 37.94 million tonnes.

24.9.21   Criticality and synchronizing of KK NPP – 2

KK NPP – 2 attained first criticality on 09-07-16. It was synchronized with the Southern regional grid on 29-08-16 at 11.17 hours and was operating at a capacity of 245 MW. The generation capacity would gradually be increased to the rated 1 GW(1000 MW)

24.9.22   Full Power capacity attainment of KK NPP- 2,the 22nd nuclear power unit

The unit attained the full rated capacity of 1 GW on 21-01-2017, raising installed capacity of nuclear power from 5.78 GW to 6.78 GW.

24.9.23   KK NPP – 2 Commercial Operation

The unit was declared for commercial operation on and from 31-03 -17.

24.9.24   KK – 1 and KK – 2 electrical power system

We had in paras 24.8.11.17 and 24.8.11.18 dealt with two important electrical system activities .We had at that point of time been narrating events with a sense of chronological importance and hence continued with erection of reactor and associated systems. Electrical power system is one of the most important and crucial ones in any nuclear power plant and are particularly so Kundankulam project which envisages a total capacity of 6 GWE.We now proceed with the due detailed narration of this system.

24.9.24.1   *Design Concept*

The concept has followed the gradually evolving and improving features of electrical system in Indian nuclear power plants. The system has adopted the now familiar four classes of power to ensure optimum construction, commissioning and operational requirements.

Kudankulam when its planned 6 Nos VVER (GWE) units are fully commissioned will be India's highest total generating capacity station. This being the case improvements in electricity system which would result in an order of safety of 2 orders of magnitude higher than in existing systems have been brought into effect. The details of these improvements will be touched upon in our continuing narration in relevant context.

24.9.24.2   *Interchange of power between 220 KV and 400 KV systems – Safety enhancement*

There are 220 KV and 400 KV systems both in RHPP 7 & 8 and KAPP 3 and 4 projects.However there is no interconnection between them .There are 2 numbers interconnecting 400kv/220 kv auto – transformers in KAPP 1 & 2.This enables to and fro transfer of power between the two EHV systems and thus contributes to more stability and hence safety in case of any regional grid disturbances. This contributes to an order of magnitude higher relating to safety in Kundankulam project.

24.9.24.3   Class III Diesel Generator Sets(DGs)

6.3 MW 6Kv DGS have been provided for providing Class III power to the station.These are 1000 RPM 0.8 PI brushless excitation units with ungrounded neutral. They are designed for start up time of not more than 12 secs following Class IV power failure.10 Nos DG units are provided for the station (4 nos. for each KK-1 and KK-2) and 2 nos. common spare for both stations.All are 100% duty units. This improved provision in KK project has provided another order of magnitude higher in station safety measure.

KK – 1 and KK – 2 24 KV Turbo-Generator(TG)

TGs in Kundankulam are 1000 MW capacity generating at 24 KV voltage ,50 H2,0.9 PI,3 phase.It is provided with a breaker.

24.9.24.4   Generator Transformers

These are single phase 24/400 KW 417bMVA capacity .50 cycles OFAF with off – circuit tap changer and solidly grounded neutral.7 nos. units are provided for the KK 1 and 2 units, 3 nos. for each unit and 1 no. common spare for the 2 units.

24.9.24.5   24 KV circulated phase bus duet(IPD)

24 KV rated IPD with 30 KA current capacity have been run from the T.G terminals to the primary terminals of the three 24/400 KV 417 NVA single phase transformers.

24.9.24.6   - 400 KV and 220 KV bus duct

2 KA capacity 400 KV Gas insulated bus duct (GIBD) connects the generator transformers and the 400 KV gas insulated switching gear through 3,300 metre long tunnels.This bus duct is the longest 400 KV installation in India and one of the largest in the world.

24.9.24.7   315 MVA 400/220 KV interconnecting auto transformers(iats)-

Shunt reactors 2 nos. "iats" one no. each for KK 1 and 2 and 3 nos each for KK 1 and 2 of 27 MVA shunt reactors have been provided.The shunt reactors ensure required power factor (0.8)and other satisfactory operating conditions for the station,particularly stability of the station electricity system , as part of the Southern Regional Grid(SREB).

24.9.24.8 *Station Unit auxiliary transformers*

2 nos. each for KK – 1 and KK – 2 of 3 phase 24/6.3/6.3 KV. 63 MVA unit auxiliary transformers(UAT) with on load tap changer provision (- 15 % to + 10%) in steps of +- 1.25% of normal secondary voltage (6.3 KV), short circuit voltage/impedance value of 11.5 +- 0.5 %,oil natural air natural cooled (ONAN)/ONAF(Oil natural air forced) – Neutral resistance grounded (NGR) are provided.

24.9.24.9 *Reserve/Common station auxiliary transformers (RAT/CSAT)*

3 phase 220 KV/6.3/6.3 KV,63 MVA (RAT/CSAT) transformers ,2 nos each for KK – 1 and KK – 2 on-load tap changer provision of +7.5 % to – 15 % with steps of +-1.25 % short circuit voltage/ impedance value of 10.5 % vector group Yn/Yno/Yno, Oil natural air forced(ONAF) with Neutral resistance grounded (NGR).

24.9.24.10 *6 KV switchgear (SWG) panels*

4 nos SF6 filled 6KV SW Gr rated 3/50/2000/630 amps 40 KA S.C capacity (I second) Sw Gr panels,2 nos reliable and 2 nos common/ station panels for each of KK – 1 and KK – 2 units.

24.9.24.11 *6KV buses – bus panels- 6 KV Sw Gr cabinets – circuit breakers*

Each of KK – 1 and KK – 2 is provided with 2 nos reliable buses and 3/50/2000/630 amps circuit breakers SF6 filled and 40 KA S.C capacity (I second) and 2 nos common station buses ."SIPROTEC" (brandname) circuit breaker relays are provided for normal systems and station relays for emergency power supply systems. 2 nos buses are for class IV and 2 nos for Class III loads.500 nos. – 6.6 KV cabinets have been provided for each unit.

24.9.24.12 *400 volts/220 volts switchgear panels,Motor Control Centres(MCS)*

Current ratings of above panels and MCS vary from less than 630 amps,630 amps and 1600 amps for meeting needs of various motors /other electrical equipment including lighting and 3 pin/2 pin 220V receptacles are available.101 nos panels of above have been provided for each unit.

24.9.24.13   Rectifiers/Inverters

132 KVA 380 V Input voltage and 220 D.C output voltage with a charger current rating of 100 Amps to 630 Amps Inverter rating 20,40 and 60 KVA.220 V D.C Input voltage and A.C 220 V/380 V output voltage .Normal Power supply is from 2 rectifiers for each of KK – 1 and KK – 2.Reserve power supply system is from 4 rectifiers and 8 inverters for each of KK – 1 and KK – 2 .Emergency Power supply is from 12 rectifiers and 24 inverters for each of KK – 1 and KK – 2.Stand by power units – 2 nos rectifiers and 2 nos inverters for each of KK – 1 and KK – 2.

24..9.24.14 110V/220 V D.C batteries

24.9.24.14.1   For Reserve Power Supply System (RPSS) and Emergency Power Supply system(EPSS)

24.9.24.14.2   RPSS: 220 VDC 1500 AH,2 hours period.4 nos battery banks for each unit of KK – 1 and KK – 2.

24.9.24.14.3   EPSS: 220 VDC 1500 AH,2 hours period.12 nos battery banks for each unit of KK – 1 and KK – 2.

24.9.24.14.4   NOPS – Normal Operations System

2 battery banks for each unit of KK – 1 and KK – 2.

24.9.24.14.5   Stand by Battery : 2 nos. banks common for each unit of KK – 1 and KK – 2.

24.9.24.14.6   Total battery banks : 38 nos for both KK – 1 and KK – 2.

24.9.24.14.2   110 VD Control Power Supply(CPS)

Normal Backup 110 VDC, 1500 AH 2 hours period.

24.9.24.14.3   Stand by battery

1 no. 220 VDC 1500 All 2 hours period, common for both RPSS/EPSS/CPS.

24.9.24.14.4   KK-1 and KK-2 Cable pans

Galvanised metal structures/painted metal structures both open and closed types, cable trays/with supporting consoles/racks ,cable boxes, routing the power and control cables over trestle in high areas.

KK- 1 – 1600 metric tons

KK-2 - 1600 metric tons

KK-1 and KK- 2 – 1500 metric tons

Total   4,700 metric tons

24.9.24.14.5   Power Cabling

We have dealt extensively in Chapter 6 covering designed erection of RAPP 1 and 2 the type, grade and sizing of both power and control cabling. A similar approach to provision of cabling has been made in KK-1 and KK- 2 with the observation that KK-1 and KK-2 are vastly upgraded unit capacity projects and cabling has been adopted to provide for suitably larger sized power cabling to meet the requirement of larger motors and other electrical equipment and services. No large change in control cabling was called for.

The sizing of 6 KV, 400 Volts and 220 V power cables has varied from 0.5 sq. mm to 500 sq. mm and covers 110 types .Details are as below:

HV covers 1.1 KV upto 22 KV – HV Power – 300 kms and LV Power – 1500 kms

LV covers 220 volts to 1.1 KV – Lighting/communication cables (LDC) – 1500 kms

Total power cabling load for KK-1 and KK- 2 including common requirement is 3,300 kms.

24.9.24.14.5.2   Control cabling (mostly upto 110 volts)

Total quantity including instrumentation and communication cables for both KK - 1 and KK- 2 is 13,500 kms.This includes common requirement for KK-1 and KK- 2.

24.9.24.14.6   Sighting KK-1 and KK- 2

Extensive and precise requirements for a typical nuclear power station have been described in chapter no. 6.Same practice has been adopted in KK-1 and KK- 2.

A separate 400 KVA 3 phase 50 V 6KV/415 Volts lighting transformer has been provided to meet the lighting loas of KK-1 and KK- 2.

50,000 lighting fixtures manufactured by "Philips India" have been provided in KK-1 and KK- 2.

24.9.24.14.7   Grounding

We have, as in other specific requirements for a generic nuclear power station extensively dealt with grounding.This is one of the most critical requirements and hence was dealt with in great detail in Chapter No. 6.The same practice has been adopted in KK-1 and KK- 2.

We mention here the specific requirement adopted for outdoor grounding grid network covering the entire of KKNPP-1 and KKNPP- 2.70,000 metres of 70 sq. mm tinned copper conductors have been laid for this purpose.

24.9.24.14.8   Auxiliary Transformers – Motor Feed Details –

24.9.24.14.8.1   Auxiliary Transformers

Dry type 6KV/400 Volts 1000 KVA/400 KVA transformers, DyM11,8% and 5 % short circuit impedances respectively for 1000 KVA and 400 KVA units with F type insulation ,neutral solidly grounded 101 in numbers are provided for meeting class IV and class III loads. Separate buses for these two classes have been provided.

24.9.24.14.9   Feeds for s and other loads rated voltages

24.9.24.14.9.1   Rated voltages for the above are 6KV,380 V
and 220 V Ac rated 50 H2,voltage variation +/- 10% frequency variation +/-5% combined voltage and frequency variation +/ - 10%. For 220 DC motors voltage variation + 10% to – 15%.

24.9.24.14.9.2   Voltage Levels/Motor Capacity

Motors above 200 KW unit capacity are fed from 6 KV SWGr. Motors between 11 KW and 200 KW unit capacities are fed from 380 V SWGr Motors below 11 KW unit capacity are fed from 380 V mecs.

Heaters of the pressure compensation systems are fed from 380 volts SwGr/mecs.

Advanced Heavy-Water Reactors (AHWRs)

25.0   Introduction

This chapter sets out to describe the advanced heavy-water reactors (AWRS) that are now (2003 to 2004) being designed in BARC. This reactor utilises Thorium, of which India possesses the largest quantity in the world; the reserves are estimated to be about 3,00,000 metric tonnes.

25.1   The design concept

The design concept has been set to achieve the following objectives:
1)   The utilization of Thorium as a nuclear fuel for commercial power-generation.
2)   The demonstration of Thorium nuclear fuel as a viable option.
3)   The sustenance and improvement of the expertise and technology already developed for the HWPTRs; the utilization of this expertise towards the exploitation of thorium fuel.
4)   The incorporation of advanced nuclear safety and the improvement of the economy of plant operation that is currently being set as international standards for the next generation of Nuclear Power Plants.

25.2   Similarity and differences between HWPTRs and AHWRs

25.2.1   General

As discussed earlier, the concept of AHWRs has been devised from the experience regarding the successful running of the existing HWPTRs. It is hence useful to observe the similarities and differences between the above two types of reactors.

25.2.2 Similarities in Design

1) Both the reactors are heavy-water moderated. This is a partial similarity; AHWRs also utilise Amorphous Carbon as a moderator. AHWRs are, however, mainly heavy-water moderated.
2) Both have pressurized coolant tubes for the conveyance of heat from the fuel to the coolant.
3) Both have on-load fuelling facility.
4) Both have adopted a low-pressure moderator system.
5) Primary and Secondary shut-down systems are similar for both systems.

25.2.3 Dissimilarities in the design

1) The Calandria of the AHWR is vertically positioned as against that of the HWPTR which is horizontally positioned.
2) There are no primary heat transport pumps in the AHWR; the flow of coolant is by gravity.
3) The coolant in the AHWRs is light water while that in HWPTRs is heavy water.
4) The fuel used in AHWRs is a mix of nuclear fuel while that in HWPTRs is UO2.

25.2.4 Comparative Data

S. No.	Description	AWHR	HWPTR	Remarks
1	Reactor Power (Thermal)	750 MWe	694 MWe	
2	Core Configuration	Vertical Pressure	Horizontal Pressure	
3	Fuel Cluster	Tube type Total 52 pins Th-PuO2-20 pins Th-U233-32 pins	Tube type 12 bundles of 19 elements of UO2 in each bundle	
4	Active Fuel length	3500 mm	3150 mm	
5	Linear heat rating of fuel	350 watts/cm	503 watts/cm	
6	Type of moderator	D2O and Amorphous Carbon material	D2O	

S. No.	Description	AWHR	HWPTR	Remarks
7	Reflector material	D2O	D2O	
8	Coolant	Boiling light water under natural circulation	D2O	
9	Total core mass flow rate (of the coolant)	2,576 kgs/sec	3,856 kgs/sec	
10	Core inlet temperature	271 degrees C	249 degrees C	
11	Feed-water temperature	165 degrees C	171 degrees C	
12	Average steam quality	0.86 dry	0.9975 dry	
13	Quantity of steam produced	1,303 tonnes per year	1,162 tonnes per year	
14	Steam pressure and temperature	70 bars, 285 degrees C	41 bars, 251 degrees C	
15	PAT loop height to maintain natural circulation	39 metres	Not applicable (the PHT system is pressurized)	
16	Calandria diameter	8600 mm	6050 mm	
17	Calandria height	5000 mm	5080 mm	
18	Number of coolant channels	432	306	
19	Lattice pitch	294 mm square pitch	229 mm square pitch	
20	Pressure tube ID	120 mm	82.6 mm	
21	Primary shut-down system	36 shut-off rods having B4C	14 shut-off rods having B4C	
22	Secondary shut-down system	32 poison tubes injected with Lithium Penta borate solution	12 poison tubes injected with Lithium Penta borate solution	
23	Reactor control system (i) Number of adjustor rods (ii) Number of regulating rods	2 4	4 4	

The attached table brings out the comparative data of parameters concerning the AHWR and an equivalent power-level standardised

HWPTR (220 MWe) that are already commissioned and being constructed in India.

25.2.5 Significant factors regarding the choice of AHWRs

25.2.5.1 General

We have seen in para 25.2, that the concept of AHWRs has been derived from our experience regarding the successful operation of the PHWTRS which has the bedrock of all the commercial power reactors so far commissioned in India. In the upcoming sections, we will see some of the factors highlighted by this experience which in turn have led to corresponding design changes in these areas for the AHWRs.

25.2.5.2 The coolant system

HWPTRs utilise a pressurized heavy-water system for primary heat transport. The primary pressure boundary encompasses the coolant channels, heavy-water feeder pipes, heavy-water heaters, PHT piping, pressurizing pumps and inter-connected piping system and the PHT pumps with their inter-connected system. The boundary is extensive with numerous pump glands, piping connections and associated gaskets. The D2O leakage imposes an economic penalty so an extensive heavy-water recovery system consisting of drier fans, heaters, etc. is required. The management of the heavy-water inventory becomes quite a critical item in the reactor plant operation. It is with a view of simplifying the primary heat transport system and avoiding extensive heavy-water inventory management that recourse has been taken to a light-water gravity-flow heat-transport system in the AHWRS.

25.2.5.3 Emergency core cooling system (ECCS): the need for a longer form of ECCS

The experience regarding the operation of the ECCS in PHWTRs has pointed out the need for a longer-term ECCS. Both the proposed AHWR and the existing PHWTRs utilise a gravity-feed coolant system to remove the residual heat from the fission products in case of loss of pressurized or non-pressurized coolant within the coolant boundary. The volume of this feed is 14met^3 under hot conditions in the case of PHWTRs.

25.2.5.4 *The diversion of moderator D2O to the PHT system in PHWTRs*

PHWTRs also have the provision of a moderator system—D2O being pumped into the PHT system on sensing a drop in the PHT pressure.

25.2.5.5 *Longer-term ECCS in AHWRs*

The ECCS in AHWRs is designed to remove the core heat by passive means in case of a postulated loss of coolant accident (LOCA). The core cooling is achieved initially by a flow of borated light water from the advanced accumulators; the latter cooling is achieved through a gravity-driven borated water pool (GDWP). The inventory of the GDWP is adequate to cool the reactor core for three days without operator intervention.

The ECCS consists of four accumulators of a total capacity of 260 Met^3 and a GDWP of 6000 met^3, both connected to the ECC system heater.

25.2.5.6 *The reactor plant and associated equipment*

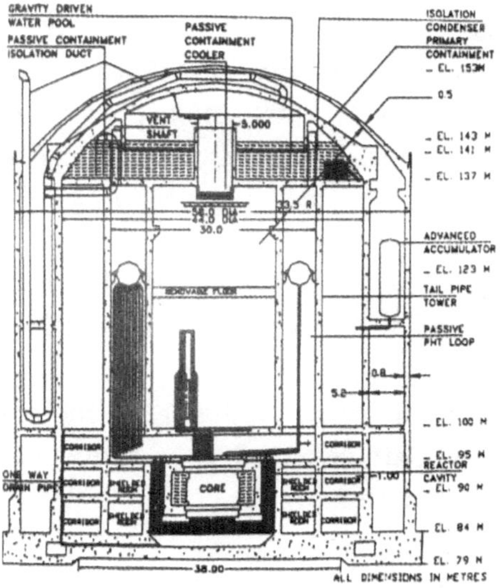

General arrangement of AHWR building

25.2.5.6.1   The reactor building and containment structures

The reactor building is a circular one that is 58 metres in diameter outside the containment wall which has both inner and outer containment walls. The walls are capped by inner and outer containment domes (ICD and OCD). ICD is 1 metre thick and OCD is 0.8 metres thick. The mud-mat foundation of the RB is 5 metres in thickness. The overall height of the building from the bottom of the mud-mat is 74 metres.

The attached sketch shows the general arrangement of the AHWR building.

25.2.5.7   *The containment domes*

The height of the OCD from the floor of the dome is 16 metres. This floor accommodates the borated gravity-driven water pool with a depth of 6 metres and a volume of 6000 met^3.

25.2.5.8   *The reactor vault containment bearing plates for the end shields*

The reactor vault containment is 2.2 metres thick and made up of heavy concrete. The vault is filled with light water which provides thermal and biological shielding against neutrons and gamma rays.

25.2.5.9   *The reactor vault containment: revolving and removable floors*

A deck plate is provided at an elevation of 100 metres for accessibility to the top of the reactor to facilitate on-power fuelling and other maintenance operations. The deck plate consists of inner and outer revolving floors. The deck plate serves as a platform for the removal of fuel assemblies and supports the shielding shirt of the fuelling machine. These floors are supported on special bearings to enable alignment to any lattice position by suitable rotation of the revolving floors.

A removable floor is provided at 120+ metre elevation. This floor will facilitate maintenance and outage work on the reactor unit and reactor auxiliaries.

25.2.5.10 *Bearing plates on the reactor heavy concrete vault*

Bearing plates are provided on the reactor's heavy concrete vault for the support of the top and bottom end-shields.

25.2.5.11 *Reactor pile and associated nuclear equipment and systems*

25.2.5.11.1 Reactor Pile

The reactor pile block structure houses the vertical Calandria top and bottom end-shields, pipe vault and feeder pipe vault.

25.2.5.11.2 Calandria

The Calandria is a vertical vessel with a main cylindrical shell connected by a flexible annular plate to a sub-shell at each end. Vertical Calandria tubes are arranged in square lattice pitches and rolled to the lattice tubes of the end-shields at the top and bottom of the Calandria.

Nozzles and penetrations are provided in the shell and sub-shell surfaces for circulation and level control of the heavy-water moderator and Helium cover gas. Vertical penetrations are also provided for housing the following:

(a) Primary and secondary shut-down systems

(b) Reactivity mechanisms

25.2.5.11.3 The moderator system

The moderator system is a twin system consisting of Amorphous Carbon and heavy water in a 4:1 volume ratio, respectively. The moderator system is flexible, and several moderator features including the dispensing of an Amorphous Carbon-based scatterer are feasible without any essential changes in the engineering design.

25.2.5.11.4 End-shields

The end-shields are provided at the top and bottom of the Calandria. The end-shields are welded to the Calandria sub-shells in the site.

25.2.5.12 *Heat transport system*

The system is a passive one with gravity-driven light water from the water pool removing the fission heat from the fuel. Fuel heat is removed by natural circulation caused by thermal siphoning. Primary circulation pumps are thus eliminated.

Heat-removal paths and primary-flow systems have been indicated.

25.2.5.13 Safety systems

Other passive systems facilitate reactor operation and shut-down, and contribute to safety functions such as residual heat removal, emergency core cooling and the confinement of radioactivity.

25.2.5.14 Nuclear steam service system

Four steam drums are located in the centre of the drum proper, 39 metres above the bottom of the coolant channels. The steam water mixture from each coolant channel is led through 125-mm nominal diameter (nd) tailpipes to the four steam drums. The steam in the drums is generated at a pressure of 70 kg/cm^2 and is led to the steam turbines through two 400 mm nd pipes.

Conventional Systems are provided at the points of entry to the coolant channels to ensure thermal hydraulic stability and also to ensure uniformity of steam quality in all the channels.

25.2.5.15 Fuel assembly

There are 424 fuel clusters, one in each of the 424 coolant channels. The 384 channels are loaded with clusters, each having twenty Th-PuO2 pins and thirty-two Th-U233O2 pins. The remaining 76 channels are loaded within fuel clusters, each of which houses 52 pins of Th- U233O2.

The fuel cluster has a length of 420 cm and is suspended inside the coolant channel. The design of the cluster enables the separation of the shield plug from the spent-fuel, joining the same to a new fuel cluster inside the fuelling machine. This design simplifies the re-fuelling operation.

25.2.5.16 The coolant channels

There are 432 channels, in all, spaced at a square pitch of 294 mm. The coolant channels are designed with a view of shop assembly and ease of installation and replacement. The pressure tube along with the bottom-end fitting can be removed by utilising the rolled joint detachment technology without disturbing the top-end fitting, tailpipe, feeder connections or any other parts of the channel. This considerably simplifies the replacement and maintenance of the coolant channel. A sketch showing the coolant assembly is placed below.

25.2.5.17 *Reactor protection and regulatory system*

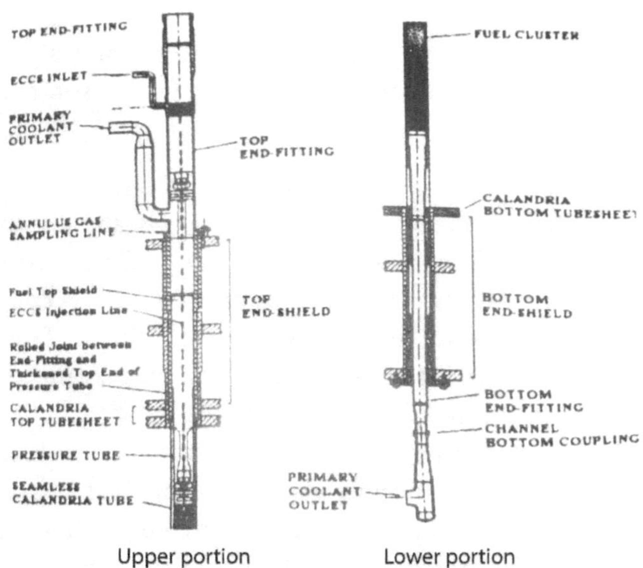

Coolant channel assembly of AHWR, fuel and portions of end-shields

25.2.5.17.1 Reactor protection

The reactor is provided with two independent fast-acting shut-down systems. The primary shut-down system (PSS) consists of absorber rods having boron carbide as the neutron-absorbing material. Boron carbide is filled in an annulus formed by stainless shells of suitable thickness.

The secondary shut-down system (SSS) consists of a liquid poison injection system in which lithium penta borate solution with suitable Boron content is injected into the poison tubes.

25.2.5.17.2 Regulatory system

Four adjuster rods and four regulator rods are provided for carrying out the regulatory functions.

25.2.5.17.3 Reactivity control system

The reactivity control of the reactor is achieved by the following methods:

1. Re-fuelling takes care of the loss of positive reactivity arising out of fuel depletion.
2. Moderator-level control takes care of power changes.
3. Adjuster rods are utilised for improving the poison over-ride operations.

25.2.5.18 *Emergency core cooling system (ECCS)*

The longer-term emergency cooling system was already touched upon in para 25.2.5.3. Further details of the ECCS are furnished herein. The ECCS heater referred to in para 25.2.5.5 is connected to the individual coolant channels above the tailpipe. The ECCS coolant enters the core through 8 perforated tubes arranged in the fuel cluster to ensure the welding of fuel pins by spray action. The ECCS coolant, after emerging from any ruptured coolant pipe, accumulates in the reactor cavity along with the PHT coolant and is re-circulated through heat-exchangers to ensure longer-term cooling. The design ensures that all the ECCS water from the accumulators and gravity-driven water pool (GDWP) reaches the fuel core irrespective of the location of the break in the coolant tube.

25.2.5.19 *Isolation Condensers (ICs)*

Eight isolation condensers have been provided in the reactor's primary heat transport system. Each set of four ICs is connected to its inlet and outlet heaters. The function of the ICs is to remove the fuel-core decay heat during the normal routine shut-down using steam condensation in the natural circulation mode. The ICs are immersed in a gravity-driven water pool (GDWP).

25.2.5.20 *Isolation of the reactor building ventilation scheme under LOCA: removal of containment heat*

The isolation of the reactor building ventilation system under LOCA is achieved by a water seal which gets activated due to pressure rise in the containment building following LOCA.

The heat in the containment system is removed by passive containment coolers (PCCS) with the GDWP serving as a heat sink. Please refer the sketch given below:

25.2.5.21 *Conventional systems*

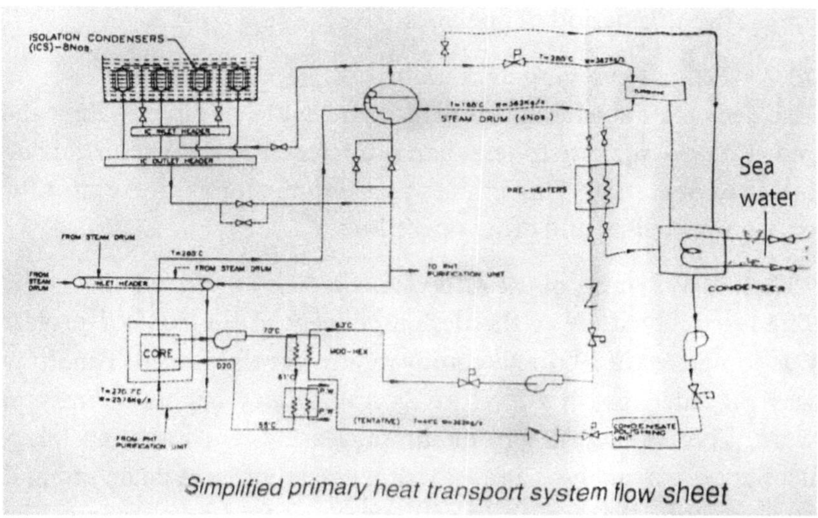

Simplified primary heat transport system flow sheet

25.2.5.21.1 The condensing system

The condenser has been designed for sea-water cooling; it is capable of condensing the steam exhausted from the turbine in operation or by pass mode. The full flow of steam condensate is purified in the condensate polishing unit and is pumped back to the steam drums through pre-heaters and feed-water pumps.

25.2.5.21.2 The steam quality re-heater system

The condensate from the condenser is purified, re-heated and pumped back to the steam drums at a temperature of 150 degrees C. About 99.5% of dry steam is generated at a temperature of 285 degrees C in the steam drums.

The feed water mixes with the water separated from the steam-water mixture. The water level in the drum is related to the reactor power and is maintained at a pre-set level during reactor operation.

25.2.5.22 *Maintenance Facility*

25.2.5.22.1   The accessibility of reactor areas after the reactor shut-down limitation of radioactivity

The end-shields have been designed to limit the dose rate to less than 5 microseimers per hour within one hour of reactor shut-down. This limitation will enable the access of personnel to these areas after this period of one hour.

25.2.5.22.2   Revolving deck plate

The deck plate shielding is provided above the tailpipes to limit the radiation dosage rate to less than 6 microseimers per hour during full-load operation. This enables access to the top of the reactor for on-power fuelling and other operations.

25.2.5.22.3   Status of the AHWR project

The feasibility study of the design concept of the AHWR project was completed in 1997 and a study report was also issued. The study was preceded by first-level analytical studies and experimental work. The detailed design of the nuclear system was then taken up. Supportive analysis and experimental work were taken in hand after the completion of the detailed design of nuclear systems. The drawing up of detailed specifications for the non-nuclear systems was also taken up.

The Nuclear Training Centre (NTC)

26.0   Introduction

The need for a training programme to train selected staff, depending on the critical and specialised nature of work carried out by these staff in any organisation carrying out large-scale production work, does not call for any emphasis.

This need becomes even more crucial in any plant involving extensive and complicated process systems, high levels of automation, radioactivity control and Health Physics surveillance such as Nuclear Power Plants.

There are 15 nuclear power-generating units operating in India as of 2005. These units employ about 15,000 to 18,000 personnel at all levels. The operating staff have to carry out sophisticated pre-laid operating procedures for operating and maintaining the generation of power at the required levels and shutting down these units, as part of their day-to-day activities as per the requirements that may arise from time to time. These sophisticated activities require the station personnel to be pre-trained adequately and possess appropriate certifications and licenses before being deployed on station activities.

26.1   The Genesis of Nuclear Power Plant Training

The rudiments of a Nuclear Power Plant could be said to have commenced in 1968-1969, when familiarisation courses were conducted for Engineers engaged in the construction of the Rajasthan Power Project, at Rawatbhata near Kota. The course was conducted over several weeks. There were no examinations conducted or any certification given. The familiarisation course provided a very good opportunity for the Engineers to get acquainted with the principles governing the generation of nuclear power.

26.2 Setting up of the NTC

A nuclear training centre (NTC) was set up at the RAPP site in 1968. This was the first centre where engineers, supervisors and technicians, who would man the operation and maintenance of the HWPTRs along with the related duties, were trained.

26.3 Goals of the NTC

The goals of the NTC are as follows:

1. Training of personnel to ensure the hassle-free operation and maintenance of PHWTRs.
2. Sustaining and improving plant performance through induction training including on-the-job training, refresher training and exposure to good international practices.
3. Lowering of heavy-water loss and man-rem consumption through effective training on heavy-water management and radiation protection measures.
4. Enhancing the public perception of nuclear power programmes, and promoting awareness that nuclear power is a safe, reliable and non-polluting source of energy—by organising public-awareness programmes and active participation in workshops, seminars, etc.

26.4 Methods adopted by the NTC to achieve its goals

The NTC adopted certain measures to achieve the goals mentioned in the previous section. They are listed below:

1. Ensuring the availability of competent, qualified and licenced manpower through effective management of qualification and licencing schemes.
2. Enhancing the quality of training by improving and strengthening infrastructure and training documentation.
3. The use of effective training, re-training and simulator training to achieve the elimination of human errors, and enhance the

capability to handle abnormal and emergency operations successfully.

4. Fine-tuning the training programmes based on the evaluation and feedback from various training programmes.
5. Involving line-management while conducting the training.
6. Developing trainees with a positive attitude towards the imparting of training.

26.5 Role of training in the preparation of the personnel

The integration of plant procedures and programmes with plant human resources in the preparation of the personnel is important and hence necessary. Thus, the role of the above-mentioned training involves the following:

1. The successful integration of system programmes and procedures with the human resources that would be required to run these programmes and procedures.
2. The qualification of personnel should match the established and proven performance standards. The qualification comprises the selection, training and evaluation with the view of meeting these performance standards.
3. The end goal of training is to prepare the personnel to achieve competence in meeting the expected standards.
4. The training system seeks to make available: qualified personnel on an on-going basis, through standardised programmes for the deployment duties of Nuclear Power Plants such as operational, maintenance, technical, administrative accounts, materials management and training duties, among others.

26.6 Competent authorities for the licensing and training regimen rates of various authorities

So far, we have covered the aim objectives and methods of the NTC. Let us now look into the regimen for training and licensing of the personnel for manning the nuclear power stations.

26.6.1 Essential players

There are three essential players involved in the training and licensing regimen. These are listed below:

(1) NTC
(2) NPCIL
(3) AERB

Let us now go into the specific roles of the above players.

26.6.1.1 *The NTC*

The NTC is a grass-roots organisation which is located at the RAPPs and MAPPs sites. The NTC is fully empowered to qualify personnel for the operation, maintenance, technical and training positions in the nuclear power stations.

26.6.1.2 *The NPCIL*

The licensing of employees for operational duties at all nuclear power stations is controlled by the NPCIL. The qualifying examinations are also centrally administered by the NPCIL.

26.6.1.3 *The AERB*

The AERB, as the regulatory body for all nuclear activities, endorses the licensing administered by the NPCIL. The AERB has a special commission of experts who conduct the final certification of the Operating Engineers and award the licence.

Twice a year, the AERB audits all the administrative procedures involved in the selection, induction, on-the-job training, simulator training and verification examinations conducted by the individual nuclear training centres.

26.7 The standard training programme: the different levels to gain a license

We will now go into the training programme carried out by the NTC in more detail.

26.7.1 The staffing of new operating stations: the induction programme

It is a general practice that for new operating stations, about 30% of the required personnel are those who have had working experience in the existing stations; the other 70% are recruited afresh.

The pre-employment training for the recruits varies from one year for executive positions, one-and-a-half years for supervising positions and two years for tradesmen.

The foundation training comprises both classroom and workshop training. The classroom training seeks to impart a strong grounding in the fundamentals of the reactor system, auxiliary equipment and safety systems.

The personnel are then deployed and on-the-job training is conducted in the construction and commissioning phases of the project.

Simulator training, structured checklists of the tasks, written examinations, walk-through tests and interviews follow for qualification and licencing purposes.

There are nine streams in operation and maintenance that the fresh trainees are inducted in. Each stream is structured in such a way that about 60% of the inputs are inducted through classroom training while the remaining 40% is done through on-the-job orientations.

26.7.2 Qualification training

The pre-job training for the employees is followed by on-the-job training which is required for obtaining a job qualification. The trainee employees are assigned to work with a qualified employee. The senior employee plays the role of a coach for the training employees. The tasks learned by these employees are verified by the coach by way of job-related checklists. Once these checklists are completed satisfactorily, the employees take the written examination and undergo the walk-through tests. The final step is the interview for the evaluation of competency required for the job. In the case of serving employees, the above-mentioned process determines the suitability for promotion to the next higher level in the organisation.

26.7.3 Licensing criteria levels and the systems of licensing

Let us now look into the criteria and details of the systems of licensing for induction into operational positions. There are three levels of licensing and three systems to qualify in. The three levels are I, II and III while the systems to be qualified for licensing are as follows:

(1) Nuclear systems

(2) Secondary system

(3) Fuel handling

While (1) is common and compulsory for all licences, (2) and (3) are optional. Operating personnel who opt to be trained to work for the main plant are trained and tested for (1) and (2), while licenses for re-fuelling operations require (1) and (3).

The three-level licenses are valid for three years at a time. Level-III is an entry level. A level-III Maintenance Engineer is qualified in 50% of the plant systems whereas a level-II Maintenance Engineer is qualified in 100% of the critical plant systems.

A level-III Operations Engineer is a Control Engineer who can operate and manipulate plant controls. A level-II license holder works in a supervisory position and coordinates and controls the work of the level-III Engineers.

Level-I Engineers are shift charge Engineers and are responsible for all the plant functions within their jurisdiction. A typical progression plan for the licensed engineers is shown in the sketch attached below:

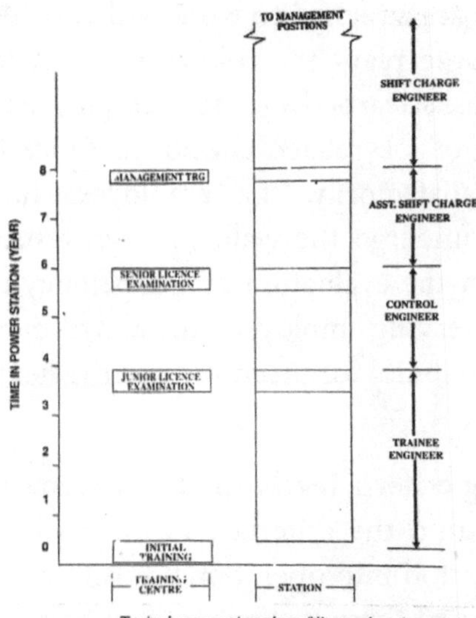

Typical progression plan of licensed engineers
(Similar concept applicable for other positions too)

26.7.4   Management and Functional Levels

There are seven functional levels in every nuclear power station organization. Of these, six levels are in each of the operation, maintenance, technical and training functions for which qualification and training are carried out by the NTC. The Pyramid Station organization is headed by the station Director/Chief Superintendent. The organization chart is shown in the attached sketch.

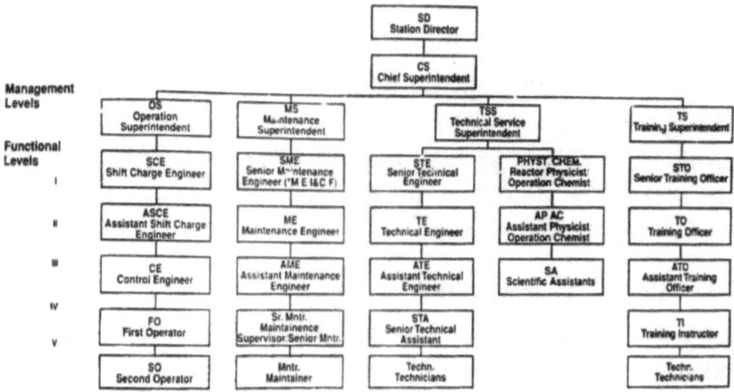

26.8   Nuclear Training Centres: Simulators

26.8.1   The NTC at TAPPs

TAPPs has a self-contained nuclear training centre catering to the requirement of licensing personnel for the operation and maintenance of the boiling-water reactors.

26.8.2   General

The NTC at RAPPs is the oldest and biggest nuclear training centre of the NPCIL.

The NTC at RAPPs is equipped for imparting training in all disciplines (nine disciplines) of the PHWTRs as well as the PWRs and the BWRs. It has well-developed workshops in mechanical, control, instrumentation and electrical disciplines. It has the latest generation simulator as well. It further has shops for Health Physics and industrial safety. These facilities are backed up by well-qualified faculty with a collective experience of nearly one thousand years of reactor operation.

26.8.2.1 *Details of the Personnel trained at the RAPP training centre*

The NTC at RAPPs had trained a total number of 1125 personnel till 1984. It has trained, on average, about 120 new personnel every year in its early years. The rate of training has been enhanced depending on the requirement. For example, the centre has trained about 1200 fresh recruits for various stations during the period 2000-2003. The centre has imparted training to the extent of 2 lakh man-days to 19,518 personnel, during the above-mentioned period, with the following break up:

(1) 2000 to 2001: 58,705 man-days to 4,843 personnel

(2) 2001 to 2002 : 48,673 man-days to 6,175 personnel

(3)   2002 to 2003 : 85,216 man-days to 8,500 personnel

26.8.3 The NTC at MAPPs

This centre is fully staffed in general for purposes of centralized core training for the PHWTRS.

26.8.4 The NTC at the Kaiga Nuclear Power Station

A full-scope simulator has been set up at the Kaiga 1 station. This simulator will help Licensed Engineers to 'experience' events such as station black-out, loss of coolant and loss of reactor regulation. The simulator would help the trainee personnel diagnose such situations and take corrective measures. This simulator experience would be of immense help to them in responding to actual events in the operating reactors.

26.9   Station Training Centres (STCs)

Station training centres have been set up at all nuclear power stations to impart station-specific on-job training and re-training to serving personnel. There are four major station functions namely operation, maintenance, technical and training; six qualification levels are built into these functions. The six qualification levels are management levels I to VI. As we have seen earlier, licensing for employees in operational functions is administered by the NPCIL.

Job-specific licensing is followed for such functions. Employees in maintenance, technical and training functions are qualified by the stations locally.

26.10 NTC at RAPPs: awards and foreign mandate

26.10.1 Awards
The following awards were secured by the NTC at RAPPs
1. Golden Peacock National Training Award : 2001
2. Golden Peacock National Training Award : 2002
3. Golden Peacock National Training Award : 2003
4.   Best house-keeping by Station Director of RAPPs : 2001
5.   Best exhibit award for RAPPs : 2002 and 2003

26.10.2 Foreign mandate
The NTC at RAPPs was invited by the Electric Co., Japan, and WANO Tokyo Centre to help them evolve their system of training and qualification.

26.11 Enhanced goals of the training programme
It was the expectation of the NPCIL corporate management that a sound training programme should be able to attain the following goals:
1. Limit the reactor outage to 1 per month
2. Minimise inappropriate actions that cause plant disturbances

It is a sobering thought that the International Safety Advisory Group (INSAG) of the IAEA, in its 1988 report, has concluded that abnormal events ranging from minor events to serious accidents have been caused by incorrect operator actions, among others.

One remedy to this problem is through automation which does away with function-specific operator action. The other is through a training programme that draws upon human ingenuity while dealing with unusual or abnormal operating circumstances.

Self-checking techniques that help in identifying potential human errors go a long way in minimising inappropriate operator actions.

26.12 Overview of the training programme

We have seen how the NTC at RAPPs, from its small beginning in 1968, has grown into the largest NTC in the NPCIL. We have seen its pivotal role in training personnel for all the nuclear stations. Its state-of-the-art simulators are beginning to play a big role in the training of personnel.

The other NTCs are also evolving and would, in due course, play a bigger role in the training of personnel for the nuclear stations which are now under construction or planned for the future.

RAPP Units 5 and 6

27.1 RAPP Units 5 and 6: Estimated Cost

End - shields before erection - RAPP-5

27.1.1 Factors for setting up the project parameters

It was decided by the NPCIL and the Government of India to locate RAPP units 5 and 6 near the site where RAPP units 3 and 4 have been set up and commissioned. There are several advantages to locating multiple units at the same site. Common services—such as administration, accounts, security, housing and other infrastructure facilities—can be shared with resultant financial advantages. Specialised and skilled manpower, already available at the site due to tailing off of work on earlier units, may be deployed on the latter units. On-site power, available at the site, may be drawn for construction and commissioning. All design parameters are identical to that of RAPP units 3 and 4.

27.1.2 Estimated cost: Financial sanction

Financial sanction for the two units, at an estimated cost of Rs. 30.72 crores, was accorded by the Government of India's Cabinet Committee on Economic Affairs (CCEA) during the last week of March 2002. The Debt/Equity ratio for the project was planned at 2:1. Dates for the commercial operation of RAPP units 5 and 6 were set up in August 2007 and February 2008, respectively.

27.1.3 Award of Civil Work

HCC was awarded the Civil work for the project early in 2001. The excavation work started on 8.11.2001 in the presence of Sir. S.A. Bohra, Executive Director (Planning and Projects) of the NPCIL, and Sri. D.K. Goyal, Project Director of the RAPP site. The intervening period from early 2001 to early November 2011 was used by the contractors for the mobilisation of workforce and the establishment of site construction facilities.

27.1.4 The laying of the formal foundation

The foundation stone for the project was formally laid on 17.1.2002 by Sri. Ashok Gahlot, CM of Rajasthan. Sri. Anil Kakadkar, Chairman of the AEC and Sri. V.K. Chaturvedi, CMD of the NPCIL, were also present at the function.

27.2 Award of contract for the erection work

27.2.1 Mega supply/contract

The construction works of RAPP units 5 and 6 were grouped into 25 mega supply and erection packages and awarded to reputed contractors.

27.3 Award of contract for the electrical system work

The work was awarded to Bombay Suburban Electricity for Rs. 100 crores.

27.4 Civil work of RAPP units 5 and 6

27.4.1 Status as of December 2002

The excavation of 8,00,000 cu. metres in hard rock by using line blasting was completed in 9 months, about half the time taken for RAPP units 3 and 4. The foundation for the RBs was 18-metre deep and the diameter was 52 metres. The first concreting for the raft of RB-5 commenced on 17.9.2002. Four pours were completed by November 2002. The 6th and the last pour was completed by the end of December of the same year. The first pour of concrete for RB-6 commenced by the end of December 2002.

The excavation for ancillaries like natural and induced draft cooling towers, emergency make-up water pools, circulating cooling water pump house, etc. commenced shortly thereafter.

27.4.2 RAPP units 5 and 6: progress of Civil work as of March 2004

27.4.2.1 *Reactor buildings of RAPP units 5 and 6: control building*

The concreting of the reactor raft of RB-5 and RB-6 was completed in a record time of 100 days and 52 days, respectively. The quality of the concrete of the RB raft was of grade M-45.

About 1,00,000 cu. metres out of a total of 2,06,000 cu. metres of concreting was completed.

The concreting of the raft of the safety-related control building was completed in only 2 pours, the second pour size being 3664 cu. metres which is the longest so far.

The adoption of state-of-the-art engineering in the design and placement of bigger-sized pours helped in considerably cutting down the time of completion of concreting. The concreting of slabs up to 100-metre elevation was completed.

The casting of the base slab of the Calandria vault of RB-5 was completed, the longest single pour being 2240 cu. metres which is the longest so far in any containment structure of the nuclear projects in India.

The concreting of the Calandria vault walls of RB-5 was taken up. A target date of the end of December 2003 was set for releasing the vault for the erection of the Calandria and the end-shields.

Internal structures up to the elevation of 95 metres of RB-5 were completed, including the slope on all sides except the West side. The outer containment wall of 99.2 metres elevation was completed as well. The parameters are the same as for RAPP units 3 and 4 except for the quality of concrete which is as follows:

a)   OCD : MB-5
b)   ICD : M-45
c)   OCW : M-35
d)   ICW : M-45

Four lifts out of 5 of the Calandria vault concreting were completed in RB-6. All the internal structures up to 91 metres elevation were completed by July 2003.

The concreting of the Calandria vault was completed within 15 months of the first concreting, that is by the middle of December 2003. This period may be viewed against the period of 36 months taken for RB-3. Similarly, the concreting of the fuelling machine vault was completed within 18 months of the first concreting, that is by mid-March 2004. The period taken for RB-3 was 36 months.

The concreting of the slab up to 100 metres of the control building was completed.

A View of Reactor Building No.5 of RAPP 5&6

27.4.2.2 Connected buildings of RAPP 5 and RAPP 6: the 220-KV switchyard and the Civil work status as of 2004

The concreting of rafts of the basement of various connected buildings was completed on the turbine building and it reached 100 metres elevation. The excavation in the 220-KV switchyard commenced thereafter.

27.4.2.3 Administration and canteen buildings

The architectural design for these buildings was finalised.

27.5 Nuclear system work: status as of mid-March 2004

27.5.1 Nuclear piping

The contract was awarded to the contractors who began mobilization at the site.

27.5.2 Nuclear service steam generators (NSSG): end-shields

Both the end-shields and all four NSSGs for RAPP 5 were received at the site, and unloaded preservation of the NSSGs was completed.

27.6 Conventional equipment work:

As of mid-2004, the contracts for the conventional equipment work like the common services piping, circulating pump-house mechanical works, among others, were awarded to the contractors. Mobilization at the site commenced thereafter.

27.7 Progress of RAPP 5 as of mid-2004: reactor system work

Both the end-shields and all the SS steel balls required for the shield filling were shifted to the end-shield ball-filling steel on the RB-6 side of the project by making use of the 650-tonne Crawler Crane. The ball-filling was in progress.

27.8 RAPP 5 and 6: overall progress as of mid-2004

About 35% of the overall work was completed.

27.9 RAPP 5: status of work as of late 2004 and 2005-2006

27.9.1 Civil work

The Civil construction work had reached up to 127 metres elevation in RB-5. The 3rd slab was completed in the control building and about 1/3rd of the concreting work in the control building was also completed.

27.9.2 The Turbine Building

About 750 cu. metres of concreting for the TG block were poured in 24 hours on the dates 22.9.2004 and 23.9.2004. Fly-ash concreting was adopted for the first time in India for the TG block. The construction of the over-head crane girder columns was expedited to enable the installation of the TB hall crane.

27.9.3 Safety-related pump-house

About 50% of the concrete slab at 100 metres was completed.

27.9.4 Circulating cooling water pump house, plant water pump and natural drought cooling towers: Common buildings

The base slab and operating floor slab were concreted. The fifth lift of the pier wall in the plant water pump house was completed.

The construction of the induced-draft cooling tower was completed up to 100 metres. The excavation for the natural draft cooling tower was also completed. Plain cement concrete work for raising the wall was completed. The work for the basin of the tower was still in progress. Civil work for common buildings was completed.

27.9.5 The 220-KV and 400-KV switchyards

The Civil work for both the switchyards was taken up. The foundations for the switchyard equipment and towers were in progress. Plain concrete work for the retaining walls of the 400-KV switchyard commenced thereafter.

Calandria for Unit II under fabrication in an engineering concern in Bombay

27.9.6   Station roads

Construction was expedited to enable the quick movement of materials and equipment. Permanent access to the TB unloading bay was scheduled for January 2005 to enable the movement of TG and condenser equipment.

27.10   Reactor plant work: progress during 2004-2008

27.10.1   Reactor and reactor auxiliary equipment: RAPP 5

Ball-filling and cover welding of both end-shields were completed in a record time of 9 days as against 2 weeks taken for RAPP 3. The Southside end-shield was lowered on 24.6.2004. The matching holes of the end-shield were aligned to the 62 stud-anchor embedded parts in the first attempt itself. The end-shields were grouted in January 2005.

The Calandria was lowered into the vault on 20.10.2004. The 650-tonne crawler crane was used for the above-mentioned work.

The Calandria tube installation was completed in 17 days in December 2005.

The work on the insulation cabinet, order installation work, etc. in the FM vaults was in hand. The work on the moderator system equipment was taken up and the primary coolant tube installation was completed in March 2006. The erection of the feeder tube was completed in a record time of 77 days during mid-August 2006 and November 2006. The PHT system hydro test was completed in August 2007. The PHT system hot-conditioning was completed in October 2007. The RB-5 leak test was completed in January 2008.

In-site welding was carried out on the end-shield of the Calandria with a modified cooling arrangement; the spent-fuel inspection bay was bored up and no leakage was observed in the fill test. The spent-fuel service bay and the suppression pool were filled with de-mineralised water.

27.10.2   Reactor Dome work in RAPP 5: ICD and OCD

Dome concreting was completed within a record time of 25 days— the time taken from the first concrete pours on the raft to the completion of the ICD and OCD concreting was 35 months and 41 months, respectively.

27.10.3   Equipment installation in the control building: progress during 2007-2008

All plate-type heat exchangers and auxiliary boiler feed pumps were installed and the cable installation was completed. About 90% of the piping work was also completed.

27.10.4   Conventional equipment: progress during 2005-2008

27.10.4.1   *The 220-MWe TG set and the TG auxiliaries*

The movement of the TG set and related equipment onto the turbine floor commenced in January 2005. The work on the condenser and related equipment also commenced during the same month.

27.10.4.2 Electrical system work on the RB-5

The work on the main power output system and station auxiliaries' system was in good progress. The work on the Class-I, Class-II and Class-III power systems, the control power-supply system, lighting cable pans and the grounding system was in an advanced stage of completion. The start-up power system was ready for operation.

27.10.4.3 The Fire protection system

The fire protection system was in the final stage of completion.

27.10.4.4 The Communication system

The internal and external communication systems and the personnel addressing and paging system were completed and ready for operation.

27.10.4.5 Emergency power transfer system (EMTR), electrical protection and metering system, station main control panel and data acquisition and recording system (DARS)

All of the above system work were completed and kept ready for commissioning and operation.

27.10.4.6 Power and control cabling

Power and control cabling terminations and connection systems were prepared for commissioning and operation.

27.11 Fuel loading

The units were kept ready for fuel loading and the approach to first-criticality

27.12 Civil work as of 2005-2008

27.12.1 The Reactor building

The concreting of the slab at 112 metres elevation was completed for both the FM vaults. The painting was in progress.

The Calandria vault was ready for Zinc metalizing. The FM vault and suppression pod area were released for mechanical and electrical work.

27.12.2 Reactor system work

The end-shield work was grouted in 2006 while the Calandria tube installation was completed in September 2007. The coolant channel installation was completed in February 2008.

27.12.3 The overall physical progress for RAPP 6 as of December 2007

Nearly 97.5% of the work was completed.

27.13 Status of work as of 2008-2009

27.13.1 The Reactor system work

The Calandria tube installation was completed in a record period of 17 days. All the four NSSGs were installed.

27.13.2 The PHT system

The hydro test of the PHT main circuit was completed in October 2008. The PHT feeder tube installations were completed in 2008. The Hydro tests on the entire PHT system were completed in February 2009.

27.13.3 The Control and instrumentation system

Independent verification and validation of the computer-based control and instrumentation system was carried out by the WANO review team.

A simulator for testing the system was designed and built.

27.14 In-service inspection requirement

The requirements were optimised, codified and issued for implementation.

27.15 The RB-6 leak test

The test was completed in May 2009.

27.16 Peer review by WANO

RAPP 6 was offered for review to the WANO.

27.17 The first-criticality of RAPP 5

The unit attained criticality at 12:51 PM on 24.11.2009. With this criticality, the installed Nuclear Power Plant capacity in the country went up to 4.33 GWe (4330 MWe).

27.18 Synchronization with the Northern Regional Electricity Board (NREB): commercial operation

RAPP 5 was synchronised with the NREB at 14:43 hrs on 22.12.2009 within 28 days of attaining first-criticality after clearance from the AERB. The unit was declared for commercial operation with effect from 4.2.2010, which is within 72 days from the date of first-criticality.

27.19 Progress of RAPP 6 as of 2008-2010

27.19.1 The Conventional systems

All the work in the circulating water systems, TG and auxiliaries' system, main and auxiliary nuclear steam supply systems, fire protection, instrumentation and control system and internal and external communication systems were completed, tested and commissioned. The start-up power system was prepared for operation.

27.19.2 Nuclear plant system

27.19.2.1 Nuclear process systems

Nuclear steam and other nuclear process systems were prepared together with reactor control and regulatory systems.

27.19.2.2 Hot-conditioning of the PHT system

This was taken up at 12:40 hours on 8.8.2009 and continued up to 15:00 hours on 11.9.2009. The Chemistry parameters of the PHT system, moderator system, feed-water system, steam generators and end-shield water cooling system were maintained within the optimum limits. A maximum temperature of 259 degrees C was obtained with the PHT circulating pumps in continuous operation. The hot-conditioning was called off after a magnetite film thickness of 0.46 0.02 microns was obtained in 50 hours and 20 minutes.

The thickness was monitored on Carbon steel coupons installed in autoclaves in the PHT system. Light water with the following PHT Chemistry was used for the hot-conditioning:

a)   pH 10 to 10.5

b)   Dissolved oxygen < 10 ppb

c)   Temperature > 200 degrees C

The system Chemistry was maintained by the addition of lithium hydroxide < 10 h through lithium, best-mixed bed ion-exchange column and hydrazine in 50 to 60 mg/litre concentration.

27.20   RAPP 6: readiness for first-criticality

All systems of RAPP 6 were prepared and kept in a state of 'Go' for first-criticality.

27.21   First-criticality of RAPP 6

RAPP 6 attained first-criticality on 23.1.2010, as a result of which the total nuclear power-generating capacity in the country went up to 4.55 GWe (4550 MWe).

27.22   Synchronization with the NREB

The unit was synchronised with the NREB at 1:53 hours on 28.3.2010.

27.23   Commercial operation

RAPP 6 was declared for commercial operation on 31.3.2010. It would be seen that RAPP 5 and RAPP 6 were declared for commercial operation within 55 days of each other, which is a very significant achievement on the part of the NPCIL.

A Panoramic View of RAPS - 5&6

27.24   Record of continuous operation of RAPP 5

The unit achieved a record with 607 days of continuous operation as on 1.4.2014 and finally achieved 765 days on 28.8.2014. It generated 4.12 billion units during this period and 8.52 billion units since its commissioning in November 2009.

The above-mentioned 765 days of continuous operation is the second highest in the world only after the Pickering Unit 7 in Canada which operated for 894 days continuously.

This 765-day operation has since been overtaken by RAPP 3 and Kaiga 1.

Kaiga Units 3 and 4

28.1 Site details: the acquisition of land

In 10.1 of Chapter 10, we have seen that the Kaiga site was chosen for housing six 220-MWe 'HWPRT' units. Of the total 120 hectares required for the 6-unit station, 70 hectares were ear-marked for housing units 3 to 6. This chapter deals with the setting up of units 3 and 4. These units, like the already established nuclear units, will lead to more local jobs.

28.2 Preliminary work: budget provision for early spending

The preliminary work at the site began during 1989-1990. An amount of Rs. 57.68 crores was spent during 1990-1991 while the provision for 1991-1992 was Rs. 70 crores.

28.3 Financial sanction: details of funding

Financial sanction for an amount of Rs. 42.13 billion (Rs. 4,213 crores) for Kaiga units 3 and 4 was received in July 2001. This amount was scaled down to Rs. 32.82 billion (Rs. 3,282 crores) due to the planned reduction in the gestation period of 18 to 24 months and lower escalation as of March 2002.

Then the current dates of expected commercial operation of Kaiga units 3 and 4 were arrived at in March and September 2007, respectively. The Government equity for the project was Rs. 10.94 billion and debt funding was Rs. 21.88 billion, resulting in a debt/ equity ratio of 2:1.

28.4 The ground-breaking ceremony

The ground-breaking ceremony for the units was held on 19.3.2000.

28.5   Permanent plant work

28.5.1   Progress as of the end of 2000

Detailed project reports (DPR) and engineering, purchase and construction contracts (EPC) were drawn up.

Notice-inviting tenders (NIT) for the preparation of DPR and EPC were issued with technical bids set for 31.12.1999 and price bids by 31.1.2000. The work involved re-engineered drawings, conversion of drawings from Kaiga units 1 and 2 to Kaiga units 3 and 4 as specified in the NIT and the re-design and re-engineering in certain specific areas of nuclear and conventional systems. About 12,500 drawings of Kaiga units 1 and 2 are applicable for Kaiga units 3 and 4. The estimated cost of work covered in the NIT is Rs. 369 lakhs.

28.5.2   Civil work

A letter of intent was issued to M/S Gammon India Ltd . The excavation of 4,60,000 cu. metres out of a total estimated quantity of 6,60,000 cu. metres was completed. Rock level was reached and the line blasting of rock for the mud-mat was to commence by mid-2001.

28.5.3   Progress as of mid-2002

28.5.3.1   *Civil work*

The main plant construction package was awarded to Gammon India Ltd. in October 2001. The excavation of 100% of the total 6,60,000 cu. metres was completed by December 2001.

The first pouring of concrete in the raft of the reactor building of Kaiga 3 was carried out on 30.3.2002 in the presence of the CMD of the NPCIL and the Chairman of the AEC. The raft is of 52-metre diameter and 5.5-metre thickness at the periphery, and 3.5 metres at the centre. About 1450 metres of Tor steel, of diameter varying from 16 mm to 36 mm, were utilised in the raft with a volume of 10,450 cu. metres of concrete. The concreting was completed in a record time of 78 days. The quality of the raft of RB was of grade M-45.

28.5.3.2 *Electrical work*

28.5.3.2.1 The contract for the 400-KV switchyard

BHEL was awarded a turn-key contract amounting to Rs. 41 crores for the design, the supply of equipment and materials and the testing and commissioning for the evacuation of power to be generated by Kaiga units 3 and 4. The contract was inclusive of the Civil work and the term for the completion was 18 months.

28.5.3.2.2 Main plant electrical system work

The contract was awarded to the contract division of Bombay Suburban Electrical Systems at for Rs. 101 crores.

28.5.3.2.3 Nuclear system work

Four Nuclear steam service generators (NSSGs) and four PHT pumps for Kaiga 3 were received at the site. The NSSGs were supplied by BHEL.

28.5.3.3 *Work packages for the construction and commissioning of Kaiga units 3 and 4*

A total number of 78 major and minor work packages for the construction and commissioning of various systems of Kaiga units 3 and 4 were identified. An advanced stage in the awarding of long-delivering packages was reached.

28.5.4 Kaiga Units 3 and 4: progress as of mid-2003

28.5.4.1 *Civil work*

28.5.4.1.1 Housing

A contract for providing a built-up residential area of nearly 50,200 square metres, comprising 11 high-rise towers and 36 independent houses for Rs. 52.20 crores, was awarded to IVR CL Infrastructures and Projects Ltd. with a time of completion of 24 months.

28.5.4.1.2 Calandria vault concreting details: FM vault construction

The Calandria vault is one of the most crucial parts of the reactor building as it houses the reactor vessel and its components. The vault walls provide safe shielding against radiation.

The vault walls are constructed with high-density (3600 kgs per metre3) hematite concrete. This compares with the density of ordinary concrete of 2500 kgs/metres3. This high-density concrete provides biological shielding.

The inner surface of the vault walls is lined with steel plates of 8-mm thickness. Plates of 2.5 mets x 10 mets size were used to limit the weld length to 120 metres. The plate was confirmed to be of ASTM-A-516 grade-70. All the embedded parts were welded to the linear plates. Grooves were provided in the inserts to enable proper welding into the linear plates.

The 28-mm diameter Tor steel re-bars were used to reinforce the concrete of the vault walls. The ends of the re-bars were threaded wherever the joining of the re-bars was necessary, and a threaded sleeve was used to join the two threaded ends of the re-bars. This was an improvement over the earlier arrangement, where the ends of the joined rods had to be lapped (lap lengths were 1400 mm and 2800 mm) for adjacent joined rods which had to be staggered for the above lap lengths. The above minimum staggered lengths had to be kept projected above each pour height. With the use of couplers, the minimum projected length above each pour was only 300 mm. No staggering of adjacent length was also necessary. The construction of the Calandria vault and FM vault was taken up simultaneously. These were sequential activities in Kaiga units 1 and 2.

28.5.4.1.3 End-shield anchor bolt grouting details

Around 200 anchor bolts for the end-shield were to be embedded in the Calandria vault around the octagonal opening of 7 metres. The bolts were to be installed within a close tolerance of + or -1.5 mm. This called for a precision work of high order. The following method was adopted to ensure the close tolerances that were mentioned above.

A support structure was fabricated and a template, along with the 200 anchor bolts, was assembled along the supporting structure. The entire assembly was placed in position in the end-shield opening. The centre lines were marked on the concreted position of the previous pore. These centre lines were matched with the centre lines of the bolts of the assembled structure template. The grouting of all the bolts was then carried out. This saved valuable time for completing the concrete pour for the grouting.

The improvements made thus far enabled the concreting of the Calandria vault of Kaiga 3 to be completed within 14 months from the first concreting of the reactor building. The scheduled time for completion was 19 months, thus a valuable time of 5 months was saved.

28.5.5 Kaiga 3 progress as of mid-2003 to 2005

28.5.5.1 *Civil work*

28.5.5.1.1 The Reactor building

The internal structures had been built up to 112.3 metres in elevation. The liner work such as the fuelling machine and fuelling machine service vaults were in the advanced stages of completion. The Calandria of Kaiga 3 was thus lowered.

28.5.5.1.2 The Reactor containment structure

The inner containment wall (ICW) had reached 111 metres in elevation. The quality of concrete was M-45 for ICW and M-35 for OCW (Outer Containment Wall). The springing level of ICW was reached by August 2004. The work of fabrication of the supporting structure (weighing 450 metric tonnes) for the domes had commenced. The parameters for the domes were the same as for Kaiga 1 and 2.

28.5.5.1.3 Progress of Civil work as of 2003-2005: control building

The building was built at a 100-metre elevation.

28.5.5.1.4 Other plant buildings: conventional equipment

All other plant-connected buildings were expected to be built up to 100 metres in elevation before the on-set of the monsoon during May-June 2004. The erection of piping spools and cable pan supports below 100 metres elevation had commenced.

28.5.5.2 *Reactor equipment work*

28.5.5.2.1 Calandria end-shield lowering and welding

Lowering of Calendria in reactor building - 3

The Calandria end-shields were lowered in November 2013. The Calandria end-shield welding was completed on 14.2.2014 within 45 days. This was four-and-a-half months ahead of schedule. The erection of Calandria tubing along with tube end rolling was taken up on 13.2.2005 and completed on 9.3.2005. The pressure coolant channel installation was taken up in October 2005 and completed on 6.11.2005.**28.5.6 Kaiga 4 Civil work in the RB**

28.5.6.1 *Progress during 2003-2005*

The building was constructed up to 100 metres elevation. The construction of the Calandria vault had commenced.

28.5.6.2 *Inner Containment Wall (ICW)*

The concreting of the ICW from 87.60 metres elevation up to 91.92 elevation was carried out in a single pour.

28.5.6.3 *Further Civil work in the RB: progress during 2003-2005*

The ICW concreting was completed on 31.5.2005 within 82 days. It may be mentioned that each lift in concreting was raised to 3.5 metres as against 2.1 metres in Kaiga 1. This resulted in less number of total pours, thus saving time for the completion.

28.5.6.4 *Kaiga units 3 and 4: total concreting quantity*

An overall concreting quantity of 1,52,192 cu. metres was carried out from 1.3.2000 to 31.3.2000.

28.5.6.5 *Kaiga units 3 and 4: overall physical progress*

A progress of 39% was achieved by 31.3.2004.

28.5.7 Kaiga 3: rear face insulation cabinet

The installation of the insulation cabinet on the rear face for the feeder and header system was taken up on 31.8.2004 and completed on 30.9.2004 within 30 days.

28.5.8 Kaiga 3: the PHT feeder tube installation

There are 612 pressure feeder tubes on both sides of the reactor. Their installation was taken up on 20.12.2005.

28.5.9 Kaiga 3 progress as of 2004-2006

28.5.9.1 *Civil work*

28.5.9.1.1 Details of work

The concreting of the outer containment dome (OCD) was completed on 28.4.2006. The control room, at 111-metre elevation of the control building, was in the advanced stages of completion. The installation of control panels in the control building was expected to commence by mid-2006.

The erection of the turbine building up to a 121-metre elevation, including the TG block, was completed by the middle of 2006.

The spent-fuel bay was completed with up to 106-metre elevation. The painting of the fuel machine vault was completed.

Kaiga Atomic Power Project (Kaiga 3&4) Kaiga - 3 on right side28.5.9.1.2 Reactor and reactor auxiliary system work

Calandria end-shield welding and grouting were completed in April 2004.

All four NSSGs (Nuclear Steam Service Generators) were placed in position and aligned. Further reactor auxiliary system work was taken up and completed in 24 days (16.1.2005 to 8.2.2005). Further fuelling machine vault activities—like the installation of the vault liners, atmospheric coolers and related work—were completed by late 2005/early 2006.

Clean room conditions were established in both the FM vaults.

Spent-fuel bay liner installation was completed on 20.1.2006

The installation of the PHT feeder tubes was completed on 26.3.2006.

28.5.9.1.3   Conventional system work

The TG and condenser erection work had commenced.

The main piping in the control building basement for the active process water system was nearing completion. The erection of 14,000-inch metres of piping was completed.

All secondary system equipment was moved to their positions and installed. Secondary system piping had commenced.

28.5.9.2   *Civil work: progress during the period March 2004 to December 2009*

<u>28.5.</u>9.2.1   Work in the RB and TB buildings

The RB-4 and containment dome work were completed. Other Civil Work in the RB, TB and connected buildings were nearing completion.

28.5.9.2.2   Nuclear system work: physical progress during 2004-2009

28.5.9.2.2.1   The end-shield and Calandria erection

The Calandria and both the end-shields were erected. End-shields were grouted in position as detailed in para 28.5.4.1.3. These were completed in November 2005. The physical progress in Kaiga 4 was 96% as of December 2007.

28.5.9.2.3   Coolant tube installation

The installation was completed in March 2006.

28.5.9.2.4   Coolant pressure channel installation

The installation was completed in May 2006.

28.5.9.2.5   The PHT feeder tube installation

Installation of all 612 feeder tubes was completed in July 2007.

28.5.9.2.6   The PHT system hydro test

The test was completed in October 2007.

28.5.9.2.7   The RB leak test

The test was completed in December 2007.

28.5.9.2.8   PHT system hot-conditioning

The work was completed in February 2008.

28.5.9.3   *Readiness for Critically Fuel Inventory*

The inventory of the nuclear fuel unit was kept ready for criticality.

28.5.9.4   *Peer review*

The unit was offered for peer review by WANO.

28.5.9.5 *Receipt of failed TG-3 unit (since its repair for use in Kaiga 4)*

It would be seen later in para 28.17 that the TG unit of Kaiga 3 failed on 26.8.2007 during field tests. The failed unit was replaced by the TG unit of Kaiga 4. The Kaiga 3 TG unit which was repaired by BHEL was received back at the site and placed in position on 3.7.2009. The heaviest lift in Kaiga 4 was the stator with a weight of 165 tonnes.**28.6 The Kaiga 4 conventional system**

The 400-KV switchyard was energised in January 2005 and the SUT was commissioned.

28.7 Kaiga 3 and Kaiga 4: overall progress as of early 2005/December 2007

A record pour of 12,000 cu. metres of concrete was made in January 2005. This was in the 34th month of the site work commencement. A total of 2,01,500 cu. metres of concreting was achieved. A total of 63,000-inch diameter of welding was completed. The overall physical progress of 63.5% was achieved as of 31.3.2005 and 98.1% as of December 2007.

28.8 Kaiga 3: status of remaining work and readiness for first-criticality as of 2006-2007

28.8.1 Civil work

Most of the work was completed in all the buildings.

28.9 First-criticality

All was a go for first-criticality.

28.10 Reactor and auxiliary system work

The fuelling machine system, spent-fuel transfer system, fuel storage bay system and all related systems were completed in all respects.

28.11 Conventional system work

28.11.1 The TG and related system work

The TG and TG auxiliary systems—including the re-heater system, TG protection and control system, static excitation system and the lubricating oil circulation and purification system—were completed in all respects.

28.11.2 Condenser and feed-water system

The condensing system work and feed-water system including the de-aerator system were completed.

28.11.3 Electrical system work

The switchyard; main power output system; system and station unit power systems; Class-I, Class-II, Class-III and Class-IV power systems; power and control cabling; lighting system and grounding systems were all completed in all respects. All electrical control equipment systems were completed.

28.11.4 Water systems

The circulating water system, high-pressure and low-pressure process water system and domestic water system were all completed.

28.11.5 Fire protection system

The fire protection system—with detectors, zonal fire indication and warning panels—was completed.

28.11.6 Telecommunication, PA and paging system work

All telecommunication systems work, including the in-station and out-of-station distance communication systems, were completed.

28.11.7 Instrumentation and Control System

All work including the installation of the main control panel was completed.

28.12 Pre-commissioning test and preparation for criticality

(1) The PHT hydro test was completed in 5 hours
(2) The RB leak test was completed within 6 days

(3) The PHT hot-conditioning was completed in 54 hours, this was a record achievement

(4) The fuel loading was completed in 70 hours

28.13 Kaiga 3: first-criticality synchronization and commercial operation

First-criticality was achieved on 26.2.2007 and the synchronization with the Southern regional grid came into effect on 11.4.2007. The station was declared for commercial operation on 26.5.2007, within 90 days from first-criticality. The unit generated 620 million units from May 2007 to March 2008 and yielded a revenue of Rs. 1,550 crores (Rs. 1.55 billion).

28.14 Chronological milestone dates

S. No.	Activity	Date
1	First pour of concrete	30.3.2002
2	RB raft completion	15.6.2002
3	Calandria vault concreting completion	22.5.2003
4	Alignment/Welding of Calandria and end-shield	30.6.2004
5	Installation of steam generators	8.2.2005
6	Calandria tubing completion	9.5.2005
7	Inner dome concreting	31.5.2005
8	Coolant channel installation	6.11.2005
9	PHT feeder tube installation	26.3.2006
10	OC Dome concreting	28.4.2006
11	The PHT hydro test	6.9.2006
12	Hot-conditioning of the PHT system	4.12.2006
13	Fuel loading	2.2.2007
14	First-criticality	26.2.2007
15	Synchronisation to the Southern regional grid	11.4.2007
16	Commercial operation	6.5.2007

28.15 Time of completion of Kaiga 3: Improvements in the construction methods

First-criticality of Kaiga 3 was achieved in 59 months from the date of first concreting. This is an improvement over the earlier record of 60 months achieved concerning TAPP 4. With improving innovations regarding the project construction, further improvement may be expected concerning the time of completion.

The following innovations in the construction of Kaiga 3 are worth mentioning:
1) The lift for ICW dome concreting was increased from 2.1 metres to 3.5 metres.
2) The number of lifts for ICW dome concreting was reduced to 9.
3) The work in the Calandria and FM vaults was carried out in parallel instead of sequentially, as in earlier reactors.
4) The assembly of the ICW dome support structures along with shuttering was done on the ground. This considerably reduced the work at the high ground of the dome site and thus saved time.

28.16 Independent verification and validation (IV and V) of the computer-based control and instrumentation system of Kaiga units 3 and 4

Independent verification and validation of all computer-based control and instrumentation systems was initiated for Kaiga units 3 and 4. A simulator for testing these systems has been designed and built for the benefit of the plant operators and maintainers.

28.17 Failure of TG of Kaiga 3: replacement of the failed TG

The TG failed on 26.8.2007 during field tests. The failed TG was replaced by the Kaiga 4 TG. The replaced TG was commissioned on 1.4.2008.

28.18 Kaiga 4: further work prior to criticality as of 2009-2010

28.18.1 Calandria man-hole cover welding

This work was carried out before the Calandria vault box-up. The following pre-welding work was done:

(1) Draining of light water from Calandria followed by Calandric drying.

(2) Intensive cleaning of Calandria internals including the over-pressure rupture disc (OPRD).

(3) Inspection of Calandria and Calandric vault by a team constituted by the site director; and the team certification of cleanliness of the Calandria.

(4) Videography of the Calandria internals was carried out.

(5) Welding of the lower-thimble housing assembly plugs of the primary shut-down system was completed and DP checks were conducted.

(6) Welding of the end-cap plugs of the secondary shut-down system was carried out and DP checks were conducted.

The over-welding was thus completed with the 100% ultrasonic and radiography tests of the welding dome.

The Calandria was then released to O and M for starting up the Helium leak test of the moderator heat-exchangers and then for the Calandria vault box-up.

28.19 Fuel loading: Kaiga 4

Fuel loading was completed in November 2010. The fuel was entirely indigenous.

28.20 First-criticality

Kaiga 4 attained first-criticality at 8:07 AM on 27.11.2010. This was the 20th reactor unit to be commissioned in the country. With the unit commissioned, India has become the 6th country globally to have 20 or more nuclear units in commission.

28.21 The synchronisation with the Southern Regional Grid: Kaiga 4

Kaiga 4 was synchronised with the Southern Regional Electricity grid on 19.1.2011. The unit was declared commercial on 20.1.11, a record one day after synchronisation.

The unit has since been supplying power to the beneficiary States of Andhra Pradesh, Karnataka, Kerala, Tamil Nadu and the Union Territory of Puducherry.

A view of KGS - 3&4

28.22 Kaiga units 3 and 4: generation during 2011-2012 and the availability factor

Kaiga units 1 to 4 generated 5.21 giga units (5.21 billion-5.210 million KW hours) during 2011-2012. The availability factor was 91.24%.

28.23 Safety Award

Kaiga units 3 and 4 were awarded the Industrial and Safety Award for the year 2011-2012 by the AERB.

28.24 Corporate Social Responsibility (CSR) activities

The construction of a mid-day meal hall at Kurnipet Higher and Primary School was initiated in July 2015. The Bhoomi Poojan for the construction was carried out under the aegis of Sri. M.P. Hanson Station Director of Kaiga units 3 and 4. Activities in the fields of education, infrastructure, health, environmental protection and skill development are also being carried out in the nearby areas of the site.

Skill-development activities have been promoted by way of providing training kits for tailoring. The training will last for 200 hours in 3 months. The training will continue for 1 year or till 200 women are trained.

28.25 Kaiga 3 record run

Kaiga 3 has created a record continuous run of over 500 days as of 23.7.2018 and was still running thereafter.

India's Interaction with the IAEA/WANO

29.1 India's interaction with the IAEA/WANO

The International Atomic Energy Agency (IAEA) was established by the United Nations during the year 1957, through a statute ratified by the required number of member countries. Over 100 Nation-States support and participate in the programme of the IAEA. The IAEA administration is based in Vienna, Austria. The objective of the IAEA is, in its own words, 'to accelerate and enlarge the contribution of Atomic Energy to peace, health and prosperity throughout the world'.

29.1.1 The IAEA

India is a member of the IAEA. India has also been designated a member of the 35-nation governing council of the IAEA, as one of the 10 countries with advanced nuclear technology.

The governing council meets five times in a year and arrives at decisions by consensus. The council prepares the decisions to be arrived at by a general conference. The general conference consists of 138 members. The IAEA maintains field and liaison offices in Toronto, Geneva, New York and Tokyo. It operates laboratories in Australia and Morrocco and supports a research centre in Fiesto, Italy, that is administered by UNESCO. The IAEA Secretariat consists of a team of 2200 multi-disciplinary professionals and support staff drawn from 90 countries. The agency is headed by a Director General and six deputy directors, the latter heading the major departments of the Secretariat.

The 138 member nations now form part of the IAEA.

29.1.2 Functions of the IAEA

The main functions of the IAEA are:

(1) To help the member countries develop nuclear applications in the fields of agriculture, medicine, science and industry. The mechanisms through which such help is rendered are conferences, expert advisor visits, publications fellowships and the supply of nuclear materials and equipment.

(2) To administer a system of international safeguards with a view to preventing the diversion of nuclear materials for military purposes. The IAEA reviews reports of individual member states on their fissionable material inventories and makes on-the-spot inspections of facilities such as reactors, fuel fabrication plants and fuel re-processing plants; such monitoring is done in respect of countries that have signed the nuclear non-proliferation treaty (NPT) on 1.7.1968, and do not officially possess nuclear weapons.

The IAEA is one of the largest science publishers in the world. It sponsors several symposia on nuclear subjects every year and publishes all the proceedings.

A total of nearly 186 countries have signed the nuclear non-proliferation treaty. Three countries—India, Pakistan and Israel—have not signed the NPT.

29.1.3 The IAEA and its role under the controversial clause of NPT: India's co-operation with the IAEA.

It is of significance at this stage to refer to clause 2 of Article-III of the NPT. This article stipulates that, 'Each State Party to the treaty undertakes not to provide (A) Source or special fissionable material (B) Equipment or material specially designed or prepared for the processing, use or production of special fissionable material to any non-nuclear weapon state for peaceful purposes unless the source or special fissionable material shall be subject to the safeguards required by this article'.

Even though India has not signed the NPT which it considers as iniquitous, the clauses, such as the above-stated, cannot prejudicially affect the development of Nuclear Energy in India and also on a global scale. The IAEA, for instance, instead of fulfilling its main function of helping member countries in the field of generation of nuclear power and other peaceful applications of the atom, has to deploy vast financial and manpower resources on the administration of international safeguards. It is of relevance to quote from the statement of the leader of the Indian delegation, during the 23rd general conference of the IAEA held in New Delhi in December 1979, 'we cannot ignore that the highest growth in the 1980 budget over the 1979 adjusted budget is under the heads **safeguards and administration**'.

The Agency seems to be steering itself into a situation where its overwhelming concern, and the major thrust of its activities, is to achieve limited non-proliferation objectives at the risk of hampering peaceful nuclear activities.

It was during the proceedings of the above-stated conference that India declared that it shall no longer be interested in receiving technical assistance from the Agency: 'since as a matter of principle, we will not be able to give our consent to any undertaking which is not in conformity to the statute'.

India has, however, attached considerable importance to the promotional activities of the Agency and has continued to provide technical assistance to other developing countries through the Agency's Technical Assistance Programme. As an instance of India's commitment, on this account, it may be mentioned that India made a payment of $76,650 as its assessed contribution to the IAEA in 1979. India has already ratified the IAEA convention on nuclear safety.

India's stand was buttressed by China and Pakistan during the meeting of the IAEA which ended on 20.9.1996 at Vienna. All these countries warned the Agency against its ambitious but controversial project to enhance the current nuclear non-proliferation goals of the Agency. India particularly laid stress on a step-by-step approach and counselled that the non-proliferation lessons learned already must be digested before embarking on more ambitious measures.

India has been co-operating with the IAEA in the field of technical collaboration, particularly regarding the peaceful development of Nuclear Energy especially in the area of generation of nuclear power. The activities in this regard are detailed in the upcoming sections.

29.1.4　The IAEA's technical committee meetings

The 3rd IAEA technical committee meeting (TCM) on 'operational safety experiences of pressurized heavy-water reactors' was held in Mumbai from 20.2.1994 to 25.2.1994. Around 44 foreign delegates from South Korea, Canada, Japan, Argentina, Romania and the IAEA took part in the TCM. From the Indian side, 75 delegates and 200 observers from BARC, NPCIL and AERB attended the TCM. An exhibition on the 'Birth and growth of heavy-water reactors' was organized for the occasion. This exhibition was opened by Mr. Rorris Rosena, Asst. Director General for Nuclear Safety, IAEA. The TCM was inaugurated by Dr. R. Chidambaram, Chairman of the AEC. The purpose of the 3rd IAEA TCM was to bring together various PHWTRs' operation member countries of the IAEA and to provide an effective forum for the interaction and exchange of views among them. It also aimed to share information and review the latest state-of-the-art developments in the field of operation and maintenance of PHWTRs. Ten technical sessions were arranged for the delegates. Cultural programmes were also organized for the benefit of the delegates. The foreign delegates were taken to Kakrapar Atomic Power Station for a technical visit.

29.1.5　The IAEA's inter-regional training

Inter-regional training on 'Strengthening Project Management' was organized by the IAEA in collaboration with the NPCIL from 1.11.1993 to 3.12.1993. This was the second such programme organized by the IAEA in India within a short period of three years. The first programme was organized at Kalpakkam (MAPP) during October-November, 1990.

Around 29 participants from Latin America, East Europe, South East Asia, China, South Korea, Turkey, Ukraine, Vietnam and Bangladesh were drawn for this course apart from India, the host. The participants visited TAPPs on 19.11.1993 and KAPPs on 20.11.1993 as part of the training programme.

29.1.6 India and OSART

An operational safety review team (OSART) was constituted by the IAEA in 1952. The OSART furnishes advice and assistance on request from member nations for enhancing the operational safety of nuclear power stations. So far, 42 OSART missions have been carried out in 33 Nuclear Power Plants, of which 27 are in operation in 22 countries.

The NPCIL has constituted a team known as the 'Internal Safety Review of Operating Stations Team' (IROS) on the lines of OSART for a systematic and elaborate review of the Indian Nuclear Operating Stations. NAPP has been chosen as the first station for the review. The review of the other stations was scheduled to follow suit.

29.2 India and WANO

The World Association of Nuclear Operators (WANO) was inaugurated in 1986. India became a member in the same year. The WANO is committed to maximising the safety and reliability of the operation of Nuclear Power Plants by promoting the exchange of information, encouraging technical communication among operators and by the comparison and emulation of accepted operational norms among its members. The WANO is headquartered in London (UK), and has four regional centres at Atlanta (USA), Moscow (Russia), Paris (France) and Tokyo (Japan). The NPCIL is represented in all these regional centres. There is a co-ordination centre in London. There are 137 nuclear operators, all over the world, who are affiliated with WANO. The NPCIL is a member of the WANO Tokyo regional centre. Other members affiliated with the Tokyo centre are KHNP-Korea, CNNC-China, PAEC-Pakistan, TPC-Taiwan and about 13 Japanese utilities. The Indian member of WANO is a Governor of the WANO Board. China has become the 15th member of WANO.

A three-day seminar on 'core management and fuel performance', organized jointly by WANO and NPCIL and hosted by the latter, was inaugurated on 27.2.1995 at the multi-purpose hall of the BARC Training School. Members of WANO from Canada,

Japan, Korea, Taiwan and the host-member India participated in the seminar. There were 18 technical papers presented at the seminar which lasted over two days; this was followed by a group discussion. A technical tour of TAPPs was organized for the overseas participants on March 1 and 2, 1995.

29.3 The convention on nuclear safety

This was ratified by India in March 2005.

29.4 The review of operating stations of the NPCIL by WANO

The first-round review of all the operating stations of the NPCIL was completed by WANO; the second round has also been started by WANO.

29.5 India's of the WANO: co-hosting of the WANO meeting

Sri. S.K. Jain, CMD of the NPCIL, took over as the president of WANO during the 2007 bi-annual general meeting of WANO which was held in Chicago in September 2007. The NPCIL co-hosted the biennial general meeting of WANO in 2010.

Around 400 inter-national nuclear executives met in New Delhi from 31.1.2010 and 3.2.2010 for the 10th WANO biennial general meeting. All the nuclear power operators of the world's 447 nuclear power reactors in over 30 countries were represented at the meeting.

The WANO has completed a peer review of every nuclear power operation unit in the world. Some of these units have been reviewed more than once. Such reviews have led to significant improvements in operational safety.

Dr. S.K. Jain's 2-year term as President of the WANO came to a close in February 2010. He handed over the charge to Mr. Qian Chinie, Chairman of the Board of China's Nuclear Power Holding Co. on 2.2.2010.

29.6 Chairmanship of WANO-TC

Dr. S.K. Jain was elected unanimously as the Chairman of the prestigious Governing Board of the WANO-Tokyo Centre (WANO-TC). Dr. Jain will also represent the Tokyo Centre on the WANO governing board. Dr. Jain retired as the CMD of the NPCIL and 'Bhavini' in May 2012, after serving for more than four decades.

29.7 India's Supplementary Compensation Agreement (SCA) with the IAEA

India signed the convention on supplementary compensation with the IAEA on 27.10.2010 at Vienna, Austria. The pact sets parameters for a nuclear operator's financial liability. The convention provides for compensation in case of trans-national consequences of a nuclear accident. Around 14 countries including India have so far signed the pact. The USA, Argentina, Morocco and Romania have so far ratified the convention. The convention would establish a uniform global regime for compensation to victims of a nuclear accident. The convention provides for the establishment of an international fund for the said purpose. The fund would serve to increase the amount available to compensate victims and allow for meeting civil liabilities including the damage from nuclear incidents occurring within a state's exclusive economic zone such as the loss of tourism or fisheries and other related income.

The pact sets parameters for a nuclear operator's financial liability and sets the time limits for legal action. It requires that nuclear operators maintain insurance or other financial security measures and provides for a single competent court to hear the claims. The convention was first adopted on 12.9.1997 and was opened for signature at the IAEA's 41st general conference in September 1997. It will come into force on the 90th day after ratification by five states that have a minimum of 400 MWe of installed nuclear power capacity.

Indo–US Nuclear Co-Operation: The 123 Agreement and Nuclear Supply Arrangements

30.1 Introduction

This chapter describes the evolving Indo–US nuclear co-operation. It traces the evolution from the time of agreement for the construction of India's nuclear power station. It also elaborates on the arrangements made by India to ensure the steady supply of nuclear fuel for India's nuclear power stations.

30.2 The unfolding of the Indo–US nuclear relationship

30.2.1 Beginning of the relationship

The Indo–US nuclear relationship has had a chequered history. The USA was the very first country with which India had signed an agreement for the setting up of the country's first Nuclear Power Plant at Tarapur in Maharashtra. The 2×190 MWe plant was commissioned and began operation in November 1969. The agreement for the plant had stipulated a 30-year tenure and further supply of enriched Uranium fuel required for the running of the units by the USA. The 1974 PNE carried out by India at Pokhran (Rajasthan) completely altered the scenario regarding the supply. The USA, acting unilaterally and against the terms of the agreement, imposed a total ban on fuel supply in 1980 after a case-by-case decision regarding the supply from 1974 to 1980. (This aspect has been dealt with in detail in paras 2.13.2 to 2.13.4 of Chapter 2 of this book.) It was for India then to make its masses to ensure a regular supply of nuclear fuel for its needs.

30.2.2   The unfolding of international nuclear relationships: domestic nuclear fuel deficit

Following the above-mentioned ban, the 45-nation nuclear supplies group (NSG), over which the USA as the only global superpower has the strongest influence, decided not to supply any nuclear equipment/materials to India. The NSG was the successor of the earlier London Suppliers Group (LSG) of which the USA, the UK, France, Germany and the USSR were the then prominent members. India was thus forced to fall back on its resources for the construction and commissioning of its Nuclear Power units. This was perhaps a blessing in disguise since India has, through its efforts, ensured a place in the high end of nuclear power-capable nations of the world in the three decades since the ban. A caveat has, however, to be entered regarding this achievement. India has enough fuel resources to operate 10 GWe of heavy-water moderated and cooled nuclear power reactors (HWPTRs). There are 54,000 tonnes of Uranium from reasonably assured resources and 23,500 tonnes from additional mine resources with the production from this commencing from 2004-2012. These resources, however, consist largely of low Uranium 238 content. Extensive and intensive mining is required to produce enough uranium oxide fuel to run these reactors. The only large Uranium mine at present is located at Jaduguda Bhatin in Bihar. This mine has been in operation for more than half a century (since 1967) and is considerably aged. It has been operating with a maximum processing rate of 2090 tonnes per day in a common mill.

The country has not been able to produce enough Uranium fuel to fulfil the capacity utilisation of existing HWPTRs and the future requirements of the new units now under construction. The requirement of uranium dioxide for the 13 HWPTRs currently in operation (2007) is 600-650 tonnes per annum at an 85% capacity factor. This requirement will double to 1200-1300 tonnes per annum once the HWPTR units now under construction have commenced operation. The requirements of TAPP units 3 and 4 which were commissioned in 2006 and of RAPP 5, RAPP 6, Kaiga 3 and Kaiga 4, then under construction, are also to be met. The above projection of required fuel will meet these requirements.

New Uranium mines in the Nalgonda and Kadapa districts of the State of Andhra Pradesh are expected to be commissioned fully during 2009.

The Uranium mines built for Rs. 270 million in Tummalapalle near Pulivendula, AP State, look very promising. An estimate of 50,000 to 1,50,000 tonnes of processed Uranium fuel has been made. The production was expected by the end of 2011. If the estimate of 1,50,000 tonnes proves to be right, then this reserve will be able to sustain 8 GWe of HWPTRs for 40 years.

Significant resources of 5 lakh tonnes of Uranium have been located in an area of 4588 km in the Srisailam forests of Kadapa in the AP state. With the above nuclear fuel resources available, India would be able to run its HWPTRs that are currently in operation (2007). Since 2007, there have been more units in the construction and proposal phases, hence India would have to go in for nuclear fuel imports. This would require negotiations with the USA which is a major member of the Nuclear Suppliers Group (NSG) and the IAEA. The IAEA has been consistently urging India to sign the Non-Proliferation Treaty (NPT), but India would never sign the NPT. We now go into more detail as to how India moved to meet this conundrum.

30.3   Negotiations with the USA

India had been continuously in dialogue with the USA over the possibility of the latter and the NSG lifting the ban on the international supply of nuclear equipment and materials to India. The USA, the only global superpower having the clout to influence any decision by the NSG, would be the prime target nation in this move by India. The consequent negotiations commenced during the tenure of late P.V. Narasimha Rao who was the Indian Prime Minister during 1991-1996. These negotiations were held away from both the countries (that is in third-party countries), to avoid media glare, public attention and resultant publicity. These negotiations continued during the tenure of subsequent Prime Ministers of India. The fact that the negotiations continued right across multiple tenures of Indian Prime ministers and the corresponding US officials would serve to bring to light the complications which these negotiations sought to resolve.

It may be pointed out that though the USA upped the ante following India's May 1998 Pokhran peaceful nuclear explosions (PNE), it subsequently relented and removed some of the restrictions it had imposed on India following the PNE. It had approved the US supply of critical nuclear equipment for ensuring the safe operation of TAPP units 1 and 2. The supplies were on a case-to-case basis and did not constitute assured future supplies. It was only during 2004-2005, which marked the beginning of George Bush II's second presidential tenure, that serious negotiations between the two countries started. The President involved himself personally and fully in promoting an agreement. He also gave a clear call to his officials that they should ensure an early agreement.

30.4 The July 2005 agreement on Indo–US nuclear co-operation

The Indian Prime Minister, Sri. Manmohan Singh, visited the USA between 17.7.2005 and 21.7.2005. An agreement was concluded between the Indian PM and the US President to enhance nuclear co-operation between the two countries during the visit. The agreement also envisaged closer and enhanced co-operation in the fields of outer space exploration and other high-tech areas. Closer commercial ties were also expected. The agreement envisaged the supply of nuclear equipment and materials to India subject to certain conditions. These conditions were as follows:

1) The separation of Indian civilian and military nuclear facilities.
2) India should conclude an India-specific agreement with the IAEA which inspects the Indian civil nuclear facilities.
3) India should join bilateral and multi-lateral non-proliferation activities.

30.5   The 123 agreement between India and the USA

The Manmohan Singh–George Bush II agreement of July 2005 was carried out by several stages of fodder. What is called a '123 agreement' was entered into between the two countries at the end of March 2006. The 123 agreements between the USA and every other country are essential before any legal steps are taken for mutual trade in nuclear equipment and materials by such a country. This 123 agreement was a historic one and envisaged the following:

a)   India would open 14 Nuclear Power units including 6 HWPTR units for IAEA inspection in perpetuity. These are RAPP units 1 and 2, TAPP units 1 and 2, Kakrapar units 1 and 2 and eight other 220-MWe units over the period ending with 2014. A schedule for such an inspection would be decided upon by India.

b)   The fast-breeder reactor units would not be open for IAEA inspection.

c)   Nuclear fuel supply for Indian power reactors would be insured in perpetuity, subject to certain conditions.

d)   India would take its decision on future reactors and would be free to withdraw from the agreement if the US Congress were to alter any terms and conditions arrived on 2.3.2006 between India and the USA

e)   The 'Cirus' reactor in BARC at Trombay, India, would be shut down permanently by 2014.

There is a 'catch' in the above 123 agreement. The agreement would not over-ride the 'Hyde Act' of the USA, according to which India would have to return nuclear equipment and materials supplied by the USA if it were to test any nuclear exclusion device. There is, however, the provision in the 123 agreement according to which, in the event of a nuclear test by India, the two countries that signed the agreement, namely India and the USA, would consult each other for a political resolution before any action by the USA for such a recall is taken.

30.6 The 123 Agreement: Political fall-out, Parliament Debate and Public Opinion

Even though the '123 agreement' was a technical one, political parties have willingly entered into debate and discussion on the agreement. This book, being a by and large technical venture, has steered clear of political factors except at places where such factors have had an impact on the Indian nuclear power programme. The '123 agreement', being a crucial one, had attracted a lot of political attention and had also led to debate in the Indian Parliament. Relevant political developments have been touched upon in our narration of the agreement and its fall-out.

A large body of retired DAE engineers, other scientists and technocrats had expressed themselves in favour of the agreement. According to them, it was the best possible solution in the prevailing circumstances, given the views of the then-current International nuclear enforcement/monitoring bodies and their clamour for harsher steps to be enforced on the international community to ensure nuclear non-proliferation.

The logical follow-up of the '123 agreement', on the other hand, would open up International co-operation and ties for nuclear power development in India. This aspect will be touched upon in the following sections.

The stand of the BJP was that the testing of a nuclear explosion device was a sovereign action and that such a venture should not attract any adverse action against India. This stand flies in the face of international sanctions against India that were introduced following the 1998 Pokhran PNE when the BJP Government led by the late A.B. Vajpayee was in power. The stand of other smaller political parties against the '123 agreement' was mere political posturing rather than arising out of a deep study of the said agreement.

30.7 Further details about the '123 Agreement'

We have seen in para 30.5 that the nuclear co-operation agreement between the US President, George Bush II, and the Indian Prime Minister, Sri. Manmohan Singh, was signed in July 2005. This was carried further and a preliminary agreement was arrived at between the countries on 2.3.2006.

A 300-working hour effort in which the people from India—S. Jaishankar, technical expert; R.B. Grover, Chief of the strategic plans division of the DAE; Shivashankar Menon, Foreign Secretary of GOI; Raminder Jaesal of the Indian Embassy, USA; Anil Kakodkar, Chairman of the AEC; Ronen Sen, India's ambassador to the USA; M.K. Narayanan, National Security Adviser to India's PM—took part in. They were joined by people representing the USA namely Richard Stratford, Richard Burns and Stephan Hendley, the USA's National Security Adviser. This effort resulted in a binding 30-page '123 agreement' that was agreed upon on 20.7.2007.

India would, as per the 123 agreement, set up a dedicated re-processing facility in which all foreign-supplied and hence Indian-used and spent-fuel would be re-processed under the IAEA safeguards. The other terms of the agreement were as follows:

(1) India would be assured of lifetime nuclear supplies for all of its civilian reactors. India would not be forced to return US-supplied nuclear fuel even if it conducts a nuclear test.

(2) The re-processing of US-supplied spent-fuel would call for a separate agreement with the USA.

(3) The USA may still withdraw nuclear co-operation with India in case of an Indian Nuclear test; this action if and when taken would be technical and not a political one.

30.8 India's talks with the IAEA

The IAEA signed the treaty but its legislative bodies are yet to ratify the same.

30.9 Debate in the Lok Sabha: the Indian Government's talks with the leftist parties and public opinion

The Lok Sabha held a debate on the '123 agreement' issue on 28.11.2007, which lasted nine-and-a-half hours. The opposition parties held the view that the sense of the house was against the agreement even though there was no voting on the issue on 28.11.2007. We may point out that certain sections of the public held the view that the future security of the country would be compromised if the LS voted for the agreement. However, a large majority held that the agreement was beneficial to the country with the current and possible future scenarios of the International Security ambience duly considered.

It will be relevant at this stage of parliamentary debate on the issue to consider in more detail the left parties' stance on the issue. The parties were not fully united, particularly on their views on the stability of the Government following the final voting on the issue. While all of them had mentioned the view that they would not let the agreement be operationalized, the most dominant party of the left conglomerate namely the Communist Party of Marxists (CPM) maintained that it would not let the Government fall on the issue. The CPM, the revolutionary socialist party of India (RSP), and similar leftist parties maintained that the left should withdraw support for the Government on the issue.

30.10   Government talks with the leftist parties

The Government held seven rounds of talks with the leftist parties as of November 2007. The talks did not succeed in the bridging of their differences. The left's stance then was 'either scrap the agreement or face withdrawal of support'.

30.11   Consideration by the IAEA

The Government of India referred the India-specific safeguards agreement to the IAEA for its clearance during the middle of 2008. The leftist parties promptly withdrew their support to the Government of India and challenged the Government to prove their majority in Parliament.

The Parliament debated the issue on 22.7.2008. The Government had moved a confidence motion before the debate started. The Government won the motion by 275 votes for and 256 votes against the motion.

The IAEA board of governors unanimously sanctioned the safeguard agreement entered into with India on 1.8.2008. The next step would be for the 45-member NSG to accord a clear (and unconditional as per India's stand) exception from the provisions of the NPT to enable India to participate in international trade in nuclear equipment and materials.

30.12   Consideration by the NSG

The 45-member NSG, at its meeting on 21.8.2008 with Germany as its then Chairman, could not decide on the waiver sought by India. Several small countries such as Austria, Ireland and New Zealand raised several queries especially as to why India was not signing the NPT, why it was not opening all of its nuclear facilities for international inspection and how it was going to control and monitor nuclear fissile materials used by it. They, perhaps, were not made aware of India's scrupulous adherence to non-proliferation and its rigorous control over the export of nuclear materials and equipment.

The USA, Russia and France convinced these countries about India's non-proliferation credentials and, as a result, the NSG at its meeting from 5.9.2008 to 7.9.2008 gave India a clear waiver from joining the NPT and CTBT (comprehensive test ban treaty), thus making India continue its international trade in nuclear equipment and materials.

The Indian Foreign Minister, Sri. Pranab Mukherjee, reiterated that India's voluntary embargo on nuclear weapons testing would help in the NSG's decision regarding China's issues. This would have cut no ice as the unanimous waiver was sent to China's delegate to the IAEA, who was not present when the waiver went through. The above-mentioned meeting, which took place in the headquarters of the IAEA at the 21st to 24th floor of Andromede towers, lasted 76 hours. It then remained for India to sign an additional India-specific protocol with the IAEA before it entered international trade in nuclear materials and equipment. The IAEA continued to retain from India the following international denials:

1)   Transfer of technology per re-processing of spent-fuel

2)   Latest technology for the production of heavy water

India is self-sufficient in both aspects. These two denials are hence not of any significance to India.

30.13   The US law body's approval of the '123 agreement'

The Foreign Relations Committee of the US Senate approved the '123 agreement' by 19 votes for and 2 against. The committee included a rider in the shape of an enabling legislation that stipulated that the transfer of nuclear technology would cease if India carried out a nuclear weapon test. The Herman bill HR 7081, which is a house version of the agreement, was passed by the House of Representatives of the USA by a two-thirds majority of 298 votes for and 117 against on 27.9.2008. The bill had to be passed by a full house of the US Senate for it to become law.

The US Senate approved the '123 agreement' by 86 votes for and 13 against on 30.92008.

30.14   India's Nuclear co-operation Agreement with France

India and France signed a nuclear co-operation agreement that assured India of a lifelong supply of nuclear fuel for Indian power reactors. The agreement gave India the right to re-process spent French-supplied nuclear fuel subject to India's observance of the IAEA safeguards concerning the end use of such re-processed nuclear fuel. Sri. Anil Kakadkar, AEC Chairman, and the French foreign minister signed the agreement on behalf of India and France, respectively. The agreement came into force on 14.1.2010 after due ratification of the agreement by both countries.

The US Senate approved the '123 agreement' by 86 votes for and 13 against on 2.10.2008 (Mahatma Gandhi's Birthday). The Senate also rejected the killer amendments to the '123 agreement' which were stipulated earlier by the Senate Foreign Relationship Committee. It then only remained for the US President to sign the bill to make it a due law.

30.15   The US President's enactment of the '123 agreement'

The US President, John W. Bush, signed the '123 agreement' into law at 2:30 PM (US time) on 9.10.2008. The law assured, among others, the re-processing of foreign fuel supplied to India subject to the setting up of a dedicated re-processing facility by India and India's observance of the IAEA safeguards.

India and the USA formally signed the '123 agreement' law on 11.10.2008. The then foreign minister of India, Sri. Pranab Mukherjee, and the US Secretary of State, Condoleeza Rice, signed the agreement on behalf of India and the USA, respectively.

30.16 Uranium fuel importing in India: agreement with other countries and safeguards agreement with the IAEA

The NPCIL of India inked an agreement with the French energy firm 'Areva' for the supply of 300 tonnes of Uranium fuel per year on 16.12.2008. This was the first fuel supply agreement that India entered into after the '123 agreement'.

India signed the safeguards agreement with the IAEA in Vienna on 2.2.2009 for the inspection of additional civil reactors to be set up in India. The agreement has cleared the way for India to enter into trade in power reactors, reactor components and nuclear materials subject to India's entry into the NSG.

It is important at this stage, as regards to India's broad agreement for foreign aid and nuclear equipment supply, to note that India and Soviet Russia entered into an agreement on 10.12.2008 for the setting up of four additional IGWe nuclear units at Kudankulam, (TN) near the present site.

India had signed an agreement by the middle of December 2008 with Mongolia, Kazakhstan, Argentina and Namibia. An agreement with Canada for the Indian import of nuclear fuel and nuclear technical co-operation was also in the pipeline.

30.17 Global Uranium producers

It would be relevant at this stage to have a look at the list of top Uranium producers. In 2009, Canada was the highest producer of Uranium in the world. The second highest producer was Australia. Other Uranium producers were Niger, Kazakhstan, South Africa, Namibia, Zambia and Uzbekistan. Of these, Australia had long held the view that it would not supply any Uranium fuel to India till the latter signed the NPT. Australia, however, changed its view in 2012 and announced that it would supply the fuel subject to necessary talks that were held in 2014. As a result, Australia would start its supply in due course.

30.18   More particulars about the '123 agreement'

The agreement would be valid for 40 years. Either party (the USA or India) has the right to terminate the agreement before its expiration with prior notice to the other party. The reason for termination shall be stated. The termination notice can be cancelled if the notice is withdrawn before the end of the notice period. The two countries shall consider relevant circumstances and promptly hold consultations to address the reasons for termination.

If the USA exercises its right to seek the return of equipment and materials in case of a nuclear test by India, it shall compensate India promptly for the costs incurred as a consequence of such removal.

30.19   The Nuclear Liability Bill

The '123 agreement' with the USA and subsequent agreements with France and Russia would have to be backed up by a law to fix the liability of various parties regarding the setting up and operation of Nuclear Power Plants and related plants and facilities. A law called the Civil Liability for Nuclear Damage Act 2010 (38 of 2010) was introduced in the Lok Sabha budget session of 2010, at the beginning of February 2010, and was referred to the relevant parliamentary committee. The committee made several changes and the Government introduced several amendments to the law based on the recommendations of the parliamentary committee.

The bill was passed by the LS on 25.8.2010 by 252 votes for and 25 against. The bill establishes the liability of the operator/supplier of equipment and materials arising out of nuclear accidents, and consequent damage to public property and life. A maximum liability of Rs. 1500 crores has been stipulated in each accident. Rules under the act were notified by the Government of India on 11.11.2011. The GOI has finalized the rules of implementation of the nuclear liability laws. The rules were notified in September

2010. The notification will help progress various contracts for the setting up of large Nuclear Power Projects, particularly, the 6×1.65 MWe 'Areva' (French) units at Jaitapur in Maharashtra.

30.20   Current Uranium production and further Future plans for production import

30.20.1   Jaduguda mines production

We have seen in para 30.2.2, that the underground mine at Jaduguda has been operating since 1967. Two other underground mines, Narwapahar commissioned in 1995 and Turamdih commissioned in 2005, have been operating along with Jaduguda with a common mill capable of processing 2090 tonnes of ore per day. A new mill with a processing capacity of 3000 tonnes of ore per day was commissioned in 2008.

30.20.2   Plans to further Uranium production

Plans were formulated during 2005-2006 to further Uranium production by opening Uranium ore mines for 700 million dollars. These plans are listed below:

1) Banduhurang in Jharkhand was the first open-cast mine in India which commenced production in 2007. The underground mine at Bagjata was commissioned in 2008 and the third one in Jharkhand commenced production in 2010. It is to be noted that all three mines are situated in Jharkhand.

2) The mine at Domiasiat-Mawthabah at Meghalaya has mill facilities attached to it. It will process 3,75,000 tonnes of Uranium ore per day.

3) Lambapur-Peddagattu in the Kadapa basin of AP is another Uranium-producing mine in India. Three kinds of Uranium minerals have been located here. The site is located 110 km South East of Hyderabad. The environmental clearance for one open cast and three underground mines was accorded.

It has been estimated that 6000 tonnes of Uranium can be processed from these mines.

The quantity of processed Uranium, as of 2010, was 1055 tonnes per year.

30.20.3 Uranium imports and annual plans

The import of processed Uranium was planned as follows:

1) From Areva of France : 300 tonnes during 2008-2010
2) From Russia (TVEL) :
   a) 2000 tonnes of natural Uranium over 10 years from 2013
   b) 210 tonnes during 2010-2011
   c) 60 tonnes of enriched Uranium for Tarapur units 1 and 2 during 2010-2011
3) From Kazakhstan:
   a) 300 tonnes of processed Uranium during 2010-2011.
   b) The State has promised the undertaking of the supply of 2100 tonnes of processed Uranium over a longer term, completing their supply during 2008-2014.

India imported 4,458 tonnes of Uranium from 2008 to 2014 from all the above-mentioned countries.

30.20.4 New fuel re-processing plants

Two new fuel re-processing plants have been planned at Kalpakkam (Tamil Nadu) and Trombay (Maharashtra) under the IAEA safeguards.

30.20.5 Indo–Canadian nuclear co-operation agreement

After 15 years, the Indian Foreign Affairs Minister, Sri. Salman Khurshid, visited Canada to hold talks with his Canadian counterpart, John Baird, about co-operation in the field of Nuclear Energy, among other subjects. This meeting took place during the last week of September 2013. The preliminary talks were held earlier in April 2013 when the two countries signed an Appropriate Arrangement Agreement (AAA) that would facilitate the supply of Uranium fuel by Canada to India. This marked a significant step ahead, considering that Canada had stopped all supplies of nuclear equipment and materials to India following the 1974 PNE by India. The September 2013 talks would also facilitate further interaction on global economic and security issues.

30.20.6 The Indo–Canadian agreement for Canadian supply of Uranium to India

During the Indian PM's official stand-alone visit to Canada, from 14.4.2015 to 16.4.2015, the two Prime Ministers, Sri. Narendra Modi and Stephen Joseph Harper agreed to the Canadian supply of 3,000 tonnes of Uranium to India over 5 years for $254 million to India. A commercial agreement was effected with the Canadian Uranium producer 'Cameco' to this effect. The first tranche of the fuel arrived in India in December 2015. Nearly 1,000 tonnes would be supplied during 2015-2017.

30.20.7 The Indo–Australian Nuclear co-operation Agreement: supply of Uranium

India and Australia finalised their nuclear deal on the sidelines of the G-20 summit meeting in Antalya, Turkey, between 15.11.2015 and 16.11.2015. The Indian PM, Sri. Narendra Modi, and the Australian PM, Malcolm Turnbull, met for the first time during the said G-20 meeting. The two announced the exchange of instruments and discussed about how the Australian Parliament had earlier approved Indo–Australian nuclear co-operation. The two courts exchanged 'note verbales' (notes of agreement) on 13.11.2015. The agreement would enable India to import the much-needed Uranium feed from Australia which is the largest producer of nuclear fuel in the world.

30.20.8 The GOI's 'Nuclear Liability Fund'

The Government cleared the decks for the setting up of a 'nuclear liability fund', with a corpus of Rs. 20 billion (2000 crores) which would pitch in if damages resulting from a nuclear plant accident exceeds the limit specified on the 'Nuclear Liability Act' of August 2010. The DAE has notified the 'Nuclear Liability Fund Rules 2015' to the above effect. The fund would go up by a levy collected from the operators of Nuclear Power Plants. The levy would be between 5 and 10 paise per unit of electricity sold. The levy will be collected and credited to the fund till the corpus reaches Rs. 220 billion. Credit to the fund will resume in the event of withdrawal from the fund to ensure the corpus always has Rs. 220 billion at its disposal. The corpus will be contained in the consolidated fund of India and transferred to the public account: 'MH 8235, general and other reserve fund'.

The Government of India will be required to obtain the Indian Parliaments' approval before making payments out of the fund, after due assessment of the cost incurred and establishing liability for the payment.

Station operators would be required to make payments to the fund every quarter. Delay in payment would attract interest at 18% daily.

30.20.9   Additional Nuclear Accident Funds

The General Insurance Corporation of India (GIC), an insurance public sector unit of the Government of India was mandated by the GOI to arrange a cover of Rs. 20 billion at a meeting held on 13.12.2011 for officials of the DAE, the NPCIL and the AERB.

The GIC made a presentation at the meeting on mortality by arranging the above cover. Potential entrants and on-going suppliers of nuclear equipment like L&T, BHEL, NTPC, IOC and NALCO were present at the meeting.

The GIC brought into the meeting, international nuclear under-writings from China and Russia. These countries whose nuclear activities are not fully in the public domain have avoided the services of international insurers for their country's nuclear damage liability. Their services are being inspected by 'pool' insurance agencies.

'Pool' is a terminology used in the international damage-cover process and denotes the underwriting capacity of the voluntary association of insurers to provide the required cover. This cover would be beyond the capacity of individual insurers.

The GIC brought to light in the meeting that inspection by pool agencies is not a cause for security concern for individual countries. It may be mentioned in this connection that the IAEA has its methodology for mutually acceptable inspection measures.

The NPCIL mentioned that WANO has been conducting peer evaluation of Indian Nuclear Power units.

It was pointed out by the 'pool' agencies to the NPCIL that pool inspection is qualitatively different from the WANO inspection which stresses on qualitative procedures in the plants.

Pool inspection aims at economically driven asset-loss prevention measures and there abides mutual economic interest with the operator in such inspections.

30.20.10   Convention on supplementary compensation for nuclear damage (CSC)

The Government of India has ratified the CSC at Geneva during the first fortnight of February 2016. This ratification opens the door for external funding for Nuclear Power Plants in India.

30.20.11   Indo–Kazakhstan nuclear agreement

During Sri. Manmohan Singh's (Indian PM) official visit to Kazakhstan between 15.4.2011 and 16.4.2011, the two countries signed seven pacts. One of them was for the supply of processed Uranium by Kazakhstan. (para 30.20.3(3) can be referred to in this regard).

The two countries agreed on the following:
(1) The use of radiation technology for healthcare in Kazakhstan
(2) The use of isotopes
(3) Reactor safety mechanism
(4) The exchange of scientific and research information exploration
(5) Joint mining of Uranium design, construction and operation of nuclear plants

Another pact envisaged the ONGC of India taking up 25% equity in the Satpayev exploration block of the country in the Caspian Sea.

30.20.12   India's undertaking of the construction of parts of nuclear equipment and the provision of nuclear services in foreign countries

In a first-of-its-kind venture for India as regards its role in nuclear power development in a foreign country, India entered into a tri-partite agreement for the supply of equipment and materials to the Rooppur $2\times 1,200$ MWe pressurised Nuclear Power Project in Bangladesh. The Soviet Federation is to construct the project which is the very first Nuclear Power Project in that country.

30.20.13   India's admission to the Australia Group (AG) and CERN

In January 2018, India was admitted into the Australia Group (AG) which is responsible for the control and regulation of Chemical and Biological Weapons (CB). It is one of the four international groups for the control and regulation of weapons of mass destruction (WMDs). The others are ATCR, Wassenaar and NSG. The Wassenaar Group ensures the regulation of exports of conventional weapons and products of dual-use (technical and weaponisation). India also became an associate member of the European Organization for Nuclear Research (CERN). India is now the only one to join the NSG (Nuclear Suppliers Group), the last of the four International nuclear control regimes.

30.20.14   Fuel Supply

India and Uzbekistan signed a deal for the long-term supply of Uranium from this Uranium-rich country to India on 19.1.2019. The deal was signed during Uzbekistan President Shavkat Mirziyoyev's visit for participation in the ninth edition of the 'Vibrant Gujarat Summit' at Gandhinagar, Gujarat, that took place between 18.1.2019 and 20.1.2019. Uzbekistan has just become the second Central Asian country after Kazakhstan to supply the much-needed nuclear fuel to India. Uzbekistan is the seventh largest exporter of Uranium fuel in the world. Canada also supplies this fuel to India. India also plans to import Uranium from Australia, the world's largest Uranium producer.

30.21   The role of the AERB

After having detailed the 123 agreement and the IAEA's role in India's exemption from signing the NPA (Non-Proliferation Agreement), we will now speak about India's own Atomic Energy Regulatory Board (AERB) that was instituted by the Government of India.

The AERB was set up by the GOI on 15.11.1983 with the mandate to ensure that the use of ionizing radiation and the production of nuclear energy does not cause undue risk to the health of the general public and the environment. The board consists of a full-time Chairman, an ex-official member, three part-time members and a secretary. The body, set up under the aegis of the Department of Atomic Energy (DAE), is headquartered in Mumbai.

The AERB fully involves itself in the activities of the NPCIL, right from pre-erection site tests, erection processes, testing, and the commissioning and maintenance of nuclear plants. It is fully involved with the erection of major nuclear equipment and services, site testing and commissioning. Its clearance is necessary at crucial points of the erection work when it deputes its personnel for inspection. It deputes senior engineers at the time of first approach to criticality, achievement of criticality and subsequent gradual increase in power-generation. Its function is critical to the task of nuclear power-generation and it has contributed immensely to the safety of Indian nuclear power plans. It is fully autonomous in its functioning.

There has been criticism from power system players that the reputation of nuclear power production should be done by a body outside of the DAE, the sole nuclear power player. This suggestion is not practical. The expert personnel, equipment and services for nuclear power reputation are not available to the degree required outside of the DAE. It has to be emphasized at this step that the AERB is fully autonomous and no occasion has arisen so far to question its autonomy.

Kakrapar Units 3 and 4

31.1　General

In Chapter 23, we dealt with the establishment and operation of the Kakrapar units 1 and 2 which were of 200-MWe capacity each. Now, it is for the first time that indigenously designed and manufactured 700-MWe units are being set.

We have seen in Chapters 15 and 16, the need for going in for higher-capacity units.

31.1.1　Criteria for and advantages of higher-capacity units

The NPCIL decided to set up units 3 and 4 at Kakrapar in line with deriving advantages from lease considerations. Apart from what has been stated in the previous section, economies of scale would accrue when large capacities are installed at one site and large quantities of power are transmitted at high and ultra-high voltages for onward transmission from ultra-high capacity receiving stations. Nuclear power stations are operated based on base generation; large generation capacity contributes to the economical rule of generation.

31.2　Preliminary work: status as of early 2009

The Ministry of Environment, Forest and Climate Change of India gave clearance for the units in March 2009. The preliminary works were then taken in hand. The plan layout was finalised and the specifications for long-delivery items of critical equipment was issued.

31.3 Civil work: ordering of NSSGs: the status as of the last quarter of 2009

The ground-breaking ceremony was held in December 2009.

Larsen & Toubro were awarded the contract for Civil work for Rs. 844 crores. The work included the reactor and auxiliary buildings, waste-management facility. The Company was also awarded another contract worth Rs. 345 crores for the manufacture of 4 NSSGs for KAPP units 3 and 4. The packages for other nuclear and conventional equipment were under finalization.

31.4 The Civil work for KAPP units 3 and 4: the status as of late 2010

The Civil work was taken up for execution after the receipt of clearance from the AERB. Excavation work was completed by the end of August 2010.

31.5 Cost estimate and financial sanction: equity stake by other Government of India PSUs

The financial sanction for KAPP units 3 and 4 at an estimated cost of Rs. 120 billion was issued in early 2010. For the first time, as per the GOI's decision to allow equity stake in the NPCIL by other PSUs/State Governments, the National Aluminium Company Ltd. (NALCO) would pick up a 49% stake of equity at Rs. 1700 crores in KAPP units 3 and 4. The remaining 51% equity of Rs. 1,770 crores would be held by the NPCIL.

31.6 The purchase of long-delivery major equipment progress during 2010-2011: the placement of purchase orders (POS)

Purchase orders were placed for two 716-MWe (842.5 MVA) 21-KV 3000-rpm Hydrogen-cooled 'THDF' type Turbo-generators to BHEL.

31.7 Flagging off of the Calandria of KAPP 3

The Calandria for KAPP 3 was flagged off at Walchandnagar Industries Ltd. (WIL) workshop at Walchandnagar on 29.3.2013.

The senior officials of BARC, NPCIL and WIL were present on the occasion.

31.8   Progress of Civil Work during 2011-2014

The rafts for nuclear building 3 (NB-3) and nuclear building 4 (NB-4) were completed. Civil work in other buildings was in progress.

31.9   Progress during 2014-2016:

31.9.1   Nuclear work of KAPP 3

The pressure vessel weighing 228 metric tonnes was lowered into the Calandria vault on 12.6.2014. The welding of the Calandria and end-shields was completed as well.

End-Shield lowering in progress in RB-3

Calandria being lowered into RB-3

31.9.2   Civil work: KAPP 4

31.9.2.1   *Concreting*

Nearly 6,20,000 (six lakhs twenty thousand) cu. metres of concreting was completed.

A view of the KAPP-3&4 main plant area under construction

31.9.2.2   *Civil structural work: KAPP units 3 and 4*

The work for several main plant buildings was completed and the finishing work was in progress. The construction of the last lift concreting of the inner containment wall (ICW) of KAPP 3 was also completed. The pre-fabricated Carbon steel liner for the wall was provided for the first time in an ICW for better leak tightness. All 16 steel panels were fabricated at the site and installed following the panel-by-panel modular installation. The installation took around 4 months. The natural draft cooling work was completed for KAPP 4.

All internal civil structural work including the steam generator areas was completed. Preparatory work for the erection of the internal containment dome for KAPP 3 had commenced. Civil structural work up to 115 metres elevation in the Calandria vault and both the FM vaults in the RB-4 and the pump-house were completed.

A bird's eye view of the KAPP-3&4 main plant area, with KAPS- 1&2 visible in the background

31.9.2.2.1   Steel liner for ICW

We have stated, in para 31.9.2.2, about the installation of a steel liner for the first time in the inner containment wall of any Indian Nuclear Power Plant. We will now go into more detail about the installation. The 16 panels that were installed were 10 metres in circular length, 5 metres in height and weighed 5.5 tonnes each. The panels were lifted by a mobile crane and kept in position one after the other at the periphery. After all the panels were positioned and the alignments chocked and fit up, tack welding and final welding were carried out.

The analysis of the ring liner structure during lifting and placement in position was carried out jointly by the technology development group of the Engineering Division and the execution team of the KAPP units 3 and 4 at the site. The AERB reviewed the study later on.

Spider wires strung from the lifting crane and an even structural beam (9 tonnes weight) were used to stabilize the liner ring during lifting.

The 7th Ring of PRL being installed at KAPP- 3&4

31.9.2.2.2   More details of the ring liner: lifting crane capacity

The ring liner is 50 metres in diameter with a height of 5 metres and a weight of 80 metric tonnes. The spider was load-tested for 140 metric tonnes. The ring assembly was lifted by the 'Liebherr' crane LR 11350. A total weight of 172 metric tonnes was handled by the crane during the erection. The crane load of the liner ring marked one of the dimensionally largest engineering structures to have been lifted and placed in position in India. The effort was the culmination of a well-coordinated erection teamwork. The total weight to be lifted and placed as a single structure in the ICD was 340 metres. The LR 11350 crane was tested for 415.7 metric tonnes before the lifting of this structure.

31.9.2.2.3 Turbine building (TB) for units 3 and 4

The civil construction work of the turbine generator concrete deck for both units was completed. Nearly 75% of the steel column fabrication work for both units was completed and the erection was in progress. Most of the turbine plant and auxiliaries were received at the site. The manufacture of balance auxiliary was in the advanced stage. The construction of both induced and natural draft cooling towers was taken up and was in progress. The condenser for unit 3 was received at the site and the erection was in progress.

31.9.2.2.4 Piping and other nuclear and conventional equipment work during 2014-2016

Nearly 75% of the pipe fabrication work in the common services system and the primary piping system was completed. Around 55% of the piping erection for common services was completed. The erection of equipment such as the ECCS, air accumulators, chiller units, air compressors, air dryers, moderator heat-exchangers, moderator storage tanks, moderator pumps, shut-down cooling pumps, fuelling machine bridges and columns were completed.

The primary coolant pumps, fuelling machine supply pumps, five reactor headers and feeder pipes were all received at the site and were being progressively erected. The coolant channel installation was completed in all respects for KAPP 3 on 4.6.2016.

The pre-assembly of the end-fittings, a pre-requisite for the coolant channel installation, was also completed.

31.9.2.2.5 The construction of the inner containment dome of KAPP 4: progress of the form-work

31.9.2.3 The Electrical work

31.9.2.3.1 The 400-KV and 220-KV switchyards

The 220-KV switchyard work was completed and both the main buses were charged. Civil work for the 400-KV switchyard was completed as well. The tower and tower yard erection work were nearing completion.

31.9.2.3.2 Other electrical work

Cable tray, cable conduit and lighting installation work were nearing completion in priority areas and progressing well in other areas. The installation of the 6.6-KV Class-IV switchgear was completed and the Class-IV 415-volt switchgear was commissioned.

All Class-III DG sets were received at the site and were being installed.

31.10   Progress during 2015-2016: KAPP 3

31.10.1   Nuclear services system: steam generators and Calandria

The first and second nuclear steam service generators (NSSGs) were received at the site on 16.6.2015 and 6.7.2015, respectively. Each NSSG weighed 225 tonnes and was 25 metres in length. The first NSSG was installed in the North SG vault on 22. 6.2015 with the help of LR-1350 and LR-1650, the two heavy-duty cranes at the site. The 2nd NSSG was installed on 7.7.2015 one day after receipt at the site and was lowered into the vault.

31.10.2   Conventional equipment: KAPP 3

31.10.2.1   The Feed-water system: the erection of the de-aerator and the natural draft cooling water tower (NDCT)

The de-aerator storage tank, of 40.097-metre length, 4.93 metres diameter and 715 metre capacity, was fabricated as a single unit for the first time in India. It was lifted and erected in the TB-3 on 26.8.2015 by Dodsal. The lifting structure for the storage tank was fabricated at the site and load-tested with a locally fabricated cradle.

LR-11350 cranes, with an operating radius of 81.5 metres and a load capacity of 206 metric tonnes, used for the lifting had to meet a combined total of 198 metric tonnes—comprising storage tank lifting structure, tackle and crane hook.

The tank was erected at an elevation of 127 metres between grids 1 and 2 and AG. The natural draft cooling tower (NDCT) was in the advanced stages of completion.

31.10.3   Charging of 220-KV cables for SUTs of KAPP units 3 and 4

It was the first time in the history of the development of nuclear power in India that 220-KV power cables were used for the charging of start-up transformers.

The 220-KV power-supply through the underground/trench route was extended from the existing 220-KV switchyard (of the KAPP units 1 and 2) by adding two 220-KV bays for feeding SUTs 3A and 3B (of KAPP units 3 and 4). The SUTs were of 70-MVA capacity and double winding 220 KV/6.9 KV - 6.9 KV. Nearly 3 km of 500-mm2 cables were used from the switchyard of KAPP units 1 and 2 to the 220-KV bay 3 of KAPP units 3 and 4.

The control cable for the bus bar coupler protection was laid from the control room of KAPP units 1 and 2 to the SUT protection panel in control building 3 at an elevation of 116 metres.

A specialized team from South Korea terminated the 220-KV cables. The 220-KV cables were charged on 19.9.2015 at 20:04 hours, the SUT 3A was charged on 20.9.2015 at 12:15 hours and the SUT 4 was charged on 21.9.2015 at 12:50 hours successfully.

31.10.4   The TG Equipment: positioning of the TG stator of KAPP 3

A 2-pole 3000-rpm TG stator of 842.9 MVA and 716.6 MWe weighing 325 metric tonnes, with 9.8 metres length and 4 metres diameter, was placed in position at 118 elevation between grids 7 and 2 of the TG block of KAPP 3 on 29.10.2015. The LR-11350 crawler crane used for the handling was configured to make it capable of handling this heavy lift with the required lifting radices. The crane was tested with a weight of 402 metric tonnes on 24.10.2015 to establish the lifting capability before stator handling.

Generator rotor

31.10.5   Conventional Equipment of KAPP 3: the LP turbine bottom casing and positioning

BHEL-Alstrom supplied a 700-MWe LP turbine bottom casing for KAPP 3. The Construction—of size 8 metres×3.2 metres×4.6 metres, weighing 29 MT and having 5×5 stages of double-flow and double-steam entry—was erected on the foundation of the TG-3 turbine deck on 28.4.2016.

The pre-erection work began a month after the completion of the erection of the condenser up to neck level. The equipment was erected using a 125-tonne TB-3 EOT crane.

31.10.6   Generator rotor Positioning

A generator rotor—of 13.135 metres in length, 1.216 metres in diameter and weight of 75.8 metric tonnes—was successfully threaded into position on 9.1.2017. Shid shoes were assembled on the rotor, skid plates were inserted inside the stator, IR and PI values of the stator were taken and all stator internals were checked before the threading.

Generator Rotor Insertion in Stator

Generator rotor

31.10.7   Progress of KAPP 3 as of early 2018: the D2O moderator system

The Moderator Storage Tank 3–3200-TK2, with a capacity of 160 metres, was commissioned in January 2018. The D2O addition station in the reactor auxiliary building at 94 metres elevation was commissioned on 21.4.2018. D2O was transferred from the D2O drums to the tank with the use of instrument air. 3-3200-TK2 is made up of Carbon steel with both the inside and outside surfaces painted with epoxy paint. D2O would be transferred to the moderator system in due course.

31.11   Progress during 2016-2018

31.11.1   Conventional Equipment: natural draft cooling towers (NDCT)

The structural construction of NDCT-3A and NDCT-3B was completed on 11.5.2017 and 4.9.2017, respectively. The hyperbolic-profiled towers are 166 metres high. Each tower is designed to dissipate a heat load of 790 MWe from the condenser water cooling system and the auxiliary water systems. The foundations of the NDCTs and the cooling water basin are independent. Each NDCT has a diameter of 142 metres at the bottom. The tower is 1.6 metres thick at the bottom, 0.3 metres at the throat and 0.75 metres at the top. The tower shell concreting was completed in 107 lifts with each lift being 1.5 metres. About 17,200 cu. metres of concrete and 2,580 tonnes of re-inforcement steel were utilised for each tower.

Natural draught cooling tower

31.11.2 Nuclear service system of KAPP 3: Calandria tube and feeder tube installation

The Calandria tube installation, comprising 392 channels, was completed in 42 days on 30.8.2017. A total of 784 rolled joints between the tubes and the Calandria vessel were involved. The feeder tube installation in KAPP 3 commenced thereafter and about 35% of the feeder tubes were completed as of September 2017.

Feeder Installation at KAPP-3

31.11.3 Conventional Equipment of KAPP 3: charging of SUTs

The start-up transformers SUT-3A and SUT-3B, each of 70-MVA capacity 220 KV/6.9 KV/6.9 KV of KAPP 3, were charged on 22.3.2017 from the main control room of KAPP 3. All the tests for the protection system of circuit breakers and isolation were completed before charging. A full-scale mock-up test of the fire water-deluge system was completed successfully before charging. The deluge system has since been kept poised. The statutory clearance from the Western Region Load Despatch Centre (WRLDC) was obtained before charging.

Start-up Transformer yard of KAPP-3

31.11.4 Conventional Equipment of KAPP 3: charging of 6.6-KV Class-IV and Class-III buses and the commissioning of the service water pump

All the 6.6-KV Class-IV buses were charged on 11.4.2017. The commissioning of the first 6.6-KV Class-III bus was also carried out. Class-III service water pumps 3-718-1 PM 5 and 6 were commissioned and no-load tests on these pumps were carried out on 27.4.2017 and 3.5.2017, respectively. Service water pumps were continuously operated for closed-loop flashing of different circuits of the service water system and primary loop flashing of the chilled water systems.

31.11.5 Completion of work

All the various construction systems were completed and tested on schedule. The unit was kept ready for fuel loading and the first approach to criticality.

KAPP-3 in foreground

31.11.5.1 Fuel loading

Fuel loading was completed in mid-March 2020 after obtaining clearance from the AERB. The AERB also cleared the unit's first approach to criticality.

31.11.5.2 *First approach to criticality and achievement of criticality*

The first approach to criticality was initiated, and all the Physics and safety parameters were duly monitored. The unit achieved criticality at 9:35 hours on 22.7.2020.

31.11.6 Current number of nuclear units and total installed capacity: new projects

KAPP 3 has the distinction of being the first 700-MWe capacity HWPTR to be fully designed, constructed and commissioned indigenously. It is the 23rd Nuclear Power unit to be commissioned in India and with this unit commissioned, the total installed nuclear power capacity goes up to 7,480 MWe. As of now, there are eleven other 700-MWe units in various stages of construction. Six 1.65-MWe units are also in the preliminary stage of construction at Jaitapur in the State of Maharashtra.

RAPP Units 7 and 8

32.1 General

The Government of India has decided to establish two 700-MWe units (RAPP 7 and 8) at the Rawatbhata site. The State of Rajasthan was once considered to be part of the 'Bimaru' group (Bihar, Madhya Pradesh, Rajasthan and Uttar Pradesh), the least developed States of India. It is no longer the case. We have seen in Chapter 7, how the establishment of the very first indigenously constructed and commissioned 220-MWe units in Rajasthan triggered the industrial and agricultural development of the State and the city of Kota in particular. The State has been on the fast track of development in the last few decades; there need be no other sign of this development that the island of Nuclear Power units on the Rawatbhata site would boast of an installed capacity of 2600 MWe once RAPP 7 and RAPP 8 (700 MWe each) are commissioned, the largest so far at a simple site in India. The GOI decided on the construction of RAPP 7 and 8 in 2009.

32.2 The design, plant layout, provisional safety analysis and progress during 2009-2010

Identical activities for KAPP units 3 and 4 (700 MWe each) were initiated. The plant layout and the reactor core design were both finalised. The provisional safety analysis report was also made.

32.3 Financial sanction: preliminary work during 2010-2011

RAPP units 7 and 8 could boast of lead advantages over other nuclear sites in that several common facilities were already available at the site. However, preliminary work specific to RAPP units 7 and

8 was necessary and they commenced early in 2010, right after the receipt of the financial sanction amounting to Rs. 120 trillion (Rs. 12,000 crores) which is the same amount as for KAPP units 3 and 4.

Excavation for the RB-7 site commenced in early 2010 and the work was completed by August 2010. The first concreting commenced on 22.11.2010.

32.4   Formal official commencement of concreting: the emergency cooling water system

The formal inauguration of the M-45 concreting, for the base floor of the emergency cooling water system (ECC) building, took place on 18.7.2011 at the hands of Sri. Srikumar Banerjee, Chairman of the AEC, and Sri. S.K. Jain, CMD of the NPCIL. Senior officials of RAPP units 7 and 8 were also present during the function. The formal inauguration of the Civil work of RAPP 8 took place on the same day.

(L to R) Mr. O.P. Arora, Project Director, RAPP - 7&8, Mr.C.P.Jhamb, ite Executive Director, RR Site, Dr. S K Jain, CMD, NPCIL and BHAVINI, Dr. S. Banerjee, Chairman, AEC and Secretary, DAE, Mr. S. A. Bhardwaj, Director(Technical), NPCIL and other senior officers at the brief inauguration ceremony of FPC

A view before concreting

32.5   Long-delivery major equipment
Purchase orders were placed for all the required materials.

32.6   RAPP units 7 and 8

32.6.1   Civil work during 2011-2012
The concreting of the raft of RB-7 was completed by August 2012 and that of RB-8 was nearing completion; the concreting of the raft of the station auxiliary building of RB-7 (SAB-7B) was completed and the concreting of columns above the raft was taken up. The wall-liner grid work for the South Fuelling Machine Area (FMSA) of RB-7 was completed. The concreting of the bottom slab of the de-mineralised water tank of RAPP 7 was completed.

The water tank (DM) of RAPP 7 was completed.

It is to be pointed out that during the concreting work in 2011, a quantity of 5,757 cu. metres was poured into RB-7 on 3.11.2011 which was the highest amount to be carried out on a single day so far, according to the DAE and the NPCIL.

A view of RAPP NB-7 and NB-8 rafts

32.6.1.1   *Other Civil work*

New guard houses for RAPP units 7 and 8 were constructed and made functional.

32.6.2   Component assembly infrastructural work

The construction of the reactor component assembly (RCA) workshop and the electrical workshop was completed.

32.6.3   Manufacture of critical equipment

The manufacture of major critical equipment like Calandria, steam generators, end-shields, primary circulating pumps, diesel generators, transformers and conventional and nuclear service pumps was in progress in various stages. L&T was awarded the contract worth Rs. 732 crores for the supply of the remaining turbine packages for RAPP units 7 and 8.

'L&T power', a subsidiary owned completely by L&T will supply, install and commission the packages.

32.7   Progress as of 2013

The Calandria vessel of RAPP 7 reached the site on 10.7.2013. It was received by the site project Director J.P. Gupta. It was earlier flagged off by the CMD of the NPCIL in Mumbai on 24.3.2013. The vessel was transported by sea up to Kandla Port and thereafter by road to the RAPP site. The vessel is made up of SS 304L material of 32-mm thickness. The diameter of the vessel is 7.8 metres and the weight is 40 metric tonnes. The vessel was manufactured by Godrej at their workshop in Mumbai.

Thus, the Calandria vault construction was completed.

A view of under construction units-7&8 of Rajasthan Atomic Power Project.

32.8 Industrial Safety Award

RRC Rawatbhata, Rajasthan site, of which the RAPP units 7 and 8 form a part, received the Atomic Energy Regulatory Board (AERB) award for industrial safety for the year 2011.

32.9 Progress during 2014-2015

32.9.1 End-shield for RAPP units 7 and 8

The first end-shield of RAPP 7 was transported to the site on 30.9.2014 in 60 days from L&T's nuclear equipment manufacturing facility at Hazira in Gujarat. The transport covered a distance of 808 km distance and was carried out in the face of heavy monsoons and poor roads on the way to the RAPP 7 site. An innovative, hydraulically operated, rotating steel fixture was mounted onto the transport vehicle and used to clear various obstacles on the way which helped in cutting down the delays. The second end-shield for the RAPP 7 was received at the site and the ball-filling activity, an essential pre-requisite for the end-shield erection, was completed. Photo P-35 shows the dispatch of the first end-shield for RAPP 7 from L&T's Hazira plant in Gujarat on 1.8.2014.

End-Shield being dispatched from M/s. L&T`s Hazira Plant for RAPP-7

32.9.2   FM work

The erection of the fuelling machine bridge and columns in the RB-7 was completed in November 2014.

32.10   RAPP units 7 and 8: progress in Civil work during 2014-2015

The construction of the inner containment wall (IC) and the internal structures of RB-7 were progressing well. The concreting of the Calandria vault of RB-8 was nearing completion. The construction of other structures in RB-8 was also progressing well. The Civil works in the control building of RB-7 and RAB-7 were in progress.

RAPP- 7 is seen on the left and RAPP- 8 on the right in the foreground

32.10.1   Auxiliary plants

Work on the waste-management plant, fire water pump house, plant water pump house, the DM water plant and the chlorination plant was in progress.

32.10.2   The TG block

The construction of the turbo-generator decks in TB-7 and TB-8 was in progress.

32.10.3   Natural Draft Cooling Towers

The excavation work for the natural draft cooling towers was nearing completion while the plain cement concrete work was in progress. The construction of induced draft cooling towers was also in progress.

32.10.4   Total concreting

Around 3.45 lakh cu. metres of concreting was carried out at the site while the excavation of 19,03,368 cu. metres was completed.

32.11   Electrical work

The construction of the 220-KV switchyard was completed and the switchyard was energised. Civil work and the erection of structures in the 400-KV switchyard were in progress.

32.12 Nuclear and conventional equipment: progress during 2014-2015

32.12.1 Equipment delivery

The delivery of equipment is progressive. Most of the critical equipment has already arrived at the site. The manufacture of the remaining equipment is in progress. The Calandria vessel for unit 7, the SUTS, the DG sets and fuelling machine bridge column assemblies were all received at the site. Low-pressure feed-water heaters for RAPP 7 were dispatched from the manufacturers.

Low-Pressure Feed Water Heater(LP-3) at the manufacturer's workshop

32.12.2 The erection of the construction equipment

650-MT and 1350-MT cranes were erected, commissioned and made functional.

32.12.3 Coolant channel mock-up facility

The facility was completed to enable the coolant channel installation at the site.

32.13 Progress of RAPP 7 during 2015-2016 and 2016-2017

32.13.1 Nuclear Equipment Installation

32.13.1.1 Coolant channel installation

Outer face of Calandria showing one of the End-Shields with the installed calandria tubes

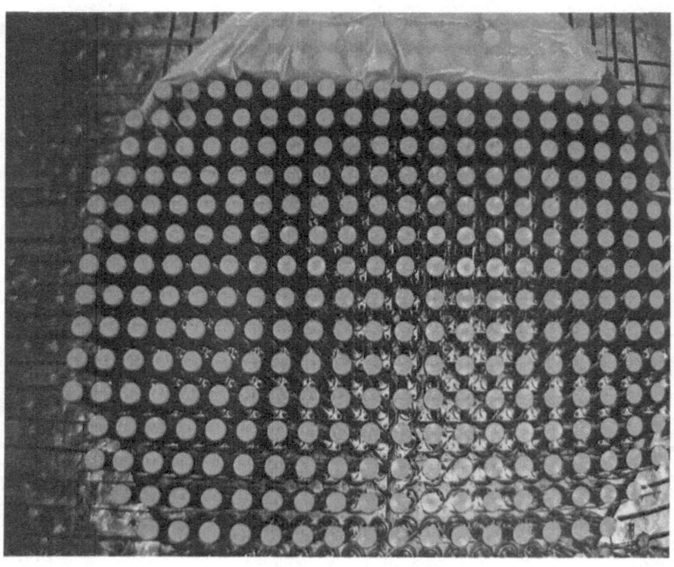

Outer face of calandria showing one of the end-shields with installed coolant tubes.

With the completion and trial of the channel mock-up facility, the installation of 392 coolant channels was taken up on 19.6.2016. The installation was completed on 27.7.2016 at 21:54 hours in a record period of 39 working days. The work comprised the following:

1) Rolling of tubes at both ends with the tubes.
2) Prior work of trimming to size of the tubes—cleaning, insertion and rolling under clean conditions. Helium leak testing was also carried out.

32.13.1.2 *Conventional Equipment Installation*

A view of RAPP-7 Reactor Building - Foreground

32.13.1.2.1   The 400-KV switchyard at RAPP units 7 and 8

The electrical lighting installation was successfully charged at 18:15 hrs on 29.3.2016 after its interconnection with the existing 400-KV switchyard of RAPP units 5 and 6. The new switchyard is an extension of the existing 400-KV switchyard of RAPP units 5 and 6 and is situated next to the latter. The excavation for the new switchyard structures and blasting were carried out in a controlled manner to avoid any damage to the existing switchyard structures.

The switchyard now stood connected to the National grid and was ready to evacuate 1.4 GWe of power to the National grid after both RAPP units 7 and 8 were commissioned. The 400-KV Shujalpur 1 and 2 lines along with the line reactors were completed, charged and synchronized with the National grid.

The 400-kV switchyard of RAPP-7&8

32.13.1.2.2   Unit 7 Civil work

The erection of the liner panels of the ring beam was completed and the concreting of the ring beam had commenced.

The construction of the vault of one of the two SG units was completed while that of the second one was still in progress.

Further Civil work in the control building, turbine building, electrical building, pipe and cable bridge and other utility buildings was also in progress. The GT and SUT foundations were completed and the construction of four natural-draft and four induced-draft coating towers was in progress.

32.13.1.2.3   Units 7 and 8: the erection and testing of conventional equipment

Hydro-testing of the condenser cooling water inlet and outlet lines (CCW) was completed. The laying of further CCW pipes was in progress.

The erection of cable trays and electrical equipment in the control building of RAB-7, TB -7 and TB-8 was still in progress.

The erection of two DG sets and related piping work in SAB-7B was completed.

32.13.1.2.4 Nuclear and Common Services System in RB-7

The erection of piping for the nuclear and common services system and equipment like the PHT shut-down pumps, the ECCS accumulators and the LZe system equipment in the RB-7, RAB-7 and CB was in progress. Low-pressure feed-water heaters were received in March 2007

32.13.1.2.5 The overall progress of the RAPP units 7 and 8 as of 31.1.2017

The overall progress of the work was 70% in RAPP 7 and 57% in RAPP 8.

32.13.1.2.6 RAPP 7 Progress since 31.1.2017: the conventional system

Low-pressure feed-water heaters LP-2 and LP-3 were dispatched from the manufacturer's workshop in March 2017.

32.13.12.7 RAPP 7 Nuclear service system (NSS)

32.3.12.7.1 The passive-decay heat-removal system and heat-exchangers (PDHRS exchangers)

PDHRS Heat Exchangers at the manufacturer's workshop

Four PDHRS heat-exchangers were flagged off from the manufacturers' workshop on 15.2.2017.

The PDHRS heat-exchangers are one of the most critical equipment of the NSS. These ensure the continued availability and non-circulation of the secondary-side steam generator for decay heat removal during a hypothesised condition of station black-out. The PDHRS heat-exchangers were manufactured as per the ASME-SEC III-NE design section criteria.

The Jaitapur Nuclear Power Project (JAPP)

33.0   Introduction

In Chapter 2, we have seen how the nuclear power generated at TAPP units 1 and 2—with 190-MWe unit capacity and 380-MWe total capacity—was synchronized with an Indian Regional Power Grid (WREB) for the first time in 1969. The above-mentioned unit capacity was raised to 200 MWe in the RAPP, MAPP, KAPP and Kaiga nuclear power stations. The capacity was further increased to 540 MWe in TAPP units 3 and 4 and later to 1000 MWe in the Kudankulam units 1 to 4. This progressive increase in unit capacity was dictated by the emerging need to achieve economies of scale, and gradually increase the share of nuclear power in the total installed electrical power in the country. In this regard, unit capacities of 1.6 GWe and above were sought after.

The unit capacity, being quite large, the Government of India decided to go in for negotiations with the Government of France for technical collaboration in a project. France has the largest share of total installed capacity of nuclear power units, globally accounting for more than 80% of the share.

Both the governments signed an agreement on 6.12.2010 for the construction of two 1.65 GWe 'Areva' units at the Jaitapur village in the Ratnagari district of Maharashtra. Maharashtra is the most industrialised state in India; hence there is a great demand for augmenting its total electrical capacity to meet the increasing power demand.

The above-mentioned agreement cast an obligation on the part of the French Government to supply enriched nuclear fuel for a period of 25 years from the date of commissioning of each unit.

The Prime Minister of India, Sri. Narendra Modi, further pushed the French PM, Francois Hollande, for the project's progress during the course of his official visit to France from 9.4.2015 to 11.4.2015.

The NPCIL and the makers of the 'Areva' brand of Nuclear power units signed an agreement on 10.4.2015 for the assessment of licensability, adherence to the Indian Atomic energy codes/laws and the clarification of the technology used for the 'Areva' brand nuclear power units. The NPCIL would later take up the licensing of the project by the 'AERB', which is India's nuclear regulatory authority. The agreement envisaged the maximisation of the manufacture of nuclear components for the project indigenously, as part of the 'Make in India' Campaign of the Government of India.

A second agreement was signed by 'Areva' and 'Larsen and Toubro' on 10.04.2015. India's giant industrial manufacturer, L&T, was brought into the project to reduce the cost of the project by augmenting indigenous technology and manufacture for the equipment and materials of the project.

It can be pointed out that one such 'Areva' 1.65-MWe unit was built and commissioned for the UK's nuclear power programme by the 'EDF'—which is France's principal technology output. The unit has been functioning satisfactorily.

33.1   Environmental clearance: scope of the project

Former Minister of State (Environment and Forest). Mr. Jairam Ramesh(left), hands over the Environmental Clearence to Dr. Srikumar Banerjee, Chairman AEC, secratary DAE (center - right), in the presence of Honourable Chief Minister of Maharashtra, Mr.Prithviraj Chavan(right)

33.1.0   Site Clearances: initial work at the site

The Ministry of Environment, Forest and Climate change in India, GOI, gave clearance for the project at a function held at Mumbai on 28.11.2010. The 'Areva' brand project comprises advanced pressurised light-water reactors fuelled by enriched nuclear fuel. 'Areva' of France would supply help with the erection and commissioning of the project.

33.1.1   Initial work at the site: rehabilitation

Initially, the work was stalled due to opposition from local stake holders, particularly farmers. The project would require a land area of 968 hectares. The Government of Maharashtra held protracted negotiations with the said parties and secured their concurrence for the purchase of the required land area.

The land for the plant was acquired by the Government of Maharashtra and handed over to the NPCIL with the help of the local administration of the Ratnagiri district. An agreement was signed between the NPCIL and the Government of Maharashtra for the rehabilitation of the project-affected people (PAPs).

33.1.2   Project enabling works: the project construction agreement

The geo-technical investigation of the site, the construction of property cum boundary wall of the plant site and the preparation of the master plan for the residential complex were in progress. The general framework agreement, site-work agreement and the construction agreement between the NPCIL and 'Areva' were in advanced stages of negotiation.

33.1.3   The progress during the first half of 2015

The site office; quality-assurance complex and fire, industrial safety and health buildings were inaugurated on 28.5.2015. The work of trenching for geo-technical studies was also inaugurated.

33.1.4   CSR activities

Most of the previously discussed nuclear power plants started their CSR activities in the nearby villages only after the plants were commissioned and declared for commercial operation. The Jaitapur project, on the other hand, forged a new trail for starting CSR activities prior to plant construction and commissioning.

Several improvement activities were commissioned for the nearby villages which focussed on education, health, infrastructure and other social/cultural developments. Some of these activities are listed below:

(1) Three-wheeler vehicles for the physically challenged people were distributed.

(2) The project supplied school books and school bags for the Zilla Parishad schools

(3) Play items were distributed for the Anganwadi School students in the vicinity of JAPP on 7.9.2018.

The above-mentioned social-improvement activities were carried out on the request of the authorities from the respective schools.

33.1.5   Further outlook on the project

The Jaitapur project is the largest nuclear project that India has embarked upon with France's technical collaboration. The only other minor nuclear transaction was when India imported 300 tonnes of processed uranium, the French party again being 'Teresa' during 2008-2010. In recent years, India and France have vastly expanded the horizon of economic trade, defence, security and cyber agreements. Though the Jaitapur project negotiations were initiated way back in 2013, progress has not been significant thus far. This is due to the following reasons:

(1) The reluctance of the agro-rich farmers of the Konkan region to part with their land

(2) France has been battling with frequent terrorist attacks

(3) The French President Emmanuel Macron's scaling down of social security measures and his tax measures

(4) The Covid-19 pandemic which affected France very badly; India also struggled a lot in its battle against the disease

The above-mentioned points discuss the negative aspects in regard to the project. The positive aspect is that both India and France are determined to forcefully go ahead with their technical collaboration.

We may, therefore, look forward to more rapid progress in the JAPP project in the coming years.

Kudankulam Units 3 and 4 of IGWe Capacity

34.0   The Genesis of the units

In Chapter 24, we have seen the extensive description of the construction of Kudankulam units 1 and 2, each with IGWe capacity. We saw the enormous efforts made at both the GOI level and the NPCIL level to get these going. These efforts ultimately proved successful and the two units are now operating successfully.

In this chapter, we will look into more details on why this large capacity unit was chosen, considering the fact that 200-MWe unit capacity reactors were already in operation. This decision requires further justification. When the decision to go in for the construction of KK units 3 and 4 was taken, two boiling light-water type reactors of 160 MWe each (TAPP units 1 and 2) and four units of 200-MWe reactors (RAPP units 1 and 2, and MAPP units 1 and 2) were operating satisfactorily. Four more reactors of 200 MWe were waiting to be commissioned. Two units of 540-MWe capacity each (TAPP 3 and 4) were also under construction; all of these were fully indigenous in scope. Around 200-MWe units were considered to be the standard prototype. Even with 500-MWe units under construction, the quantum jump to 1000-MWe units was considered to be a giant leap.

Apart from the advantage of large-sized units providing the NPCIL with a quantum increase in revenue, the need for meeting the increasing power demand in Tamil Nadu and other Southern States were also paramount. The state of Tamil Nadu is one of the foremost generators of renewable energy sources (RES) of power in India. Even with this distinction, the share of RES in the state of

Tamil Nadu as of 2017 was hardly 6%. The share would have been much less when the decision to go in for KK units 3 and 4 was taken during 2018.

34.1   The contract for the design and construction frame work of the Inter-Government agreement

The contract for the preparation of initial design within the framework of the Inter-Government agreement, between the Soviet Federation and the Government of India, was signed by the NPCIL and 'Atomstroyexport', which is the Russian Federation's exporter of nuclear power equipment. Techno-Commercial discussions between the two countries continued after December, 2008.

34.2   The cost of the project

The protocol for the Kudankulam units 3 and 4 was signed by Shri. A.P. Joshi, special secretary of the DAE, on behalf of the Government of India and S.A. Storchak, Minister of finance, on behalf of the Russian Federation on 17.7.2012. The Russian Federation would extend credit, amounting to $3.4 billion US dollars, for financing 85% of the work supplies and services provided by the Russian agencies. The Russian Federation would provide a State credit of 0.8 billion US dollars for financing 85% of the cost of fuel and fuel control assemblies. The interest on the credit would be 4% per annum. The project credit would be repayable in 14 years, starting one year from the date of commission. The credit to fuel and fuel assemblies would be payable in four years, commencing two years after the receipt of fuel at the site. The project cost of Kudankulam units 3 and 4 is expected to be Rs. 320 billion as per the prices of 2012, Rs. 170 billion of this would be met by Russian credit.

34.3   The progress of construction activities at the site of KK units 3 and 4 during 2016-2017

34.3.1   Preliminary work

Geo-tech tests were carried out at the site following the signing of the protocol. Site roads, site offices and preliminary enabling works were also carried out. The excavation for the plant and auxiliary structures was undertaken.

34.3.2   The first pouring of concrete

This was carried out on 29.6.2017.

First Pour of Concrete at KKNPP-3&4

34.3.3   The role of 'Make in India' in the KK 3 and 4 project

The NPCIL and with its Indian industry partners strongly took up the initiative, as part of the 'Make in India' drive, to carry out the design and facilitation of the construction of structures, systems and equipment for the Kudankulam units 3 and 4. Initially, two out of the 40 large-diameter stainless steel tanks were dispatched from the manufacturer's shop to the site on 8.7.2017. These tanks would go on to be installed in the lower-most elevation of the reactor auxiliary building.

34.3.4 Common services system of KK units 3 and 4

Reliance Infra has emerged as the lowest bidder at the cost of Rs. 1000 crores for the design, engineering, procurement and manufacturing as required for all the structures and components; as well as the erection, testing and commissioning of the common services system. The contract was accordingly given to them. The contract period for the completion of the project was 56 months, starting from the end of 2017 to early 2018.

Kudankulam Units 5 and 6

35.0 The Genesis of KK units 5 and 6

In para 34.0, we have seen the genesis of the construction of KK units 3 and 4 at the Kudankulam site. The same argument holds good for the construction of the KK units 5 and 6 at the site. The construction of IGWe size units each at the Kudankulam site will hugely boost the total capacity of the Southern Regional Electricity grid. It will also help meet the growing power demands of the states of Tamil Nadu, Karnataka and Kerala. Spade work for the construction began in the middle of 2016.

35.1 The general framework between India and the Russian Federation for the construction of units 5 and 6

The general framework for the construction of units 5 and 6 were expected to be signed between the NPCIL and its Russian counterpart by the middle of 2017. The units are expected by 2022-2023.

35.2 Further progress on the framework agreement

The agreement covering the design and supply of the main equipment for units 5 and 6 was formalized between the NPCIL and the ASE group of the Russian Federation on 01.06.2017. The ASE group consists of Atomprockt, Atomstroyexport and the ASE Joint Stock Company

(Formerly Atomenergy prockt). Following this development, the NPCIL and 'Atomstroyexport' signed contracts on 31.7.2017 for the design and supply of equipment and systems for the two units. The agreement provides for reviews of the design of the equipment and systems, and for obtaining clearances and licensing of the units from India's nuclear regulator, namely the AERB. The

NPCIL would also be responsible for the construction of buildings and structures, the erection of equipment and systems as well as the commissioning and operation of the two units.

The contracts signed on 31.7.2017 followed the broad agreement signed between India and Russia during PM Narendra Modi's visit to Russia on 2.6.2017.

The four Kudankulam units—KK 3 to 6 inclusive—would cost Rs. 820 billion and 30% by equity. The 30% equity would be met with by the NPCIL's own resources and by the Government of India.

Mahibanswara Raj Power Project (MBRPP)

36.0   Introduction

In the previous chapters, we dealt with the setting up of progressively higher-capacity nuclear units ranging from 220-MWe, 520-MWe and 700-MWe units up until 1 and 1.2-GWe units. The progressive higher-capacity units were made possible due to the absorption of indigenous technology by the NPCIL, who are responsible for setting up Nuclear Power Projects in India. In July 2018, the Government of India accorded administrative sanctions for setting up ten 700-MWe nuclear power stations in India, six of which are now under construction—namely KAPP units 3 and 4, RAPP units 7 and 8 and KGPP units 5 and 6. The aim of these sanctions is to attain a total nuclear installed capacity of 12.98 GWe by 2024-2025.

36.1   The MBRPP

Of the remaining four 700-MWe capacity units, two are proposed to be constructed in the district of Banswara in the Southern part of the State of Rajasthan.

36.2   CSR work

Even at the construction stage, MBRPP had started its CSR work at the nearby villages; hitherto CSR was started after the commencement of commercial operation of the stations. The project provided twenty-five patient stretcher trolleys to the Mahatma Gandhi Hospital at Banswara on the request of the Chief Medical officer at the hospital, Sri. Halian, on 1.11.2018.

36.2.1   CSR work at MBRPP

The Mahatma Gandhi (M.G) hospital is the largest hospital in the district of Banswara with bed strength of 540. Around 1000 to 1200 patients come here daily for out-patient (OP) treatment. The sub-divisional Registrar and Land acquisition officer of Banswara, Smt. Pooja Parth (IAS); the Chief Engineer of MBNPR, Sri. S.B. Joshi; the Assistant. Director of the Social Welfare Dept. of the Government of Gujarat and other senior officials of M.G hospital were present during the presentation ceremony.

Mithi Virdi Bhavnagar Atomic Power Project (MVAPP)

37.0   Introduction:

The MVAPP is the 5th in the series of ten 700-MWe Atomic Power Projects that were planned to be set up by the GOI in July 2018. The proposed Atomic Power Project is to be constructed in Mithi Virdi in the district of Bhavnagar, Gujarat.

37.1   CSR work at MVAPP

37.1.1   Contribution to education

37.1.1.1   *Trapaj village*

MVAPP became the second Atomic Power Project of the NPCIL to start its CSR even before the construction activity had commenced. It began its work at the Trapaj village which is about 12 kms from the project site. The said village has a population of 6,500. The Trapaj Kendravarti School (K.S) in the village has 450 students.

On the written request of the Principal of 'K.S', the MVAPP supplied the following for the school:

(1) Twelve sets of desktop 'Lenovo' computers with UPS, Intercom and tables for educational development

(2) A P.A system from Ahuja Co. for prayer and academic and cultural programmes.

The P.A system comprised of one amplifier, fourteen wall-mounted classroom speakers and a microphone and a mike stand with 10 metres connectivity.

37.1.3 Navgam Nana village

Navgam Nana Village has a population of 1600 and a primary school with 230 students. It is located about 7 kms from the MVAPP site.

At the request of the Principal of the above-mentioned school, MVAPP provided an over-head projector of the model type Epson EB-X05 and one projector screen and tripod of the Liberty Make at the cost of Rs. 40,680. The equipment provided will meet the educational needs of the school.

37.1.4 Goriyali village

Goriyali village has a population of 1770 and is situated about 5 kms from the MVAPP site. The village has a primary school with 175 students.

At the request of the Principal of the school, the MVAPP authorities supplied the following to the school:

(1) Bead counters and puzzles; charts on the alphabets, numbers 1 to 20; 3D charts on the human organs; reading and writing charts, etc.

(2) Geographical Maps portraying the World continents, Indian rivers, Indian crops, Indian political maps and the Bhavnagar district map

(3) Science activity kits

(4) Green boards and dusters

These materials were provided at a total cost of Rs. 48,053.

Comic books titled 'Have Badloi Gayo Mansukh' Part-III were distributed among the students.

The handing over of the above-mentioned materials was carried out at a dedication programme held on 05.10.2018

37.1.5 Further CSR work of MVAPP

MVAPP further carried out its CSR work by providing a pre-cast State bus stand at the Kaked village at a cost of Rs. 3, 60,000. Kaked village is about 6 kms away from the project site with a population of 2,200. The bus stand was dedicated on 22.07.2019 in the presence of Sri. Kumar pal, Sarpanch of Kaked village; Sri. N.P Gandhe, the Additional Chief Engineer of the NPCIL and the villagers of Kaked.

MVAPP built a borewell with 60 metres depth and 2.5 metres diameter and handed it over to the Bhavanipur primary school situated at about 8 kms from the project site at a cost of Rs. 85,450.

Gorakhpur Haryana Anu Vidyut Pariyojana (GHAVP)

38.0 Introduction

The GHAVP is a 2700-MWe Atomic Power Project situated to the west of the Gorakhpur Village in the Fatehabad district of Haryana. The project is one among the ten 700-MWe units being constructed by the NPCIL as per the sanctions of the GOI. There were several concerns about the fossil-fuel power units and their distance from coal mines, mostly situated in Central India, on which they were very much dependent. The GHAVP was thus an optimal choice to meet the growing power requirement of Haryana's rapidly growing economy which was already gaining the attraction of new industrial enterprises. The GHAVP is the first Atomic Power Project to be set up in Haryana.

38.1 CSR work

Similar to previous projects, the GHAVP also started its CSR projects even before the construction work had begun.

The project organized free dental medical camps on 28.8.2018 at the Gorakhpur Govt. Girls Senior Secondary School (GGPS) for the students of classes I to XII. About 450 students and staff members participated in the camp.

38.2 Topographic Survey

The work on initial surveys commenced on 6.9.2012. The officers of the NPC, Fatehpur district administration and farmers were present on the occasion of the topographic survey of the villages of Gorakhpur, Badopal and Kajal Heri.

Ceremony marking the beginning of the survey work

Kovvada (AP) Atomic Power Project (APNPP)

39.0   Introduction:

The Government of India, in association with the DAE and the Ministry for Science and technology, has decided to establish 2*700 fully indigenous nuclear power units at Kovvada (A.P) in line with its enunciated policy of constructing more nuclear power stations.

39.1   Relief and Rehabilitation for the PAP (Project-affected People)

The Kovvada (APNPP) authorities, the Srikakulam district authorities and the National Information Centre (NIC) have come together to develop a first-of-its-kind software called 'Punarvas'—which is a web-based software that can be used to expedite the processes involved in the acquisition of land, and the rehabilitation and resettlement of people who live in the Kovvada project site.

'Punarvas' has earned the gold award in the area of governance. Dr. Jitendra Singh, the Union Minister of State with independent charge of the Ministry of Science and technology, presented the award to the representative authorities of the Kovvada Project at New Delhi on 27.02.2019.

39.1.1   Salient features of 'Punarvas'

Some features of the software are listed below:

1.   It can track new structures and encroachments on Government land
2.   It can be operated through desktop or mobile units
3.   The 'Cartosat-2' satellite of ISRO can be used to capture spatial data for RDR purposes

4. 'Drones' can be used extensively to pinpoint the nature of land and encroachments
5. Direct payment to land owners can be done through 'Adhaar' or payment agencies using their bank account

'Punarvas' is a landmark scheme which may be used extensively in the significant task of e-governance.

39.2   Engagement of APNPP with the public

The Kovvada APNPP participated in a 6-day public-awareness programme organized by the administration authorities of the Srikakulam District as part of the Kalinga-Andhra Utsav Celebrations. The Kovvada APNPP exhibited part-dynamic exhibitions of the 700 PHWR and the 'AP' 1000-MWe nuclear units. The exhibits were visited by college students, Govt. employees, farmers and prominent businessman.

Epilogue

I sincerely hope that the readers found this book to be interesting, reader-friendly and informative about the development of nuclear power in India. I would be happy to hear from you. Your suggestions for the improvement of this book are most welcome and they would be incorporated in future editions of the book. You can reach me by mail at **anamayrama@gmail.com**

My sincere gratitude to all the readers who chose to read this book.